INTEGRATED PRODUCT AND PROCESS DEVELOPMENT

T0324855

WILEY SERIES IN ENGINEERING DESIGN AND AUTOMATION

Series Editor
HAMID R. PARSAEI

GENETIC ALGORITHMS AND ENGINEERING DESIGN
Mitsuo Gen and Runwei Cheng

ADVANCED TOLERANCING TECHNIQUES
Hong-Chao Zhang

INTEGRATED PRODUCT AND PROCESS DEVELOPMENT: METHODOLOGIES, TOOLS, AND TECHNOLOGIES
John M. Usher, Utpal Roy, and Hamid R. Parsaei

INTEGRATED PRODUCT AND PROCESS DEVELOPMENT

Methods, Tools, and Technologies

Edited by
JOHN M. USHER, Ph.D., P.E.
Mississippi State University
Mississippi State, Mississippi
UTPAL ROY, Ph.D.
Syracuse University
Syracuse, New York
HAMID R. PARSAEI, Ph.D., P.E.
University of Louisville
Louisville, Kentucky

A WILEY-INTERSCIENCE PUBLICATION
JOHN WILEY & SONS, INC.
New York · Chichester · Weinheim · Brisbane · Singapore · Toronto

Library of Congress Cataloging-in-Publication Data:

Integrated product and process development : methods, tools, and
    technologies / edited by John Usher, Utpal Roy, Hamid Parsaei.
        p.   cm.
    Includes index.
    ISBN 0-471-15597-7 (cloth : alk. paper)
      1. Computer-aided design. 2. Computer integrated manufacturing
    systems. I. Usher, John (John M.) II. Roy, Utpal III. Parsaei, H.R.
    TA174.I467 1998
    658.5—dc21   97-29338

10 9 8 7 6 5 4 3 2 1

To Our Families:
Rita, Matthew, Kelly, and David Usher
Moushumi and Urmi Roy
Farah, Shadi, and Boback Parsaei

CONTENTS

PREFACE

In the last decade, many companies have experienced immense pressure to provide a greater variety of more complex products in smaller batches. This need, coupled with a desire to lower costs and shorten the product development life cycle, has forced manufacturers to reevaluate their approach to product and process development. These challenges have prompted interest in the integration of product and process design activities employing various tools and technologies such as rapid prototyping and stereolitography, as well as manufacturing philosophies including concurrent engineering to support the required integration.

This book is intended to bring together articles that present ideas and applications that support product design, process design, or the integration of these design activities. This volume contains 14 reviewed articles divided into three sections.

The first section consists of five chapters that address ideas and methods applicable to product design. The specific chapters in this section address the following topics: using quality function deployment to identify customers' needs, issues that arise under the constraint of time-driven product development, methods for incorporating functional design to enhance automated design systems, development of data models for capturing the intent of the design, and the use of the worldwide Web to enhance design.

The second section of this volume includes four chapters. These chapters are intended to address process design and manufacturing. This section includes a review of various approaches to rapid prototyping along with their limitations and advances, a prototype implementation of a case-based process planning system, presentation of a methodology for the design of control systems of automated manufacturing systems, and a new approach for manufacturing diagnosis.

The last section of this volume consists of five chapters. These chapters present several case studies that discuss the application of various tools and techniques of integrated product and process development including a review of how integrated product and process development applies to manufacturing in the defense industry, a survey-based study of how best a manufacturer can take advantage of product planning, an application of quality function deployment, an effective use of design optimization, and a system aimed at integrating design and process planning.

We are indebted to our authors and reviewers for their outstanding contributions and assistance in preparing this volume. We would like to express our appreciation to Mr. Robert L. Argentieri, Senior Editor at John Wiley & Sons, for providing us the opportunity to edit this book. Finally, we would like to thank Shadi Parsaei, Kiran Fernandes, and Athisan Wayuparb for their exceptional help during the course of this project.

JOHN M. USHER

UTPAL ROY

HAMID R. PARSAEI

February 1998

ABOUT THE EDITORS

JOHN M. USHER, Ph.D., P.E., is Associate Professor of Industrial Engineering at Mississippi State University. He holds graduate degrees in both chemical and industrial engineering. Prior to his academic career, Dr. Usher worked with Texas Instruments in the area of process development for the manufacture of printed circuit boards. His research interests focus on concurrent engineering, as well as the application of artificial intelligence to manufacturing with special emphasis on systems that support and enhance the design and process planning functions. Dr. Usher serves as technical director for the MSU Computer Integrated Manufacturing Laboratory. He is a registered professional engineer in Mississippi and a member of IIE, SME, and AAAI.

UTPAL ROY, Ph.D., is Associate Professor of Mechanical, Aerospace, and Manufacturing Engineering at Syracuse University, Syracuse, New York. He is also Director of Syracuse University's Multi-Disciplinary Analysis and Design Laboratory (MADLAB) and Knowledge-Based Engineering Laboratory (KBELAB). He holds a B.S. (1978) and an M.S. in mechanical Engineering and a Ph.D. (1989) in industrial engineering. Dr. Roy's research interests are in computer-aided design, computer-integrated manufacturing, and artificial intelligence applications. He teaches computer-aided design and manufacturing related courses.

HAMID R. PARSAEI, Ph.D., P.E., is Professor of Industrial Engineering and Director of the Manufacturing Research Group at the University of Louisville, Louisville, Kentucky. He has been actively involved in conducting research and teaching courses in computer-integrated manufacturing systems, robotics, concurrent engineering, feature-based design and manufacturing, economic justification of advanced manufacturing systems, and productivity analysis. Dr. Parsaei has served as Editor-in-Chief, Associate Editor, and Area Editor to the *Journal of Engineering Design & Automation, Journal of Engineering Valuation and Cost Analysis, Engineering Economist, IIE Transactions, International Journal of Computers and Industrial Engineering, and International*

Journal of Environmentally Conscious Design and Manufacturing. He has also been a member of the editorial board of several journals, including *International Journal of Industrial Engineering, International Journal of Intelligent Automation and Soft Computing,* and *Industrial Engineering Research Journal,* and has served as guest editor for several international journals.

Dr. Parsaei has published over 270 articles in refereed journals and conference proceedings. He has also authored and edited over 24 books (including those in progress) since 1984. Dr. Parsaei's research projects have been funded by the National Science Foundation, Department of the Navy, National Occupational Safety and Health Administration, General Dynamics, and IBM. In 1993, Dr. Parsaei received the **Wellington Award.** This award is presented annually by the Institute of Industrial Engineers to an individual who has made the most significant contributions to the field of engineering economy. Dr. Parsaei is a registered professional engineer and Fellow of the Institute of Industrial Engineers.

CONTRIBUTORS

Ang Cheng-Leong, Gintic Institute of Manufacturing Technology, Nanyang Technological University, 71, Nanyang Drive, Singapore 638075

James E. Bailey, Industrial and Management Systems Engineering, Arizona State University, Tempe, AZ 85287

Stephen M. Batill, Aerospace and Mechanical Engineering, University of Notre Dame, Notre Dame, IN 46556-5637

Tor Shu Beng, School of Mechanical and Production Engineering, Nanyang Technological University, Nanyang Avenue, Singapore 639798

John M. Britanik, Department of Electrical and Computer Engineering, University of Arizona, ECE Building, Room 230, Tucson, AZ 85721

Graeme Britton, School of Mechanical and Production engineering, Nanyang Technological University, Nanyang Avenue, Singapore 639798

Jay B. Brockman, Computer Science and Engineering, University of Notre Dame, Notre Dame, IN 46556-5637

Muthu Chandrashekar, Systems Design Engineering, Environment and Resource Studies, University of Waterloo, Waterloo, Ontario, Canada N2L 3G1

Shing I. Chang, Department of Industrial and Manufacturing Systems Engineering, Kansas State University, Manhattan, KS 66505

Frank S. Cheng, Industrial and Engineering Technology Department, Central Michigan University, Mount Pleasant, MI 48859

Chua Chee-Kai, Centre for Engineerng and Technology Management, Nanyang Technological University, MPE, Nanyang Avenue, Singapore 639798

Ernest L. Hall, Department of Mechanical, Industrial, and Nuclear Engineering, University of Cincinnati, Cincinnati, OH 45221-0072

Kiran Hebbar, Bentley Systems, Incorporated, 690 Pennsylvania Drive, Exton, PA 19341-1136

Jeffrey W. Herrmann, Department of Mechanical Engineering, University of Maryland, College Park, MD 20742

Eng Shwe Sein Aye Ho, Department of Industrial and Manufacturing Systems Engineering, Kansas State University, Manhattan, KS 66505

Mahendra S. Hundal, Department of Mechanical Engineering, University of Vermont, 201 Votey Building, Burlington, VT 05404-0156

Eric W. Johnson, Electrical and Computer Engineering, Valparaiso University, Valparaiso, IN 46383

Khoo Li-Pheng, School of Mechanical and Production Engineering, Nanyang Technological University, Nanyang Avenue, Singapore 2263

Young-Jou Lai, Strategic Planning, Phillips Petroleum Company, 15B4 Phillips Building, Bartlesville, OK 74004

Leong Kah Fai, Centre for Engineerng and Technology Management, Nanyang Technological University, MPE, Nanyang Avenue, Singapore 639798

Todd Maher, Kellstadt Graduate School of Business, DePaul University, 1 East Jackson Boulevard, Chicago, IL 60604

Michael M. Marefat, Department of Electrical and Computer Engineering, University of Arizona, ECE Building, Room 230, Tucson, AZ 85721

Ioannis Minis, Department of Mechanical Engineering, University of Maryland, College Park, MD 20742

Dana S. Nau, Department of Computer Science, University of Maryland, College Park, MD 20742

Madeleine Pullman, Cox School of Business, Southern Methodist University, Dallas, TX 75275

John E. Renaud, Design Automation Laboratory, Aerospace and Mechanical Engineering, University of Notre Dame, Notre Dame, IN 46556-5637

Robert H. Rucker, Industrial and Management Systems Engineering, Arizona State University, Tempe, AZ 85287

Gerald C. Shumaker, Manufacturing Systems and Integration Division, The Manufacturing Technology Directorate, Air Force Research Laboratory, Wright-Patterson Air Force Base, OH 45433-6533

Stephen Smith, Department of Mathematics and Computer Science, Hood College, 401 Rosemont Avenue, Frederick, MD 21701-8575

Richard E. Thomas, Texas A&M University (Professor Emeritus), P.O. Box 340727, Beavercreek, OH 45434-0727

K. van der Werff, Technical University of Delft, The Netherlands, Department of Mechanical Engineering

Roberto Verganti, Dipartimento di Economia e Produzione, Politecnico di Milano, Piazza L. Da Vinci 32, 20133 Milano, Italy

Rohit Verna, Kellstadt Graduate School of Business, DePaul University, 1 East Jackson Boulevard, Chicago, IL 60604

Ng Kek Wee, Object Oriented, Toa Payoh North, P.O. Box 0778, Singapore 9131

Brett Wujek, Engineous Software, Inc., 1800 Perimeter Park Drive, Suite 275, Morrisville, NC 27560

D. Zhang, Department of Mechanical Engineering, City University of Hong Kong, Tat Chee Avenue, Kowloon, Hong Kong

Zhang Jian, Hewlett-Packard Singapore (Pte.) Ltd., 20 Gul Way, Singapore 629196

W. J. Zhang, Department of Manufacturing Engineering & Engineering Management, City University of Hong Kong, 83 Tat Chee Avenue, Kowloon, Hong Kong

SECTION I
PRODUCT DESIGN

1

IDENTIFYING CUSTOMER PREFERENCES IN QUALITY FUNCTION DEPLOYMENT USING GROUP DECISION-MAKING TECHNQIUES

YOUNG-JOU LAI
Phillips Petroleum Company

ENG SHWE SEIN AYE HO and SHING I CHANG
Kansas State University

1.1 INTRODUCTION

Today, many companies are interested in improving their competitive edge in the global marketplace where rapid changes occur due to technological innovations and changing customer demands. These companies realize that in order to bring product innovations and value-added services to the market in a timely fashion, they must know the wants (like-to-have), needs (must-have), and desires (wish-to-have) of their customers and quickly fulfill these wants, needs, and desires as completely as possible. Quality function deployment (QFD) is then an impressive planning methodology to help reach this end.

The QFD idea was first introduced by Y. Akao in 1966, and then used at the Kobe Shipyards of Mitsubishi in 1972. Toyota has been applying this concept since 1977. Basically, QFD translates the wants, needs, and desires of

Integrated Product and Process Development, Edited by John Usher, Utpal Roy, and Hamid Parsaei
ISBN 0-471-15597-7 © 1998 John Wiley & Sons, Inc.

customers into quality characteristics, technical requirements, and engineering characteristics of a product and subsequently into its parts' characteristics, process plans, and production requirements by integrating information from marketing, engineering, research and development (R&D), manufacturing, and management. Each translation of customer "voices" and the subsequent processes uses a chart or matrix to relate the variables associated with QFD phases. Depending on the complexity of the product development and production processes, the translation and deployment processes may require a number of matrices. Details of QFD can be found in Bossert[4] and Day.[6] It is best known by a matrix called the first house, the house of quality, as presented in Figure 1.1. This matrix contains: information on what to do (e.g., what customers want), how to do it (e.g., how customer requirements can technically be achieved), and the relationships between each of these aspects; an evaluation of the technical requirements for the company's and competitors' products; and what the company's target levels should be. Other information, such as a history of customer complaints, the degree of technical difficulty in implementation, and so forth, can also be added. By carrying "how to do" into a house of quality as "what to do," quality functions are deployed.

Recent emphasis on concurrent engineering illustrates the need to integrate all aspects of design, marketing, finance, and production simultaneously into engineering decision-making processes. Most concurrent engineering efforts focus in one way or another on improving communication by collocating and dedicating development teams. This further confirms the use of QFD to facilitate joint work among different parts of the team.

Although QFD has been used by manufacturers of electronics, appliances, clothing, and construction equipment, and by firms such as General Motors, Ford, Mazda, Motorola, Xerox, Kodak, IBM, Procter & Gamble, Hewlett-Packard, AT&T, and so on, [6] it is still not as widely used as initially expected in the United States. Major causes of this problem may be the differences that exist between the main QFD characteristics and American culture, as well as the complexity and time-consuming nature of its processes. These revealed the need for effective group decision-making techniques to achieve correct customer requirements and their rankings in dealing with QFD's team-oriented characteristics while allowing for an individual's willingness to take pride in being unique. Therefore, we believe that issues of successfully applying QFD lie in communication among team members in order to reach consensus, assigning importance levels that reflect each individual member's preferences, and mutual interaction of these two factors.

To resolve the first issue, we propose a modified nominal group technique (NGT) to obtain customer requirements. As we will discuss later, NGT is one of the best techniques to achieve group consensus without creating conflicts or hindering individual thinking among team members. For the second, we extend Hwang and Lin's[13] concepts to calculate important levels of customer requirements by using voting and linear programming techniques. The proposed agreed and individual criteria methods are used to solve situations where an agreed criteria set is reached and where no agreed criteria set can

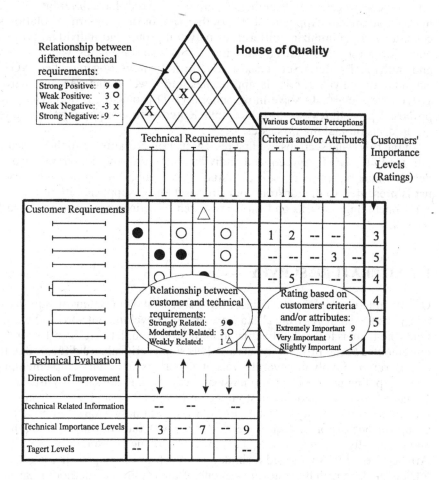

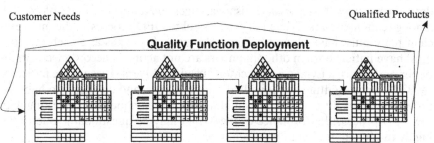

Figure 1.1 House of quality and quality function deployment.

be reached, respectively. Note that none of the researchers in QFD literature attempt to aggregate team members' opinions when they have different criteria in their judgments. Applying both together cannot only confirm a solution's consistency by minimizing differences between group and individual preferences (even for cases with agreed criteria), but also solve cases where agreed and individual criteria sets coexist. By further integrating modified NGT, agreed, and individual criteria approaches into a well-structured interactive system, we are able to solve more general QFD problems. If necessary, the proposed procedures can also be implemented with computers to expedite group decision-making processes.

In the following sections, we first review current literature, and then discuss our modified NGT, agreed and individual criteria, and integrated group decision-making approaches. A case study of developing a "light-saver" product is presented to demonstrate our proposal and its applicability in Section 1.4. Concluding remarks and future research directions are given in the last section.

1.2   LITERATURE SURVEY

Quality function deployment is based on a process involving teams of multidisciplinary representatives from all stages of product development and production. Once the team is formed, QFD first identifies the complete set of customer requirements for the product or service. This step is critical and difficult due to considerations involving several levels of customers and their complementary and competing needs. The team must collectively select the list of customers' demands based on a series of customer contacts and the compilation of supplemental information. Bossert[4] mentioned using an affinity diagram approach to group various customer requirements starting with the brainstorming method and eventually leading to a tree diagram with screened customer preferences. Armacost et al.[1] also applied a similar approach to develop their customers' attributes. Although brainstorming is one of the traditional methods of group decision making, it has several disadvantages including the issue of conflict among team members. To be fully successful in QFD processes, the most efficient team must have a good balance of personality types and each member must have patience when other members are not acting in accordance with his or her preferences. Lyman and Richter[21] used the Myers–Briggs type indicator (MBTI)—representing an individual's nature, likes, and dislikes—to help improve communication and understanding among team members working with the QFD process. Performance of the MBTI may be questionable due to the highly random nature of human behavior.

Once customer requirements are obtained, their importance has to be determined to continue the QFD process. This stage is essential because consequent technical level determinations are based on the importance of customer requirements. Team members usually have difficulty assigning measures of priority to a list of customer preferences. Given a list of preferences, they often easily lose objectivity and thus the proposed relative importance is biased.[6] Even with a smaller matrix, debates over the importance ratings of customer requirements and the strength of the relationship between customers and technical requirements are at times difficult to resolve without systematic rating and ranking procedures.[12]

Armacost et al.[1] applied an analytical hierarchical process (AHP) approach to generate and prioritize customer requirements in a case study of industrialized housing by a judgmental group of seven experts. The analytical hierarchical process was found to be successful when applied to obtain customers' priorities in their study; however, many controversies exist on the issue of rank-reversal problems while adding or deleting an alternative. Even though these problems were overcome by Lai[16] and Saaty[27] in some specific situations, Carlsson and Walden[5] and Perez[24] pointed out other practical weaknesses of AHP, such as a heavy reliance on an individual's expertise and intuitive judgment, elimination of interdependence in an ongoing decision-making process during an AHP session, and so forth.

Instead of using simple normalization, Wasserman[30] proposed a decision model of prioritizing technical requirements with consideration of their dependencies in a QFD planning process. Moreover, a linear programming model was developed to select required technical levels that optimize customer satisfaction over cost constraints corresponding to technical requirement levels. He also mentioned using AHP to generate the importance of customers' requirements. However, he did not explain how the correlation between a pair of technical requirements and the relationship between customer and technical requirements can be methodically obtained.

In the QFD planning process, the house of quality is flexible enough to be conveyed through different environments in a manufacturing organization. Belhe and Kusiak,[2] Glushkovsky et al.,[9] and Locascio and Thurston[19] have applied the house of quality in various application areas. Belhe and Kusiak[2] exploited the house of quality in an analysis of product development processes in which design process attributes are treated (e.g., average lateness of design activities, risk of violating due dates) as customer requirements and design process variables (e.g., average resource levels, number of design projects undertaken) as counterpart characteristics. Corresponding relationships between design process attributes and process variables were captured qualitatively, and finally functional relationships were established in linear regression forms by quantifying the qualitative relationship to determine design attribute values, based on design process variables. Locascio and Thurston[19] introduced a multiattribute design optimization approach with QFD processes. Customer attributes were first transformed into design attributes, and then a multiattribute utility (objective) function was constructed. The objective function was optimized under the constraints of relationships between design attributes and engineering characteristics, and their feasible bounds. Neither approach suggested how to systematically obtain qualitative relationships between attributes and design variables and thus the quantification of these relationships seems unjustified.

Although QFD has gained popularity and related works have been developed throughout the world, it has not been as widely used as initially expected. One of the main problems may be the lack of efficient and effective tools or models. Even though a number of research methods have been developed, systematic methods of finding the most fundamental and important items of QFD such as rating of customer requirements and relationships between customers and technical requirements are still deficient. In this study, the nominal group technique (NGT)—developed by Delbecq and Van de Ven[7]—is modified to obtain customer requirements, and then group decision-making techniques, based on agreed and individual criteria, are proposed to attain importance ratings of customer requirements. This approach provides a solid basis for obtaining the technical requirements, as well as for modeling the relations between customer and technical requirements.

1.3 GROUP DECISION MAKING FOR QFD

In order to implement QFD, at least one team has to be formed to collect data on customers' wants, via interviews, mail questionnaires, product clinics, and other methods. It may also need to determine target markets, demographics, and geographical distribution, and survey people external to the organization, with or without samples of the current product. However, in practice, companies are usually unable to undertake a complete customer study due to insufficient resources. In such cases, team members can research only product areas where customer requirements may not be fulfilled.[20]

Depending on different customer groups, geographical situations, and individual team members' focus, the information collected in specified product areas is different and diverse. To determine satisfactory group decisions from such information is not a trivial task, but a problem consisting of (1) aggregation of individual, conflicting preferences and (2) communication between group members during the QFD process. Group decision-making techniques should then be utilized. We propose a modified NGT to collectively and concurrently obtain important customer requirements in this study. By extending Hwang and Lin's[13] concepts, we then develop systematic group decision-making techniques by integrating agreed and individual criteria to attain importance levels of customer requirements. This approach is presented in the following section.

1.3.1 Generating Customer Requirements by NGT Process

In order to obtain a complete, comprehensive description of the product to be developed, participants must reach a common understanding about which directions to take for and what customer requirements are to be fulfilled.

These needs can be gained through the identification and resolution of conflicts arising from the cross-functional team's different perspectives. Unfortunately, team processes too often leave members exhausted and discouraged because of seemingly endless meanderings into unfruitful byways from two extremes—a search focusing on initial responses rather than a continuing creative flow and interaction among members in overly long debates. Researchers have been developing and applying a number of different group decision-making processes to help conduct various team processes.

These group decision-making techniques can basically be categorized into unstructured and structured group processes. The former involves brainstorming techniques and the latter consists of NGT and the Delphi technique. The nominal group and Delphi techniques are special techniques useful for situations where group members must elicit and combine the judgments of individuals in a particular procedure at group meetings. The major differences between the NGT and the Delphi techniques are: (1) NGT groups meet face to face around a table, whereas Delphi respondents are physically distant and never meet face to face; and (2) all communications between respondents in the Delphi process occur via written questionnaires and feedback reports from the monitoring team, whereas communications occur directly between members in NGT. Furthermore, the NGT process brings individuals together in name only, but does not allow individuals to communicate verbally. Recent research reveals that the nominal group process is significantly superior to group interaction in generating information relevant to a problem based on performance measures of uniqueness and quality of information produced. Thus, NGT is used here as a participative data collection method, a consensus-forming device, and an important component of systematic, participative, and team-oriented programs, which are the major characteristics required by QFD practitioners as mentioned by Hales.[10] It is also flexible in that it can be modified to fit various manufacturing organizations.

The nominal group technique was derived from social-psychological, management-science, and social-work-related studies and developed by Delbecq and Van de Ven[7] in 1968. It integrates elements of brainwriting, brainstorming, and voting techniques to balance participation among team members in group decision-making processes. The team size for NGT is between five to ten members. Despite the fact that NGT can only deal with simple problems one question at a time, it has been widely utilized in complex situations in various industries in the last several years.[7] Beruvides,[3] Kolano,[15] and Sink[29] have proven its effectiveness and efficiency in idea generation and group consensus-building processes. The modified NGT proposed here basically follows the procedure of the original NGT. Modifications to be made are in terms of setting the goal of obtaining a set of customer requirements, appropriate techniques to elicit related requirements, and suitable voting methods in each corresponding NGT step.

The proposed modified NGT is described in the following step by step procedures to generate a total of m customer requirements (Q_1, Q_2, . . Q_q, . . . , Q_m) (see Fig. 1.2 also):

Step 1: Generation of Customer Requirements. Each team member independently generates a list of customer requirements by writing silently. This step is very similar to nominal group brainstorming of the unstructured group process where ideas are individually generated. Nominal group brainstorming outperforms interactive group brainstorming in most social and psychological literature and recent research in engineering design.

Step 2: Reporting of Generated Lists. Use round-robin recording of ideas (RRRI) to record ideas obtained from the first step. This process is conducted by a facilitator or leader going around the table and asks for one of the listed customer requirements from each member one at a time and writes each down on a blackboard or a flip chart visible to the whole group.

This original reporting step can be modified if the issue discussed is sensitive; if there are many participants or requirements, the facilitator collects the ideas and records them individually and anonymously for the team.

Step 3: Clarification and Discussion of Customer Requirements. Each requirement is serially discussed by all members. This stage clarifies and elaborates the main ideas of all team members. Combine similar customer preferences generate the final list of customer requirements. An affinity diagram, tree diagram,[4] or hierarchical structure[1] of essential customer requirements can be generated if necessary.

Step 4: Ranking of Requirements. The goal of this step is to reach a consensus on the customer requirements under consideration. This final list of customer requirements is screened by a voting process, which aggregates group judgments into a compromise. A number of voting methods such as simple majority, Borda count, Condorcet method, approval voting method, and so on, are to be looked into in order to select the suitable method(s) for a particular team. However, it should be noted that different voting methods may generate different solutions (see Hwang and Lin[13] and Saari[26] for details).

Step 5: The team may further discuss results from the preliminary voting process and may conduct voting again until a final satisfactory compromise of customer requirements is obtained.

It is crucial to limit the amount of time in every discussion during the process, but the limitation should not make discussion unfruitful and voters feel rushed. The trade-off between time and precision is usually a good open question for decision makers and engineers.

Although eliciting customer requirements for developing new products is focused so far, QFD can actually be used in various kinds of product and process developments. For example, in Belhe and Kusiak's[2] house of quality, design process attributes (e.g., average lateness of design activities and risk of violating due dates) are equivalent to the customer requirements mentioned previously. In such cases, design project engineers and other team members should generate the essential design process attributes in their environment by using appropriate techniques such as the proposed NGT. Also, when a manufacturing firm wants to repair or buy new facilities for internal use, houses of quality can be used. In this case, the customers become the internal customers, the company's engineers, and decision makers. The generation of customer requirements should not be limited to only the information collected from customers. At times, good ideas and product features, which delight customers, may come from design team members and these need to be considered as unspoken customer wants. During an NGT session, it is important for the facilitator to frequently trigger team members to take into account unspoken customer wants in addition to what they have heard from customers before.

1.3.2 Selection of Importance Levels of Customer Requirements

After a list of satisfactory customer requirements for QFD is obtained, the next step is to determine how important each customer requirement is and thereby determine engineering requirement levels. Each team member may have different criteria for a customer requirement, and, based on these, the proposed approach is used to find their importance levels. Among many ranking and rating procedures in the literature, we use ordinal scaling to represent levels of importance such as 1 for "not important," 3 for "imporytant," 5 for "very important," 7 for "strongly important," 9 for "extremely important," and maybe even numbers (2, 4, 6, 8) for importance levels in between. This ordinal scaling system is chosen for its simplicity and applicability in describing the relative importance of customer requirements in practice. In determining the priorities of customer requirements by AHP, individual members must have a set of criteria/attributes that belong to the group and preferences are given based on the same criteria/attributes. In practice, individual members usually use different criteria/attributes in their judgments and it is difficult to come up with a set of criteria that are agreed on by all individuals. To cope with such situations, we present a technique that allows an individual to keep his or her own set of criteria during the group decision-making process.

In the following section, we extend Hwang and Lin's[13] concepts and develop an agreed criteria approach to deal with a situation where a set of criteria is generated by group consensus, and an individual criteria approach when an individual has his or her own set of criteria to establish importance levels of customer requirements. We then integrate both approaches into solving general QFD problems.

1.3.2.1 Agreed Criteria Approach This approach establishes group decision making in three phases: (1) finding a set of criteria based on which importance levels for a customer requirement are to be evaluated, (2) voting on the criteria on a pairwise comparison basis and obtaining the preference (weight) of each criterion, and (3) aggregating judgments of each team member's preference of optional customer importance levels. Thus, a group of individuals, each with his or her own set of values and preferences, must identify a set of criteria and deduce the alternative importance levels of customer requirements and then search for the level that is the best for the entire group. Each stage involves corresponding critical issues: The central issue of the first stage is the difficulty of defining criteria for combining the expressed preferences of individuals to determine the optimal team choice; the latter concerns the problem of combining conflicting individual preferences into a measure of overall group preference. Obtaining consensus in building an agreed set of criteria may be tedious, but once achieved, it provides for team solidarity and a common ground in evaluating various importance levels.

The following procedure will detail how to select a customer's importance level for a given customer requirement. By repeatedly applying this same procedure, one can evaluate all customer requirements.

Phase 1. Find a set of criteria, $C_1, C_2, \ldots, C_p$, to use for judging a customer requirement agreed on by all team members. Note that the criteria may be different in nature or number for different customer requirements while being selected with a view of the company's competitiveness, sales points, and other company objectives. Each criterion can be classified as either quantitative or qualitative. The set of criteria may simply be constructed by all team members who collaboratively include all criteria or attributes anyone suggests, including those they think their opponents may want them to include. In this case, a long list of criteria will probably be generated, but this can be alleviated by using NGT, as discussed in Section 1.3.1, except that we now try to define a set of criteria, instead of customer requirements, generally agreed on for each customer requirement by some or all members.

Phase 2. Although all team members agree to have a set of criteria, individuals may differ in their willingness to trade one criterion off against another. This is, the relative importance levels between two specific criteria are different among individuals. Here, we use an eigenvector function and least squares approach to obtain the relative importance or weight w_i for each criterion C_i, $i = 1, 2, \ldots, p$.

In this phase, each member first gives a nonranked vote for preferred criteria on each ordered pair (C_i, C_j), $\forall i, j$, and $i < j$, and then we have a total number n_{ij} of votes representing that C_i is considered as more important than C_j. The following pairwise comparison matrix **V** summarizes this information:

$$V = \begin{array}{c} \\ C_1 \\ C_2 \\ \vdots \\ C_p \end{array} \begin{array}{cccc} C_1 & C_2 & \cdots & C_p \end{array} \\ \begin{bmatrix} 1 & \frac{n_{12}}{n_{21}} & \cdots & \frac{n_{1p}}{n_{p1}} \\ \frac{n_{21}}{n_{12}} & 1 & \cdots & \frac{n_{2p}}{n_{p2}} \\ \vdots & & & \\ \frac{n_{p1}}{n_{1p}} & \frac{n_{p2}}{n_{2p}} & \cdots & 1 \end{bmatrix} = \begin{array}{cccc} C_1 & C_2 & \cdots & C_p \end{array} \\ \begin{bmatrix} \pi_{11} & \pi_{12} & \cdots & \pi_{1p} \\ \pi_{21} & 1 & \cdots & \pi_{1p} \\ \vdots & & & \\ \pi_{p1} & \pi_{p2} & \cdots & \pi_{pp} \end{bmatrix} \approx \begin{array}{cccc} C_1 & C_2 & \cdots & C_p \end{array} \\ \begin{bmatrix} 1 & \frac{w_1}{w_2} & \cdots & \frac{w_1}{w_p} \\ \frac{w_2}{w_1} & 1 & \cdots & \frac{w_2}{w_p} \\ \vdots & & & \\ \frac{w_p}{w_1} & \frac{w_p}{w_2} & \cdots & 1 \end{bmatrix}$$

where $\pi_{ij} = n_{ij} / n_{ji}$ can be interpreted as the relative strength of preference or importance level between C_i and C_j. Furthermore, π_{ij} represents the binary relationship between C_i and C_j. This binary relationship has certain properties such as homogeneity, anonymity, neutrality, monotonicity, and Pareto optimality (see Hwang and Lin[13] for details), which are essential for evaluating group decision rules.

According to the fact that any square matrix V has at least an eigenvalue λ in $Vx = \lambda x$ where x is called the corresponding eigen vector, we can first find the largest eigenvalue λ by solving $\det(V - \lambda I) = 0$, and then obtain the relative importance vector $w = (w_1, w_2, \ldots, w_p)$ by solving $Vw = \lambda w$, with an additional equation $\Sigma w_i = 1$ for normalization.

Ideally, V has three properties: (1) diagonal terms are all 1s; (2) corresponding terms above and below the diagonal of V are reciprocals, $\pi_{ij} = 1/\pi_{ji}$; and (3) for every i, j, and k, $\pi_{i,k}$, $\pi_{k,j}$ and $\pi_{i,j}$ must satisfy $\pi_{i,j} = \pi_{i,k} \cdot \pi_{k,j}$. However, only when all decision makers are very consistent, will V satisfy all three properties and the calculated w_i's will correctly reflect V. That is, $n_{ij}/n_{ji} = w_i/w_j$. In cases of inconsistencies, it is best to obtain a set of weights so that w_i/w_j is achieved as close as possible to π_{ij} for all i and j. This can be done by simply applying the least square method to minimize inconsistencies between w_i/w_j and π_{ij} and by solving the following constrained least squares problem (see Marshall and Oliver[22]):

$$\text{Minimize} \quad \sum_{i=1}^{n}\sum_{j>i}\left(\pi_{ij} - \frac{w_i}{w_j}\right)^2$$

$$\text{Subject to} \quad \sum_{i=1}^{n} w_i = 1$$

Note that by comparing the results of the constrained least squares method with those obtained from the eigenvector method, a preferred set can be identified.

Phase 3. Calculate the required importance levels of customer requirements by agreed criteria. In this stage, each team member first rates the string of importance levels $S = \{1, 2, 3, 4, 5, 6, 7, 8, 9\}$ for each customer requirement—1 being the least important and 9 being the most important—in his or her preferred order. Here, we use all nine levels for every criterion and every customer requirement. In practice, one may only choose some of these elements as candidates to distinguish importance levels for different criteria and/ or different customer requirements. We then aggregate all members' ratings r_{kj}^{s}, where $\xi_{kj} \in \{1, 2, \ .\ .\ ,9\}$ is the evaluation of the sth importance levels of the criterion j by team member k, into a combined rank r_{sj}. Finally, we can obtain a final rating for customer requirement "Q_q" by aggregating through all criteria, and the customer's importance level corresponding to the best ranking is selected.

The procedure discussed previously is completed by the following three steps.

STEP 1. Based on each criterion, each team member starts evaluating the importance levels of each customer requirement by allocating scores from the given set of ordinal numbers $\{1, 2,.\ .\ .,9\}$. We assign the scores in descending order–"1" to the importance level "s" which is most preferred, "2" to the importance level second most preferred, and so on. This ordinal scoring system is used for its simplicity and applicability when applied with voting techniques. For example, we can use a simple card to design this system without any difficulty (see Fig. 1.3). For a given criterion and all team members "n," the rating matrix is obtained as $[\mathbf{E}]_j = [r_{sj}^{s}]_{9 \times n}$, where each r_{sj}^{s} can be viewed as a preferential voting status for the importance level of the sth customer requirement by the kth team member, based on criterion j.

STEP 2. Use the Borda count to consolidate all members' ratings. Although it is the earliest preferential voting technique, recent research reveals that the Borda count is a very unique method which minimizes a wide variety of paradoxes, the likelihood of a paradox, the likelihood that a small group can successfully manipulate the outcome, the possibility of voters' errors adversely changing the outcome.[26]

We first transfer the rating of r_{sj}^{s} into Borda scores B_{sj}^{s} by $B_{sj}^{s} = 9 - r_{sj}^{s}$. Then the Borda scores of all team members are summed, $\Sigma_k B_{sj}^{s}$, for each customer's importance level. The sth customer's importance level that achieves the highest sum will be ranked as 1, the second highest will be ranked as 2, and so forth. These orderings are then summarized in the collective ordinal rank matrix $\mathbf{R} = [r_{sj}]_{9 \times p}$, where $r_{sj} \in \{1, 2, \ .\ .\ .\ , 9\}$ is the ordering of the sth customer's importance level, based on criterion j. $r_{sj} = \alpha$ if the sum $\Sigma_k B_{sj}^{s}$, $k = 1,.\ .\ .,n$, is the αth largest value among all sth importance levels for criterion j, where $\alpha = 1, 2,.\ .\ .\ ,9$.

STEP 3. Merge all criteria to obtain the final ranking of the alternative importance levels of each customer requirement such that the overall weights are maximized. This aggregation is first done by building binary (0–1) agreement matrices $[\mathbf{A}]_j = [a_{srj}]_{9\times9}$, where $a_{srj} = 1$ if the sth importance level is in position "r" for the criterion j (positions are r_{sj}'s obtained from Step 2); otherwise, $a_{srj} = 0$. Next, we combine weights (obtained from Phase 2) of criteria in each designated position of $[\mathbf{A}]_j$, $\forall j$, and obtain the collective weighted-agreement matrix $\mathbf{O} = [a_{sr}]_{9\times9}$, where $a_{sr} = \Sigma_{jar_j}w_j$, $j = 1, \ldots, p$. Finally, the sth importance level should be assigned to the rank r such that the sum of the related weights is a maximum. This is done by solving the following linear assignment problem:

$$\text{Minimize} \quad \sum_{s=1}^{9} \sum_{r=1}^{9} a_{sr} \cdot x_{sr}$$

$$\text{Subject to} \quad \sum_{r=1}^{9} x_{sr} = 1 \quad s = 1, \ldots, 9$$

$$\sum_{s=1}^{9} x_{sr} = 1 \quad r = 1, \ldots, 9$$

where $x_{sr} = 1$ if the rth position has been assigned to the sth importance level; otherwise, $x_{sr} = 0$. After solving the preceding equations, the final result is an assignment vector. The customer's importance level assigned to the first rank (rank 1) is selected to further proceed in the QFD process.

1.3.2.2 Individual Criteria Approach

As mentioned in Section 1.3.2.1, the major issue in the agreed criteria approach is to have a set of agreed criteria. In the QFD environment, members from engineering departments are usually most concerned with the quality of finished products, whereas marketing representatives are often more interested in maintaining or increasing market share, and accounting personnel regularly concentrate on quarterly profits. Therefore, it is extremely time consuming, if possible, to reach a consensus in generating agreed criteria. In such cases, each individual must elicit a small set of the most important criteria based on his or her own preferences. Each member first determines individual relative importance weights for each criterion and consolidates all criteria to obtain individual preference orderings of importance levels for each customer requirement. Individual preference orderings are then merged into a final selection of importance levels for each customer requirement. This individual criteria approach is similar to the agreed criteria approach, except that calculation steps for each customer requirement are different as follows.

Phase 1. Like the agreed criteria approach, each team member k derives a criteria set $\{C_1^k, \ldots, C_{pk}^k\}$ for the customer requirement under consideration. Note that C_{pk}^k accounts for different criteria sets for each individual. To complete this phase, for example, individuals can list all potential criteria for a given customer requirement, arrange them into a tree diagram or hierarchical relational structure, and then extract the significant criteria. It is highly recommended that, to obtain a consistent and comparable rating, all members have the same set S of importance levels for every criterion and customer requirement.

Phase 2 Weight vectors for each set of criteria are calculated individually in this phase. Considering a particular customer requirement, an individual provides his or her own preference rating of ordinal scales for each pair of criteria. This preference information of all criteria is summarized in the following pairwise comparison matrix $[\mathbf{V}]^k$ for member k:

$$
[V]^k = \begin{array}{c} \\ C_1^k \\ C_2^k \\ \vdots \\ C_{p_k}^k \end{array}
\begin{array}{cccc}
C_1^k \quad C_2^k & & \cdots & C_{p_k}^k \\
\left[\begin{array}{cccc}
1 & T_{12}^k & \cdots & T_{1p}^k \\
1/T_{12}^k & 1 & \cdots & T_{2p}^k \\
\vdots & & & \\
1/T_{1p}^k & 1/T_{2p}^k & \cdots & 1
\end{array}\right]
\end{array}
$$

where C_i^k is the ith criterion of the kth individual, $i = 1, \ldots, p_k$, and $T_{ij}^k \in \{1, 2, 3, \ldots, 9\}$ indicates the relative strength of criterion i being preferred to criteron j by the kth individual. Here, the meanings of the priority scales $1, 2, \ldots, 9$ could be clarified by Saaty's[27] definitions as: 1 for "equally preferable," 3 for "more preferable," 5 for "strongly preferable," 7 for "very strongly preferable," 9 for "perfectly preferable," and even numbers for "preference intensity in between." Note that individuals may have different sets and different numbers p_k of criteria and can modify the terms as needed.

The relative importance vectors $\mathbf{w}^k = (w_1^k, w_2^k, \ldots, w_p^k)$ are then calculated by using the eigenvector and/or least squares methods as discussed in the previous section.

Phase 3. Determine a compromise importance level for a customer requirement by the following steps.

STEP 1. Each team member k assigns ordinal ratings of $S = \{1, 2, 3, \ldots, 9\}$ as importance levels of each customer requirement in descending order as mentioned in the agreed approach for all his or her criteria. Evaluation matrices of $[E]^k = [r_{sj}^k]_{9 \times p}$, for all k, are then obtained.

STEP 2. Transform individual evaluation $[E]^k$ matrices into binary (0–1) agreement matrices $[A]_j^k = [a_{srj}^k]_{9 \times 9}$, where $a_{srj}^k = 1$ if the sth importance level is in position "r" for criterion j evaluated by the kth member; otherwise, $a_{srj}^k =$

0. Note that each $[A]_j^k$ represents an agreement matrix for criterion j of an individual.

As in Step 3 of the agreed approach, we calculate the individual weighted-preference matrix $[O]^k$ by using the weighted average of agreement indices for all criteria. That is, $[O]^k = [a_{sr}^k]_{9 \times 9}$, where $a_{sr}^k = \Sigma_j a_{srj}^k \cdot w_j^k, j = 1, \ldots, p_k$.

Finally, calculate individual ranking matrices $[R]^k = [r_s^k]_{9 \times 1}$ by solving the following linear assignment problem for the individual k, $k = 1, 2, \ldots, n$:

$$\text{Minimize} \quad \sum_{s=1}^{9} \sum_{r=1}^{9} a_{sr}^k \cdot x_{sr}^k$$

$$\text{Subject to} \quad \sum_{r=1}^{9} x_{sr}^k = 1 \quad s = 1, \ldots, 9$$

$$\sum_{s=1}^{9} x_{sr}^k = 1 \quad r = 1, \ldots, 9$$

where r_s^k represents the sth importance level in the rth position of the kth individual.

STEP 3. Combine all individual ranking matrices $[R]^k$ into a collective ranking $R' = [r_s^k]_{9 \times n}$ and aggregate all members' rankings with the Borda count method as discussed in the previous section. Hence, we calculate $B_s = \Sigma_k B_s^k$, $k = 1, \ldots, n$, where $B_s^k = 9 - r_s^k$ and select the sth importance level with the highest B_s value for the considered customer requirement.

1.3.2.3 Integrating Agreed Criteria and Individual Criteria Approach

So far, we have studied the agreed criteria and individual criteria approaches based on an agreed set or individual sets of criteria. In reality, possible cases exist where criteria sets cannot be found explicitly as ones agreed on by all members and others disagreed on. For example, some team members (or a subgroup of members) select the same set of criteria while others have their own criteria sets. Such situations can arise when each member initially decides to find criteria sets independently and terminates sets of resembling criteria chosen by more than one members who then deduce an agreed criteria set altogether; and where all members start searching for an agreed set of criteria, but finalize with a set that is fully agreed on by some members, whereas others demand their individual criteria sets.

One way to resolve this situation is to treat the agreed criteria sets as individual sets for each person in the subgroup, and then apply only the individual criteria approach as shown in Section 1.3.2.2 to get the importance levels of customer requirements.

Alternatively, a subgroup can use the agreed criteria approach to obtain importance levels while the rest of the members apply the individual criteria approach to get another set of importance levels. If the best ranking of both sets of importance levels is the same, then the corresponding level will be selected. If not, a consensus between the two groups must be reached either by voting on these two sets of importance levels or by conducting another set of group decision-making processes from a certain stage (see Fig. 1.2) with the supportive information of previously obtained criteria and weights until

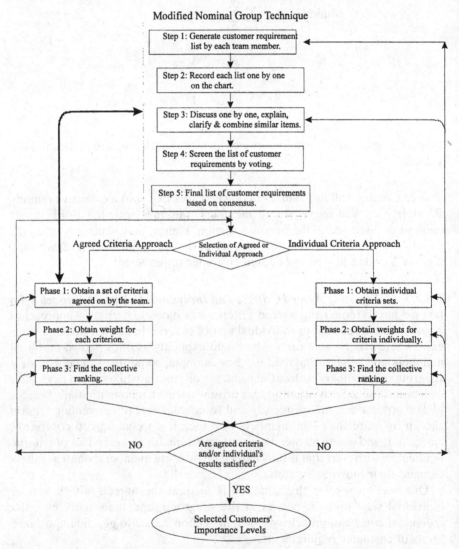

Figure 1.2 Integrated group decision-making process for selecting customers' importance levels in QFD.

a certain level of achievement is reached. Note that the proposed NGT and integrated approach can be treated as complements of each other. An appropriate NGT can be used to reconcile differences between subgroup(s) and individual members, while the integrated agreed and individual approach can also be useful in selecting critical customer requirements.

Sometimes, when a high degree of consistency between all individual results and group consensus are required, both approaches must be repeatedly used simultaneously until the required consistency is reached. Figure 1.2 provides a brief, systematic guideline to perform this process.

1.4   A CASE STUDY

A project of designing and making a LightSaver—a product that will turn lights in a room on when the room is occupied and off when unoccupied—was implemented. The LightSaver accomplishes this task by counting the number of people entering and leaving the room or upon sensing occupants. Five members were selected to serve on the QFD project team. The team had been assigned to look for information one week before the QFD analysis was implemented and was also introduced to and trained with the NGT approach, voting method, and ranking/rating systems. After collecting sufficient information, team members gathered and started to extract customer requirements based on a modified NGT approach. They then used the agreed criteria and individual criteria approaches to assign importance levels for every customer requirement and selection of other importance levels in the house of quality in order to finally decide upon the required design components.

1.4.1   Extracting Customer Requirements

The project leader arranged a meeting with all team members and a written task statement regarding a LightSaver customer requirements list was distributed to each participant along with necessary supplies.

Step 1. Each member was asked to write down phrases that represented the customer requirements of the product considered, based on collected information.

Steps 2 and 3. After about 15 minutes, the leader picked up all lists and collected a total of 26 phrases. He then wrote each member's list separately on a blackboard and serially clarified the listed ideas one by one. Note that the time needed for clarification only was allowed and any judgment of the phrases posted was discouraged. After combining similar items, 11 phrases were selected.

Step 4. To further screen out the most relevant representatives, approval voting was conducted. Each member allocated nonranked votes for as many customer requirements as he or she liked, but could not cast more than one vote for each item. The greater the numbers of votes for an item, the more essential the item is (for details of the approval voting method, see Fishburn and Little[8]).

Step 5. Voting results were posted on each item and phrases with smaller votes were further analyzed for possible elimination, combination, or rephrasing of groups of words. Finally the following eight items were selected:

1. Ability to override
2. Durability
3. Attractiveness
4. Added security feature as an option

5. Ease of installation
6. Low cost
7. Ease of use
8. Energy savings

Once customer requirements were obtained, appropriate importance levels were calculated for all items by using the agreed criteria and individual criteria approaches. Importance levels, $S = \{1, 3, 5, 9\}$, were selected to be assigned to all customer requirements in this problem. The team then tried to obtain agreed criteria sets for all customer requirements consecutively. Individuals were asked to list individual criteria sets immediately if agreed criteria sets were not brought up. Decisions on what technique, agreed, individual, or both, should be applied for which customer requirement were determined. Here, the team had achieved the agreed criteria sets for all customer requirements; however, it also applied the individual criteria approach for checking consistency of the results. In the following section, we will discuss in detail the procedures for using agreed criteria and individual criteria techniques to select customer's importance levels regarding the customer requirement "Attractiveness."

1.4.2 Agreed Criteria Approach

Phase 1. Using NGT, the criteria set of "Attractiveness" was first obtained as: companies (C_1), schools (C_2), and Households (C_3).

Phase 2. To calculate weights, the team first obtained the following pairwise matrix $\mathbf{V}$:

$$
\mathbf{V} = \begin{array}{c} \\ C_1 \\ C_2 \\ C_3 \end{array}
\begin{array}{c} \begin{array}{ccc} C_1 & C_2 & C_3 \end{array} \\
\begin{bmatrix} 1 & \frac{3}{2} & \frac{4}{1} \\ \frac{2}{3} & 1 & \frac{3}{2} \\ \frac{1}{4} & \frac{2}{3} & 1 \end{bmatrix} \end{array}
$$

The team then applied the eigenvector function approach which yielded $w_1 = 0.54$ (companies), $w_2 = 0.30$ (schools), and $w_3 = 0.16$ (households). The **V** matrix calculated from the obtained weights was found to be satisfactorily close to the original **V**, and thus the least squares approach did not need to be considered.

Phase 3. Importance levels for customer requirements were selected as follows.

STEP 1. Each member's rating (recorded in rating cards as shown in Fig. 1.3) of importance levels was collected and aggregated based on each criterion. The following individual ratings were grouped into evaluation matrices $[\mathbf{E}]_j$ for each criterion j:

Member A

$$
\begin{array}{c}
\\ 1 \\ 3 \\ 5 \\ 9
\end{array}
\begin{array}{ccc}
C_1 & C_2 & C_3 \\
\left[\begin{array}{ccc}
4 & 4 & 4 \\
3 & 1 & 2 \\
1 & 2 & 1 \\
2 & 3 & 3
\end{array}\right]
\end{array}
$$

Member B

$$
\begin{array}{c}
\\ 1 \\ 3 \\ 5 \\ 9
\end{array}
\begin{array}{ccc}
C_1 & C_2 & C_3 \\
\left[\begin{array}{ccc}
1 & 2 & 2 \\
2 & 1 & 1 \\
3 & 3 & 3 \\
4 & 4 & 4
\end{array}\right]
\end{array}
$$

Member C

$$
\begin{array}{c}
\\ 1 \\ 3 \\ 5 \\ 9
\end{array}
\begin{array}{ccc}
C_1 & C_2 & C_3 \\
\left[\begin{array}{ccc}
2 & 2 & 3 \\
1 & 1 & 2 \\
3 & 3 & 1 \\
4 & 4 & 4
\end{array}\right]
\end{array}
$$

Member D

$$
\begin{array}{c}
\\ 1 \\ 3 \\ 5 \\ 9
\end{array}
\begin{array}{ccc}
C_1 & C_2 & C_3 \\
\left[\begin{array}{ccc}
4 & 4 & 4 \\
3 & 3 & 3 \\
1 & 1 & 1 \\
2 & 2 & 2
\end{array}\right]
\end{array}
$$

Member E

$$
\begin{array}{c}
\\ 1 \\ 3 \\ 5 \\ 9
\end{array}
\begin{array}{ccc}
C_1 & C_2 & C_3 \\
\left[\begin{array}{ccc}
4 & 4 & 4 \\
1 & 1 & 1 \\
2 & 2 & 2 \\
3 & 3 & 3
\end{array}\right]
\end{array}
$$

$$
E_1 = \begin{array}{c} 1 \\ 3 \\ 5 \\ 9 \end{array}
\left[\begin{array}{ccccc}
4 & 1 & 2 & 4 & 4 \\
3 & 2 & 1 & 3 & 1 \\
1 & 3 & 3 & 1 & 2 \\
2 & 4 & 4 & 2 & 3
\end{array}\right]
\qquad
E_2 = \begin{array}{c} 1 \\ 3 \\ 5 \\ 9 \end{array}
\left[\begin{array}{ccccc}
4 & 2 & 2 & 4 & 4 \\
1 & 1 & 1 & 3 & 1 \\
2 & 3 & 3 & 1 & 2 \\
3 & 4 & 4 & 2 & 3
\end{array}\right]
\qquad
E_3 = \begin{array}{c} 1 \\ 3 \\ 5 \\ 9 \end{array}
\left[\begin{array}{ccccc}
4 & 2 & 3 & 4 & 4 \\
2 & 1 & 2 & 3 & 1 \\
1 & 3 & 1 & 1 & 2 \\
3 & 4 & 4 & 2 & 3
\end{array}\right]
$$

As only four importance levels, $S = \{1, 3, 5, 9\}$, are considered, the rating is thus $r_{sj}^k \in \{1, 2, 3, 4\}$.

STEP 2. The team transformed rating into Borda scores and summed the Borda scores for all members to aggregate the individual rating matrix $[\mathbf{E}]_j$. For C_1, the team calculated:

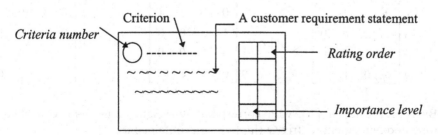

Figure 1.3 Three-by-five-inch card used in Step 1.

$$E_1 = \begin{matrix} 1 \\ 3 \\ 5 \\ 9 \end{matrix} \begin{bmatrix} 4 & 1 & 2 & 4 & 4 \\ 3 & 2 & 1 & 3 & 1 \\ 1 & 3 & 3 & 1 & 2 \\ 2 & 4 & 4 & 2 & 3 \end{bmatrix} \xrightarrow[\text{Count}]{\text{Borda}} \begin{matrix} 1 \\ 3 \\ 5 \\ 9 \end{matrix} \begin{bmatrix} 0 & 3 & 2 & 0 & 0 \\ 1 & 2 & 3 & 1 & 3 \\ 3 & 1 & 1 & 3 & 2 \\ 2 & 0 & 0 & 2 & 1 \end{bmatrix} \longrightarrow \begin{bmatrix} 5 \\ 10 \\ 10 \\ 5 \end{bmatrix}$$

Borda scores: B_{s1}^k 'S $\sum_{k=1}^{5} B_{s1}^k$

Similarly, for C_2 and C_3, the following results were computed:

$$\left[\sum_{k=1}^{5} B_{s2}^k \right]_{4\times1} = \begin{matrix} 1 \\ 3 \\ 5 \\ 9 \end{matrix} \begin{bmatrix} 4 \\ 13 \\ 9 \\ 4 \end{bmatrix} \qquad \left[\sum_{k=1}^{5} B_{s3}^k \right]_{4\times1} = \begin{matrix} 1 \\ 3 \\ 5 \\ 9 \end{matrix} \begin{bmatrix} 3 \\ 11 \\ 12 \\ 4 \end{bmatrix}$$

Then the team achieved the ordinal rank matrix **R** by assigning ranks to importance levels: rank "1" to the level with the highest Borda score, rank "2" to the level with the second highest Borda score, and so on. If there was a tie in scores, shared rankings were assigned with the following results:

$$R = [r_{sj}]_{4\times3} = \begin{matrix} 1 \\ 3 \\ 5 \\ 9 \end{matrix} \begin{array}{ccc} C_1 & C_2 & C_3 \\ \begin{bmatrix} 3.5 & 3.5 & 4 \\ 1.5 & 1 & 2 \\ 1.5 & 2 & 1 \\ 3.5 & 3.5 & 3 \end{bmatrix} \end{array}$$

STEP 3. To merge all criteria for obtaining the ranking of importance levels for each customer requirement, each column in **R** was first converted into binary (0–1) agreement matrices, $[A]_1$ ($[a_{sr1}]_{4\times4}$), $[A]_2$ $[a_{sr2}]_{4\times4}$), and $[A]_3$ ($[a_{sr3}]$ 4×4) as follows:

$$[A]_1 = \begin{matrix} 1 \\ 3 \\ 5 \\ 9 \end{matrix} \begin{array}{cccc} 1^{st} & 2^{nd} & 3^{rd} & 4^{th} \\ \begin{bmatrix} 0 & 0 & 1 & 1 \\ 1 & 1 & 0 & 0 \\ 1 & 1 & 0 & 0 \\ 0 & 0 & 1 & 1 \end{bmatrix} \end{array} \quad [A]_2 = \begin{matrix} 1 \\ 3 \\ 5 \\ 9 \end{matrix} \begin{array}{cccc} 1^{st} & 2^{nd} & 3^{rd} & 4^{th} \\ \begin{bmatrix} 0 & 0 & 1 & 1 \\ 1 & 0 & 0 & 0 \\ 0 & 1 & 0 & 0 \\ 0 & 0 & 1 & 1 \end{bmatrix} \end{array} \quad [A]_3 = \begin{matrix} 1 \\ 3 \\ 5 \\ 9 \end{matrix} \begin{array}{cccc} 1^{st} & 2^{nd} & 3^{rd} & 4^{th} \\ \begin{bmatrix} 0 & 0 & 0 & 1 \\ 0 & 1 & 0 & 0 \\ 1 & 0 & 0 & 0 \\ 0 & 0 & 1 & 0 \end{bmatrix} \end{array}$$

By aggregating all $[A]_j$ with corresponding weight values, the collective weighted-agreement matrix **O** ($[a_{sr}]_{4\times4}$) was obtained as:

$$
O = \begin{array}{c} \\ 1 \\ 3 \\ 5 \\ 9 \end{array}
\begin{array}{cccc}
1^{st} & 2^{nd} & 3^{rd} & 4^{th} \\
\left[\begin{array}{cccc}
0 & 0 & w_1/2+w_2/2 & w_1/2+w_2/2+w_3 \\
w_1/2+w_2 & w_3+w_1/2 & 0 & 0 \\
w_1/2+w_3 & w_2+w_1/2 & 0 & 0 \\
0 & 0 & w_1/2+w_2/2+w_3 & w_1/2+w_2/2
\end{array}\right]
\end{array}
$$

$$
= \begin{array}{c} \\ 1 \\ 3 \\ 5 \\ 9 \end{array}
\begin{array}{cccc}
1^{st} & 2^{nd} & 3^{rd} & 4^{th} \\
\left[\begin{array}{cccc}
0 & 0 & 0.42 & \mathbf{0.58} \\
\mathbf{0.57} & 0.43 & 0 & 0 \\
0.43 & \mathbf{0.57} & 0 & 0 \\
0 & 0 & \mathbf{0.58} & 0.42
\end{array}\right]
\end{array}
$$

Finally, a linear assignment problem was formed with the input data of $[a_{sr}]_{4\times4}$, and its solution was as follows:

$$
\text{Importance Levels}
$$

$$
\begin{array}{cc}
\text{rank 1:} & \\
\text{rank 2:} & \\
\text{rank 3:} & \\
\text{rank 4:} &
\end{array}
\left[\begin{array}{c} 3 \\ 5 \\ 9 \\ 1 \end{array}\right]
$$

Therefore, the importance level of "3" was selected.

As this problem is simple, one could solve it intuitively by observation; that is, simply maximize the total weight by assigning importance level I to position j with the maximal weight.

1.4.3 Individual Criteria Approach

Phase 1. Although all members agreed to have a common criteria set, members were eager to analyze the selection again with the individual criteria approach, based on corresponding areas where information was collected. The individual criteria sets were as follows:

Member A and B	Member C	Member D	Member E
$C_1^A = C_1^B$: Industry	C_1^C: Manufacturing people		C_1^E: Others
$C_2^A = C_2^B$: Schools	C_2^C: Servicing people	C_1^D: Industry C_2^D:School laboratories	C_2^E: Sport centers
$C_3^A = C_3^B$: Households	C_3^C: Housekeepers		

Phase 2. Each member generated a $[\mathbf{V}]^k$ matrix with respect to his or her individual criteria sets as follows:

$$
[V]^A = \begin{array}{c} \\ C_1^A \\ C_2^A \\ C_3^A \end{array}
\begin{array}{c} C_1^A \; C_2^A \; C_3^A \\ \begin{bmatrix} 1 & 1 & 1 \\ 1 & 1 & 1 \\ 1 & 1 & 1 \end{bmatrix} \end{array}
\qquad
[V]^B = \begin{array}{c} \\ C_1^B \\ C_2^B \\ C_3^B \end{array}
\begin{array}{c} C_1^B \; C_2^B \; C_3^B \\ \begin{bmatrix} 1 & 0.5 & 1 \\ 2 & 1 & 1 \\ 1 & 1 & 1 \end{bmatrix} \end{array}
$$

$$
[V]^C = \begin{array}{c} \\ C_1^C \\ C_2^C \\ C_3^C \end{array}
\begin{array}{c} C_1^C \; C_2^C \; C_3^C \\ \begin{bmatrix} 1 & 2 & 1 \\ 0.5 & 1 & 1 \\ 1 & 1 & 1 \end{bmatrix} \end{array}
\quad
[V]^D = \begin{array}{c} \\ C_1^D \\ C_2^D \end{array}
\begin{array}{c} C_1^D \; C_2^D \\ \begin{bmatrix} 1 & 1 \\ 1 & 1 \end{bmatrix} \end{array}
\quad
[V]^E = \begin{array}{c} \\ C_1^E \\ C_2^E \end{array}
\begin{array}{c} C_1^E \; C_2^E \\ \begin{bmatrix} 1 & 0.5 \\ 2 & 1 \end{bmatrix} \end{array}
$$

As an example, $T_{12} = 0.5$ in $[\mathbf{V}]^B$ indicates that member B preferred C_2 to C_1. The weight vectors were then solved by the eigenvector function and are shown in Table 1.1.

Note that the individual $[\mathbf{V}]^k$ matrices computed from the obtained weights were found to be satisfactorily consistent with the initial $[\mathbf{V}]^k$ matrices.

Phase 3. Importance levels for each criterion were selected as follows.

STEP 1. Each member obtained the following initial evaluation matrices $[\mathbf{E}]^k = [r_{sj}^k]_{4 \times 3}$ individually:

$$
[E]^A = \begin{array}{c} 1 \\ 3 \\ 5 \\ 9 \end{array}
\begin{array}{c} C_1^A \; C_2^A \; C_3^A \\ \begin{bmatrix} 4 & 4 & 4 \\ 1 & 3 & 1 \\ 2 & 1 & 2 \\ 3 & 2 & 3 \end{bmatrix} \end{array}
\qquad
[E]^B = \begin{array}{c} C_1^B \; C_2^B \; C_3^B \\ \begin{bmatrix} 4 & 4 & 4 \\ 2 & 2 & 1 \\ 1 & 1 & 2 \\ 3 & 3 & 3 \end{bmatrix} \end{array}
$$

$$
[E]^C = \begin{array}{c} C_1^C \; C_2^C \; C_3^C \\ \begin{bmatrix} 4 & 4 & 3 \\ 2 & 1 & 1 \\ 1 & 2 & 4 \\ 3 & 3 & 2 \end{bmatrix} \end{array}
\quad
[E]^D = \begin{array}{c} C_1^D \; C_2^D \\ \begin{bmatrix} 3 & 3 \\ 2 & 4 \\ 1 & 1 \\ 4 & 2 \end{bmatrix} \end{array}
\quad
[E]^E = \begin{array}{c} C_1^E \; C_2^E \\ \begin{bmatrix} 3 & 2 \\ 2 & 3 \\ 1 & 1 \\ 4 & 4 \end{bmatrix} \end{array}
$$

STEP 2. Each member then transformed the $[\mathbf{E}]^k$ matrices into binary (0–1) agreement matrices $[\mathbf{A}]_j^k$ and obtained individual weighted-preference matrices $[\mathbf{O}]^k$. For simplicity, only member A's agreement and weighted-preference matrices are listed here:

TABLE 1.1 Criteria Sets and Weights Used by Experts

Member	C_1^A	C_2^A	C_3^A	C_1^B	C_2^B	C_3^B	C_1^C	C_2^C	C_3^C	C_1^D	C_2^D	C_1^E	C_2^E
A	0.33	0.33	0.33										
B				0.26	0.26	0.26							
C							0.41	0.26	0.33				
D										0.5	0.5		
E												0.33	0.67

$$[A]_1^A = \begin{array}{c} 1 \\ 3 \\ 5 \\ 9 \end{array} \begin{bmatrix} 0 & 0 & 0 & 1 \\ 1 & 0 & 0 & 0 \\ 0 & 1 & 0 & 0 \\ 0 & 0 & 1 & 0 \end{bmatrix} \quad [A]_2^A = \begin{array}{c} 1 \\ 3 \\ 5 \\ 9 \end{array} \begin{bmatrix} 0 & 0 & 0 & 1 \\ 0 & 0 & 1 & 0 \\ 1 & 0 & 0 & 0 \\ 0 & 1 & 0 & 0 \end{bmatrix} \quad [A]_3^A = \begin{array}{c} 1 \\ 3 \\ 5 \\ 9 \end{array} \begin{bmatrix} 0 & 0 & 0 & 1 \\ 1 & 0 & 0 & 0 \\ 0 & 1 & 0 & 0 \\ 0 & 0 & 1 & 0 \end{bmatrix}$$

$$[A]^A = \begin{bmatrix} 0 & 0 & 0 & 1 \\ 1 & 0 & 1 & 0 \\ 1 & 1 & 0 & 0 \\ 0 & 1 & 1 & 0 \end{bmatrix} \quad [O]^A = \begin{bmatrix} 0 & 0 & 0 & w_1+w_2+w_3 \\ w_1+w_3 & 0 & w_2 & 0 \\ w_2 & w_1+w_3 & 0 & 0 \\ 0 & w_2 & w_1+w_3 & 0 \end{bmatrix}$$

where w_1, w_2, and w_3 are given in Table 1.1. Then matrix $[O]^A$ was aggregated to obtain individual ranking matrices $[R]^A$ by solving the corresponding linear assignment problem. Similarly, ranking matrices for the rest of the members were computed as

$$[R]^A = \begin{array}{c} 1^{st} \\ 2^{nd} \\ 3^{rd} \\ 4^{th} \end{array} \begin{bmatrix} 3 \\ 5 \\ 9 \\ 1 \end{bmatrix} \quad [R]^B = \begin{bmatrix} 5 \\ 3 \\ 9 \\ 1 \end{bmatrix} \begin{array}{c} \\ (9) \\ (3) \\ \end{array} \quad [R]^C = \begin{bmatrix} 3 \\ 5 \\ 9 \\ 1 \end{bmatrix}$$

$$[R]^D = \begin{bmatrix} 5 \\ 9 \\ 1 \\ 3 \end{bmatrix} \quad [R]^E = \begin{bmatrix} 5 \\ 3 \\ 1 \\ 9 \end{bmatrix} \begin{array}{c} (1) \\ \\ (3) \\ \end{array}$$

where the importance levels in the parentheses are tied.

STEP 3. Individual preference orderings were finally consolidated by using the Borda count:

$$R = \begin{array}{c} 1 \\ 3 \\ 5 \\ 9 \end{array} \begin{bmatrix} 4 & 4 & 4 & 3 & 3 \\ 1 & 2 & 1 & 4 & 2 \\ 2 & 1 & 2 & 1 & 1 \\ 3 & 3 & 3 & 2 & 4 \end{bmatrix} \xrightarrow[\text{Count}]{\text{Borda}} \begin{array}{c} 1 \\ 3 \\ 5 \\ 9 \end{array} \begin{bmatrix} 0 & 0 & 0 & 1 & 1 \\ 3 & 2 & 3 & 0 & 2 \\ 2 & 3 & 2 & 3 & 3 \\ 1 & 1 & 1 & 2 & 0 \end{bmatrix} \longrightarrow \begin{bmatrix} 2 \\ 10 \\ 13 \\ 3 \end{bmatrix} \longrightarrow \begin{array}{c} 1^{st} \\ 2^{nd} \\ 3^{rd} \\ 4^{th} \end{array} \begin{bmatrix} 5 \\ 3 \\ 9 \\ 1 \end{bmatrix}$$

It was found that the result of the agreed criteria approach was "3," whereas the individual criteria approach provided "5." Because three members had tied rankings, they were asked to reanalyze their evaluations for confirmation. After some changes on their initial ratings, the importance level was then obtained as "3," which is the same as that obtained from the agreed criteria approach. "3" is then a satisfactory selection for the customer's importance ranking of the customer requirement "Attractiveness." The importance rankings of other customer requirements were calculated in the same way.

1.5   CONCLUDING REMARKS AND FUTURE STUDIES

Quality function deployment has been popular in product and process development throughout the world in recent years; however, it is still not as widely used as initially expected in the United States. Major causes of this problem may be the differences that exist between the main QFD characteristics and American culture, as well as the complexity and time-consuming nature of its processes. These reveal the need for effective group decision-making techniques to achieve correct customer requirements and their rankings in dealing with QFD's team-oriented characteristics while allowing for an individual's willingness to take pride in being unique. Therefore, we believe that issues of successfully applying QFD lie in communication among team members in order to reach consensus, assigning importance levels that reflect each individual member's preferences, and mutual interaction of the these two factors. We first modify the NGT to build a structured communication system to obtain customer requirements. Unlike brainstorming, NGT possesses advantages of eliminating interpersonal conflicts and/or domination of one or more group members because of its strict allowance of verbal communications and its processes of discussion and disagreement without argumentation. It attempts to increase the efficiency of conventional brainstorming by reducing production blocking, evaluation apprehension, and social loafing.

We then develop group decision-making techniques to obtain the importance levels of customer requirements. The agreed criteria and individual criteria methods combine voting and linear programming techniques to aggregate individual preferences into group consensus in situations where an agreed criteria set is reached and where no agreed criteria set can be reached, respectively. Applying both together cannot only confirm a solution's consistency by interactively minimizing differences between group and individual preferences, but also solves cases where agreed criteria and individual criteria sets coexist. By further integrating modified NGT and the agreed and individual criteria approaches into a well-structured interactive system, we can solve various mixed QFD problems and obtain consistent solutions. The computational simplicity of the proposed methodology is also shown in our student project of developing a light-saver product.

Although we propose the integrated group decision-making approach for determining the importance levels of customers' requirements in detail here, it can be applied in customers' competitive evaluations and in determining the strength of relationships between customers and technical requirements. In determining the relationship for a given pair of customer and technical requirements, alternatives such as "strongly related," "moderately related," "weakly related," and "no relation," are to be ranked and selected, based on criteria of corresponding product and engineering characteristics. The proposed method can also be applied in setting technical requirement target values by defining some technical requirement levels in certain ranges and specifying them as alternatives, and it can be easily extended to solve a hierarchical structure, instead of a list, of customer requirements.

The NGT process suggested may be fine tuned, depending on work environments such as different voting styles and/or the methods of expert judgment and group participation in organizing customers' requirements. It will be an interesting future research area to implement and computerize the proposed integrated system shown in Figure 1.2 into a group decision support system (GDSS). As Beruvides[3] mentions, present GDSS does not yet merge with the NGT technique. Moreover, our integrated group decision-making technique is practical to support computerized documentation processes such as those proposed by Khoo and Ho[14] in complete QFD planning.

In this study, we model linguistic variables by ordinal numbers; for example, "1" stands for the linguistic value of "the least importance level." In practice, this may not work well enough to handle various situations for the customers' preferences or customer requirements containing ambiguity and multiplicity of meaning. Recently, fuzzy sets have increasingly been used to model these linguistic variables with membership functions or possibility distributions in many fields. Khoo and Ho[14] and Masud and Dean[23] apply fuzzy sets in modeling the linguistic variables of importance levels and relations between customer and technical requirements. It might also be practical to deal with linguistic weights by using fuzzy sets. In such cases, our approach can be extended without difficulties.

The proposed method may further be enhanced in determining technical target vectors by transforming them into multiple-objective optimization problems where multiple technical requirement levels, which are normally conflicting, need to be designed simultaneously to achieve nominated importance levels of various requirements within technical limitations. Lai and Hwang[17, 18] provide some guidelines for exploring fuzzy multiple objective modeling in QFD.

REFERENCES

1. Armacost, R. L., P. J. Componation, and W. W. Swart, 1994, An AHP Framework for Prioritizing Customer Requirements in QFD: An Industrialized Housing Application, *IIE Trans.,* Vol. 26, No. 4, pp. 72–78.

2. Belhe, U., and A. Kusiak, 1995, The House of Quality in Analysis of Product Development Processes, *Fourth Industrial Engineering Research Conference Proceeding,* B. W. Schmeiser and R. Uzsoy (eds.), IIE, Norcross, pp. 996–1003.

3. Beruvides, M. G., 1995, Group Decision Support Systems and Consensus Building: Issues in Electronic Media, *Computers in Industrial Engineering,* Vol. 29, pp. 601–605.

4. Bossert, J. L., 1991, *Quality Function Deployment*, ASQC Quality Press, Milwaukee.

5. Carlsson, C., and P. Walden, 1995, AHP in Political Group Decision: A Study in the Art of Possibilities, *Interfaces*, Vol. 25, No. 4, pp. 14–29.

6. Day, R. G., *Quality Function Deployment: Linking a Company with Its Customers*, ASQC Press, Milwaukee.

7. Delbecq, A. L., A. H. Van de Ven, and D. H. Gustafson, 1986, *Group Techniques for Program Planning: A Guide to Nominal Group and Delphi Processes*, Green Briar Press, Middleton, WI.

8. Fishburn P.C., and J. D. C. Little, 1988, An Experiment in Approval Voting, *Management Sci.*, Vol. 34, No. 5, pp. 555–568.

9. Glushkovsky, E. A., R. A. Florescu, A. Hershkovits, and D. Sipper, 1995, Avoid a Flop: Use QFD with Questionnaires, *Quality Progress*, June, pp. 57–62.

10. Hales, R., 1995, Adapting Quality Function Deployment to the U.S. *IIE Solutions*, October, pp. 15–18.

11. Hauser, J. R., and D. Clausing, 1987, The House of Quality, *Harvard Business Rev.*, Vol. 66, No. 3, pp. 63–73.

12. Hunter, M. R., and R. D. Van Landingham, 1995, Listening to the customer using QFD, *Quality Progress*, April, pp. 55–59.

13. Hwang, C. L., and M. J. Lin, 1987, *Group Decision Making under Multiple Criteria*, Springer-Verlag, Heidelberg.

14. Khoo, L. P., and N. C. Ho, 1996, Framework of a Fuzzy Quality Deployment System," *Internat. J. Production Res.*, Vol. 34, No. 2, pp. 299–311.

15. Kolano, F., 1991, Using the Nominal Group Technique in Value Engineering, *Society of American Value Engineers' Proceedings*, Society of American Value Engineering, Northbrook, pp. 189–195.

16. Lai, S. K., 1995, A Preference-Based Interpretation of AHP, *Omega*, Vol. 23, No. 4, pp. 453–462.

17. Lai, Y. J., and C. L. Hwang, 1992, *Fuzzy Mathematical Programming*, Springer-Verlag, Heidelberg.

18. Lai, Y. J., and C. L. Hwang, 1994, *Fuzzy Multiple Objective Decision Making*, Springer-Verlag, Heidelberg.

19. Locascio, A., and D. L. Thurston, 1993, Multiattribute Design Optimization with Quality Function Deployment, *Second Industrial Engineering Research Conference Proceedings*, D. A. Mitta et al. (eds.), IIE, Norcross, pp. 82–86.

20. Lockamy, A., and A. Khurana, 1995, Quality Function Deployment: A Case Study, *Production Inventory Management J.*, Second Quarter, pp. 56–59.

21. Lyman, D., and K. Richter, 1995, QFD and Personality Type: The Key to Team Energy and Effectiveness," *Industrial Engineering*, February, pp. 57–61.

22. Marshall, K. T., and R. M. Oliver, 1995, *Decision Making and Forecasting*, McGraw-Hill, New York.

23. Masud, A. S. M., and E. B. Dean, 1993, Using Fuzzy Sets in Quality Function Deployment," *Second Industrial Engineering Research Conference Proceedings*, D. A. Mitta, et al. (eds.), IIE, Norcross, pp. 270–274.

24. Perez, J., 1995, Comments on Saaty's AHP, *Management Sci.*, Vol. 41, No. 6, pp. 1091–1095.

25. Prasad, B., 1994, Product Planning Optimization Using Quality Function Deployment, in *Artificial Intelligence in Optimal Design and Manufacturing,* Z. Dong (ed.), Prentice-Hall, Englewood Cliffs, NJ, pp 117–151.

26. Saari, D. G., 1994, *Geometry of Voting,* Springer-Verlag, Heidelberg.

27. Saaty, T. L., 1990, *The Analytic Hierarchy Process: Planning, Priority Setting, Resource Allocation,* RWS Publications, Pittsburgh.

28. Saaty, T. L., 1994, How to Make a Decision: The Analytic Hierarchy Process, *Interfaces,* Vol. 24, No. 6, pp. 19–43.

29. Sink, D. S., 1983, Using the Nominal Group Technique Effectively, *Nat. Productivity Rev.,* Spring, pp. 173–184.

30. Wasserman, G. S., 1993, Technical Note: On How to Prioritize Design Requirements during the QFD Planning Process, *IIE Trans.,* Vol. 25, No. 3, pp. 59–65.

2

FUNCTIONAL DESIGN

TOR SHU BENG and GRAEME BRITTON
Nanyang Technological University

MUTHU CHANDRASHEKAR
University of Waterloo

NG KEK WEE
Object Oriented Pte., Ltd.

2.1 WHAT IS FUNCTIONAL DESIGN?

There are a number of serious problems with current computer-aided design/
computer-aided manufacturing (CAD/CAM) systems despite the tremendous
advances that have been made in solid modeling. First, the current systems
are generic by nature. The generic systems meet the CAD vendors' objective
of making their products attractive to a large number of customers, but they
are not ideally suited to each individual customer's needs. Most companies
perform design work within a relatively narrow domain. They do not need
the wide range of functionality provided by generic CAD systems. Consequently, there is a long learning curve to effectively use these systems in an
organizational context, and there are few commercially available CAD tools
to assist designers working in specific domains.

Second, the CAD systems are based on geometric manipulation and reasoning and cannot deal with nongeometric information effectively, if at all. As a
result, much of the design intent is not captured by the CAD system and has
to be documented separately, usually using other software. It is also difficult
to link the geometric solid models with the conceptual models developed

Integrated Product and Process Development, Edited by John Usher, Utpal Roy, and Hamid
Parsaei
ISBN 0-471-15597-7   © 1998 John Wiley & Sons, Inc.

during the early design phases as they are based on different representation schemes.

Third, the design process is not fully visible to the designer. Current computerized design management systems provide some design process visibility by tracking the status of components in assemblies. However, they cannot show how a design change in one component affects other components.

Fourth, there is increasing pressure on companies to shorten the design–manufacturing cycle lead time. In principle, this can be achieved by overlapping design and manufacturing. In practice, the design–manufacturing life cycle has to be reengineered to facilitate closer integration. How is this integration to be achieved? The design process itself needs to be scientifically studied in order to design new processes that can take full advantage of existing computer technology.

These problems have been known since the late 1980s and a considerable amount of research is being conducted to develop feasible solutions (e.g., Sata,[36] Finger and Dixon,[15] Black,[9] Andreasen and Olesen,[3] Jakobsen,[21] Tomiyama et al.,[39] Gui and Mäntylä,[17] Shah and Mäntylä.[37] The early research in this area focused on feature modeling representation schemes. However, it was soon realized that these are very limited in terms of what can be represented. Current research takes a broader perspective and attempts are being made to develop representation schemes and CAD tools that include both geometric and nongeometric data. In addition, the design process itself is being redesigned. The design processes and CAD tools being developed are referred to as "functional design."

Functional design covers the upstream phases of design. The aim is to provide computer tools to link design functions with the structural (physical) embodiments used to realize the functions. Some of the approaches researchers have adopted to tackle this problem are:

- Functional decomposition[29]
- Function-to-structure mapping[27, 28, 40]
- Functional language[22, 39]
- Graph theory[17]

Our own work in this area is based on a rigorous scientific epistemology and ontology from the systems area. We believe the systems viewpoint offers a rich and fruitful approach for redesigning the design process. The key features of functional design from our viewpoint are:

- The research and development epistemology and ontology should be based on systems theory and practice. That is, a rigorous scientific theory is needed to guide the modeling of the design process and engineering assemblies.[16]

- The design process and CAD support tools should be designed for a specific domain of design activity.
- The CAD tools should be developed to improve productivity. All design involves some creativity and this cannot be replaced by automated tools. Our aim is to help the "creative" designer work more quickly by automating routine CAD tasks. Hence the tools should:
  - Complement, not replace, the designer.
  - Assist the designer to rapidly generate geometric and nongeometric data.
  - Be able to trace the implication of a design change, display this to the designer, and then automatically update all affected components once the change is confirmed.
  - Provide a roadmap of the design process so that designers know what design decisions have been made and what decisions are yet to be made.
- A graph-theoretic representation scheme should be used to rigorously define the graphical systems models. Graph theory provides a powerful way to represent the interactions between systems and objects. This will be justified later on.
- Object-oriented (OO) modeling and software development should be used to develop the software. The OO programming languages and concepts provide an efficient means for implementing the systems and graph-theoretic models.
- A software methodology should be used to provide discipline for the software development efforts.

The remainder of this chapter explains how systems, graph-theoretic, and OO concepts can be integrated to model design processes and products.

2.2 MODELING THE DESIGN PROCESS

A number of writers divide the design process into phases (see Asimow,[6] Matousek,[31] and Shah and Mäntylä[37]). Typically, these phases are presented in sequence from general to specific as follows:

1. *General Functional Specification.* This is a statement of the user's needs in general functional terms. The output is a general functional specification of a class of products that will satisfy the user needs.
2. *Specific Functional Specification* (*Conceptual Design*). The conceptual design stage consists of defining a set of subfunctions and their interrelationships to achieve a general functional specification. It involves the generation and synthesis of ideas and performance analysis.

3. *Structural Design.* Sometimes referred to as embodiment design, this stage defines the geometric layout of the design. The design is now represented as an image, but in general terms.

4. *Detail Design.* This is a fully detailed design of a product, including manufacturing notes and bills of materials.

The traditional viewpoint of the design process is that it proceeds in some fashion from phase 1 to phase 4. That is, the process is one of continually refining a general design concept until it is fully detailed. This is a top-down view of the design process. If the general functional specification is considered to contain a set M of the total design properties and the detail design another set A, then this view can be expressed as $M \subset A$. It is recognized that there may be feedback and iterations between phases before a design is finalized.

We argue that this view of the design process is of little use for organizing and managing design teams and hence does not provide a useful framework for developing CAD tools. This is because it ignores the fact that designers use past experience in current design situations. During conceptual design, for example, a designer may develop a concept on the basis of past embodiment designs. In this case, embodiment design precedes conceptual design. In other cases, where there is considerable reuse of past designs, the conceptual design phase might be not be required at all. The design process may only require a new embodiment variant with some additional detail design. Furthermore, the phases may be performed concurrently in large design teams. If computer systems are to be used as an aid in this process, then they must be flexible enough to allow initiation of change at any point in the process, parallel processing of design activities, all feasible iterative sequences, and design ambiguity (imprecise specification) to prevent premature closure of the design concept.[3, 9, 17, 21, 30]

Cyberneticians Ashby[5] and Beer[8] offer an alternative view of the design process. They argue that the design process starts with a very large number of possibilities. The aim of the designer is to specify constraints to reduce this large number to just one: the final design. In other words, the cyberneticians argue that $A \subset M$. This is antithetic to the traditional viewpoint.

The cybernetic viewpoint facilitates incremental and iterative development of the design as the constraints can be specified in any order. The design process can be tracked by measuring the number of decisions remaining to be taken (constraints to be specified) at any point in time during the process. The model of the design process needed to track the decisions can also indicate those decisions that have been taken and show how each decision affects the other decisions.[8] The constraint-based approach for designing has been used to improve productivity in electronic design,[14] and our own research indicates that it can do the same for mechanical design.

Cybernetic decision-making models are divided into two parts: parameterised variables and data.[7] The parameterized variables represent specific objects or systems and the relationships between them. Data, stored separately, are combined with the parameterized variables to represent a particular decision. The major advantage of this approach is that a very compact model can generate a large number of actual real-world situations. This type of approach is common these days in CAD: It is called *parametric design*. However, we have referred to Beer's work because the cybernetic approach is much more general than simply parameterizing geometric shapes; for example, nongeometric data can be parameterized.

The final issue we address here is how to relate the different types of design models (functional and structural). At the geometric level, design models represent structural objects, morphologically defined structural classes of similar objects, and the structural relationships between them. The structural relationships are deterministic. Modern, commercial, solid-model assembly modelers can capture class and subclass information (hierarchical relationships) and relationships between features of objects (network relationships). They can display the hierarchical relationships in a graph tree (assembly tree), but they provide no means for graphically displaying the network relationships. The latter are normally displayed as lists of expressions.

In a functional (conceptual) design model, the physical classes are not defined by their structural properties, but, instead, by what they do: their function. A functional class differs from a structural class in three ways. First, the class contains structurally dissimilar objects, not similar objects. Second, the functions and the functional relationships between functional classes are probabilistic, not deterministic. Third, functional superclasses inherit properties from the subclasses. That is, a functional superclass can do everything its subfunctions can do, but the subfunctions cannot do what the superclass can do. This is the opposite of structural classes, where subclasses inherit properties from the superclasses.

We can see now that there are two totally different ways of viewing and classifying design models: structural and functional. How are these two views related? Gui and Mäntylä[17] have made a very good attempt to define these views, but their work does not have a rigorous epistemological basis. Fortunately, a philosophy of science has been developed that provides just such a basis: it is called *experimentalism*.[11] The experimentalists argue that two apparently conflicting scientific "realities"—a mechanical image from the physical sciences and a functional image from the life sciences—are simply different views of the same reality and hence they are complementary rather than antithetic.

In a mechanical image of nature, the properties of a bounded part of space, at any given moment of time, can be completely defined. The bounded part of space is referred to as a time slice and is taken to consist of elementary entities (e.g., point particles or aggregates of these). Changes in the values of the properties over time are given by one or more deterministic relationships (mechanical laws). The set of initial properties and the laws are sufficient to define the values of the properties at any later moment of time. The time slices and laws form a deterministic, closed system (called a natural mechanical system). The initial set of properties (the cause) is necessary and sufficient for any following set (the effect). If the first time slice does not occur, then neither will the following one. If the first one does occur, then the following one must also occur.

A producer–product (probabilistic) image is obtained by considering part of a time slice. One part of the time slice is taken to be observable (the object being studied) and the rest to be partially or completely unobservable (the environment). It is assumed that the object and its environment form a natural mechanical system. Under these conditions, the object sometimes produces a particular result and sometimes it does not. This is because the environment also plays a part in determining the result, and some, or all, of the properties of the environment are not observed (known). More rigorously, we can say that the object (the producer) is necessary but insufficient for a given result (the product). If the producer does not occur, then neither will the product. If it does occur, then the product may or may not occur. The occurrence of the result is defined by a probability distribution. The producer–product image is an intermediate image linking the mechanical and functional images.

The mechanical and producer–product images are based on one natural mechanical system. A teleological (functional) image is formed by considering sets of natural mechanical systems.

The simplest teleological image is a set of natural mechanical systems containing different structural objects (x_i) that produce the same structural class of products (y) [see Fig. 2.1]. The objects (x_i) can be considered as a functional class. In this case the function is extrinsic. Each producer has a function only because there are other different types of producers that also produce the same product. An extrinsic class can also consist of a set of events or actions. Two examples of extrinsic classes are (1) the set of time-keepers—mechanical clocks, digital watches, sun-dials, sand-glasses, water clocks, etc.; and (2) the set of meters— voltmeters, ammeters, etc.

Note that an object or event in an extrinsic functional class is neither necessary nor sufficient for the product. It is not necessary because other objects can produce the product. It is not sufficient because functionality is defined by a producer–product relationship.

Now consider a set of natural mechanical systems, containing different structural actions, that produce the same product (extrinsic function) and that are themselves produced by one object. In this case the object is also a producer of the product. It is said to have an intrinsic function because it produces all the actions in the extrinsic functional class. No other objects are necessary to define its function; functionality is inherent. Intrinsic function is defined by considering two or more producer–product relationships in sequence [see Fig. 2.2]. Some examples of systems having intrinsic function are heating and ventilating systems, auto pilots, computers, cats and people.

The definitions above can be used to develop a complete, hierarchical, functional classification of systems.[1, 2] The simplest system is one that can act in only one

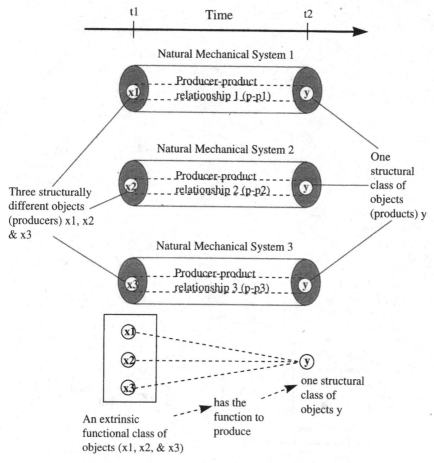

Figure 2.1 Teleological image of nature: extrinsic function.[11]

way in all environments and belongs to an extrinsic functional class. Such a system is referred to as passive functional because it does not change when the environment changes.

A more complex type of system . . . is one that has one or more intrinsic functions, can act in only one way in any given environment, and can act in different ways in different environments. The selection of the different ways of acting (initiation of action) is caused by the environment. If the system has more than one function then change of function is also caused by the environment. This type of system is said to be reactive functional because it reacts to changes in the environment. It cannot choose its actions nor its functions, but it does co-produce the execution of its actions.

The most complex type of system . . . is one that has one or more intrinsic functions, and can act in different ways in any one environment and in different

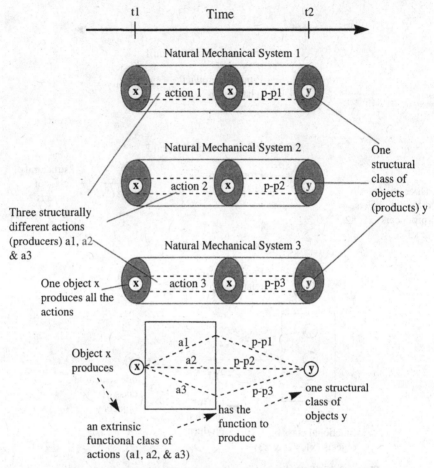

Figure 2.2   Teleological image of nature: intrinsic function.[11]

environments. The selection of the different ways of acting is co-produced by the environment and the system. Such a system is called goal-seeking if it has only one function (goal). A multi-goal seeking system can seek two or more goals, but only one goal in any one environment; the selection of goals is caused by the environment. A purposeful system can seek two or more goals in any one environment and in different environments. It co-produces the selection of goals. It has the greatest freedom of all because it can choose its actions and its goals. This freedom enables it to build extended hierarchies of functions: ends, goals, objectives and ideals.[11]

Experimentalist ontology provides an integrated, conceptual framework for modeling the design process and designs. Conceptual designs, defined within a functional image, can be related to the geometric models in the structural image. This conceptual framework can be expressed rigorously using linear graph theory. Graph theory allows us to represent the connectivity (relationships) between components in design assemblies. All types of relationships can be represented: functional, structural, hierarchical, and network.

2.3 REPRESENTATION OF DESIGN MODELS

Design models can be represented by graph-theoretic methods (GTMs) because the methods are concerned with component actions and how the component interconnection pattern affects the action at the system level. Graph-theoretic methods are applicable to lumped or distributed, linear or nonlinear, scalar or vector, deterministic or probabilistic, and continuous- or discrete-time systems.[12] Thus, they meet all of the criteria discussed previously.

The GTM approach provides a matrix-oriented methodology for modeling interdisciplinary systems. The resulting system of equations is suited for computer implementation and leads to efficient formulation and solution procedures, taking full advantage of object-oriented programming.

Graph-theoretic methods have been applied to a wide variety of problems including electrical, mechanical, hydraulic, magnetic, structural, thermal, and combinations of these systems.[4, 13, 18, 23, 25, 26, 32, 34] More recent application of GTMs to two- and three-dimensional mechanical systems, for analysis of linkages, robots, and mechatronic systems, has particular relevance to tolerance modeling and tolerance analysis for mechanical assemblies.

This section explains essential graph concepts and shows how graph trees can be used to represent the different types of connectivity encountered in mechanical design.

2.3.1 Essential Graph Concepts

A *linear graph* consists of lines and points. A line is called an *edge* (*branch, link,* or *chord*) and a point where edges are connected is called a *node* (*vertex*). A graph is denoted by $G(n, e)$, where n is the number of nodes and e is the number of edges.[38] Edges of an *oriented* graph use arrows α to define measurement conventions for quantities associated with the edges. The nodes are usually designated by letters and edges by numbers. A graph $G(4, 5)$ is shown in Figure 2.3.

A graph is used to represent the connectivity of a system. For example, consider a graphical representation of an electrical circuit. The edges represent electrical components, such as resistors, capacitors, and sources, and the nodes represent the interconnection points of these components in a circuit.

Similarly, in a two- or three-dimensional mechanical system, the edges represent mechanical components, such as rigid bodies, springs, dampers, joints, and external applied forces, torques, or displacements; and the nodes represent the points of interconnection. In a transportation system, the edges

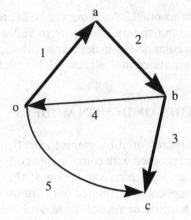

Figure 2.3 An example of a graph $G(4, 5)$.

represent the roads and the nodes represent the intersections. In a path planning exercise for a robot, a graph could be used to represent the visual three-dimensional space, where the edges represent the walls in a room and the nodes represent the corners where the walls meet.

In a tolerance analysis application, the edges represent the nominal dimensions of various elements of an assembly and the nodes represent the connectivity of the elements. For example, in Figure 2.4, a graph is used to represent the vectorial dimensions of a slider-crank mechanism in two dimensions. The edges represent the nominal dimensions associated with each member of the mechanism, for example, from the base to the first pin joint at A (edge 1 from o to a), the tolerance of the pin itself by edge 2 from a to a', and so on. We assume that the mechanism is free to move at point C, and closure is attained at edge 6, which defines the tolerance gap.

Although the graph in Figure 2.4 is drawn as a vector diagram, it is not necessary to do so. In general, the edges represent the connectivity, and the actual vector quantity is expressed separately as an equation to accompany each edge, and thus the edge together with its equation defines the tolerance. Thus, the graph in Figure 2.4 could be drawn as shown in Figure 2.5 without any loss of information.

2.3.2 Graph Definitions

The following is a list of some graph terms and their definitions:

An edge is said to be *incident* on a node if it is connected to it (positive if oriented away, negative otherwise).

A *subgraph* of a graph is defined as a subset of edges and the associated nodes of the graph.

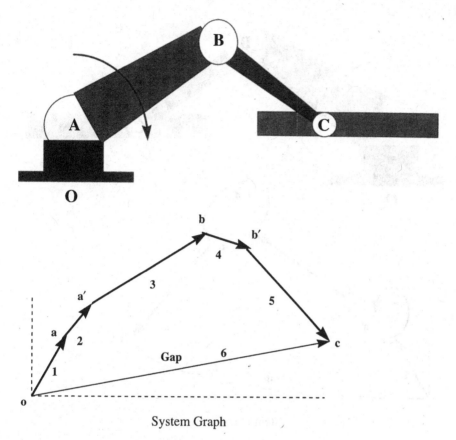

Figure 2.4 Slider crank and its tolerance graph.

A *path* between a pair of nodes p, q is a subgraph such that there is exactly one edge incident at p and q, and, at every other node, there are exactly two edges incident.

A subgraph is said to be *connected* if there is a path between every pair of nodes in the subgraph.

An unconnected graph is said to be in *parts*.

A *circuit* (or loop) is a subgraph that is connected and has exactly two edges incident at each node.

A *tree* is a subgraph that is connected and has no circuits.

A *spanning tree* is a tree that contains all the nodes of the graph. [A spanning tree of $G(n, e)$ has exactly $n - 1$ edges.]

A *forest* is a set of trees, each associated with part of an unconnected graph.

A *cutset* is a subgraph (not necessarily connected) that leaves the graph in two parts when it is removed, and no subgraph of this subgraph has this property.

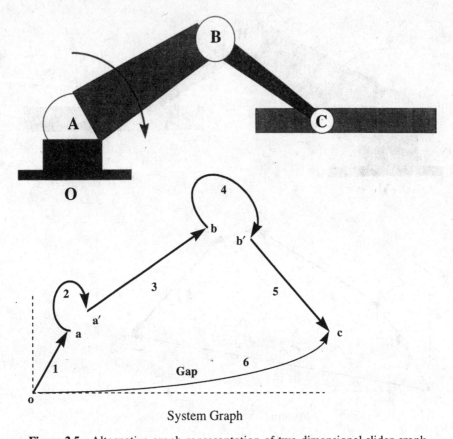

System Graph

Figure 2.5 Alternative graph representation of two-dimensional slider crank.

A *cotree* or *coforest* is the remaining edges in a graph after a tree or the forest has been removed.

The edges of a tree or forest are called the *branches*, and the edges in a cotree or coforest are called the *chords* or *links*.

2.3.3 Graphs to Matrices

The most fundamental matrix is called the *incidence* matrix A, which contains the incidence information of each edge on each node in a graph. Matrix A is an $(n - 1) \times e$ matrix whose elements are defined as follows:

$$a_{ij} = \begin{cases} 0 & \text{edge } i \text{ is not incident on node } j \\ +1 & \text{edge } i \text{ is positively incident on node } j \\ -1 & \text{edge } i \text{ is negatively incident on node } j \end{cases}$$

An edge (arrow) oriented away from a node is considered as positive. This is a matter of convention: For Figure 2.3, matrix A is defined as shown in Table 2.1.

Matrix A is also called the reduced incidence matrix because the information of node o is omitted as it can be derived, by summing each column of A and changing the sign of the result, and hence is redundant.

It can be shown that the rows of A are linearly independent, and hence matrix A is a row full-rank matrix. This is important for selecting a tree and determining some important matrices associated with the graph, as will be explained next.

Because A represents the basic connectivity information about the graph, all other relevant matrices and associated structural or topological information can be derived from A.

In order to manipulate matrix A, we need to use the elementary operations from linear algebra, which include the following three operations: (1) multiply a row (or a column) by a scalar constant, (2) add two rows (or columns), and (3) interchange two rows (or columns).

For example, multiplying a row by a constant and subtracting it from another row, a typical Gauss elimination step, is a combination of the three elementary operations. We will use these operations to manipulate A to obtain other useful matrices associated with a graph.

With respect to the graph in Figure 2.3, a tree is shown by thick lines, edges 1, 2, and 3, and the associated cotree is shown by light lines, edges 4 and 5. Edges 2, 3, and 5 make a path between nodes a and o. A circuit consists of edges 1, 2, and 4. Finally, a cutset is defined as edges 2, 4, and 5.

A fundamental cutset (f-cutset) matrix Q is defined as follows. A cutset is selected such that it contains exactly one tree edge. The collection of all such cutsets when arranged in a matrix defines Q. Obviously, Q is an $(n - 1) \times e$ matrix because there are $n - 1$ edges in a tree. Table 2.2 shows Q for the graph in Figure 2.3.

The following points should be noted about Q: The edges (columns) have been partitioned into the tree (1, 2, 3) and cotree (4, 5); there is an identity matrix defined by the tree edges; and the entries are all +1, −1, and 0.

In order to understand the sign convention, consider row 2 of Table 2.2 (010-11). Each f-cutset is defined with respect to a tree branch, in this case

TABLE 2.1 Incidence Matrix A for Example Shown in Figure 2.3

Edge Node	1	2	3	4	5
a	−1	1	0	0	0
b	0	−1	1	1	1
c	0	0	−1	0	−1

TABLE 2.2   Fundamental Cutset Matrix Q for Figure 2.3

1	2	3	4	5
1	0	0	-1	1
0	1	0	-1	1
0	0	1	0	1

edge 2, which is assumed to be positively oriented. The orientation of the chords forming the cutset is defined relative to the defining tree branch. For example, imagine a cut or line is drawn across the paper such that it cuts edges 2, 3, and 5. Using the orientation of edge 2 as the defining direction, edge 4 is oriented in an opposite direction to edge 2, and edge 5 is oriented in the same direction as edge 2; therefore, the cutset is defined as 010-11.

A fundamental circuit matrix B is defined as follows. Given a tree in a graph, it defines a unique path between every pair of nodes in the graph. If chords are added to the tree one at a time, then a unique circuit will be defined for each chord. The collection of all such circuits in a matrix defines B. Obviously, B is of size $(e - n + 1) \times e$. Table 2.3 shows the B matrix for the graph in Figure 2.3 and the tree 1, 2, 3.

The following points should be noted about B: Edges have been partitioned into tree and cotree groups; there is an identity matrix under the cotree edges; and the entries are $+1$, -1, and 0. Each row of B defines a unique circuit in the graph. For example, the first row consists of 11010 and defines the circuit consisting of edges 1, 2, and 4. Here the defining chord edge (4) has the positive orientation, and the other edges assume positive, negative, or zero values. A zero value indicates an edge is not associated with the circuit.

2.3.4   Orthogonality of Q and B matrices

A fundamental theorem in graph theory states that each row of Q is orthogonal to each row of B. As a result, the matrix product QB' (where ' denotes the transpose) is a null matrix:

$$QB' = 0$$

TABLE 2.3 B Matrix for Figure 2.3

1	2	3	4	5
1	1	0	1	0
-1	-1	-1	0	1

This results in the following: If Q is redefined as $[UQ_c]$ and B as $[B_tU]$ corresponding to the tree and cotree partitioning, then

$$Q_c = B_t'$$

In effect, once Q or B is known, the other can be obtained immediately. That is, the nonidentity portions of Q and B are the negative transposes of each other. This is a very useful and fundamental result which will be used later.

2.3.5 Across and Through Variables

In any given system it is possible to identify a type of variable X that is measured across two points or nodes (e.g., voltage, displacement, pressure, and temperature). Variables whose values are determined by two-point oriented measurements are called *across* variables. Similarly, another type of variable Y can be defined which is measured at a point (e.g., current, force, flow rate, and heat flow rate). Variables whose values are determined by one-point oriented measurements are called *through* variables. The product XY has the units of *energy* or *power* in each discipline (e.g., electric, mechanical, hydraulic, or thermal).

It is possible to associate a pair of X and Y variables with each edge of a graph because each edge corresponds to a component in a system and the variables are properties of the component. For example, an electrical resistor, represented by an edge, has the properties voltage drop (across variable X) and current (through variable Y) associated with it. The variables can be scalar, complex, or vectorial, depending on the physical domain being modeled.

2.3.6 Circuit and Cutset Equations

Each physical domain or discipline has natural laws that apply to an assembly, a collection of interconnected components, or a system. It is interesting to note that such laws pertain to the X and Y variables of the system separately. For example, Kirchhoff's current and voltage laws involve currents Y and voltages X of the system. Similarly, force balance and compatibility laws apply to the forces F and displacements r in a mechanical system.

The cutset and circuit matrices have an interesting interpretation in this context.

Circuit Postulate. If X represents the across variables associated with a system and B is its circuit matrix, then

$$BX = 0$$

Cutset Postulate. If Y represents the through variables associated with a system and Q is its cutset matrix, then

$$QY = 0$$

In relation to the graph in Figure 2.3, the circuit and cutset equations are given by

$$BX = \begin{bmatrix} 1 & 1 & 1 & 1 & 1 \\ -1 & -1 & -1 & 0 & 1 \end{bmatrix} \begin{bmatrix} x_1 \\ x_2 \\ x_3 \\ x_4 \\ x_5 \end{bmatrix} = 0$$

and

$$QY = \begin{bmatrix} 1 & 0 & 0 & -1 & 1 \\ 0 & 1 & 0 & -1 & 1 \\ 0 & 0 & 0 & 0 & 0 \end{bmatrix} \begin{bmatrix} y_1 \\ y_2 \\ y_3 \\ y_4 \\ y_5 \end{bmatrix} = 0$$

2.3.7 Derivation of Q and B Matrices from A

We will assume that A has been arranged so that the first $n - 1$ columns correspond to the tree and the last $e - n + 1$ columns correspond to the cotree. Let this matrix A be represented as

$$A = [A_t A_c]$$

2.3.7.1 Obtaining Q from A A comparison of Q with A reveals that

$$Q = [UQ_c] = [UA_t^{-1}A_c]$$

Thus, a unit matrix is created under the tree section of the incidence matrix by elementary operations. This will be demonstrated using the example in Figure 2.3:

$$A = \begin{bmatrix} -1 & 1 & 0 & 0 & 0 \\ 0 & -1 & 1 & 1 & 0 \\ 0 & 0 & -1 & 0 & -1 \end{bmatrix}$$

where the columns represent edges 1, 2, 3, 4, and 5.

STEP 1. MULTIPLY ROW 1 BY -1 (SCALAR):

$$\begin{bmatrix} 1 & -1 & 0 & 0 & 0 \\ 0 & -1 & 1 & 1 & 0 \\ 0 & 0 & -1 & 0 & -1 \end{bmatrix}$$

STEP 2. MULTIPLY ROW 2 BY -1:

$$\begin{bmatrix} 1 & -1 & 0 & 0 & 0 \\ 0 & 1 & -1 & -1 & 0 \\ 0 & 0 & -1 & 0 & -1 \end{bmatrix}$$

STEP 3. ADD ROW 2 TO ROW 1:

$$\begin{bmatrix} 1 & 0 & -1 & -1 & 0 \\ 0 & 1 & -1 & -1 & 0 \\ 0 & 0 & -1 & 0 & -1 \end{bmatrix}$$

STEP 4. MULTIPLY ROW 3 BY -1:

$$\begin{bmatrix} 1 & 0 & -1 & -1 & 0 \\ 0 & 1 & -1 & -1 & 0 \\ 0 & 0 & 1 & 0 & 1 \end{bmatrix}$$

STEP 5 AND 6. ADD ROW 3 TO ROWS 1 AND 2:

$$Q = \begin{bmatrix} 1 & 0 & 0 & -1 & 1 \\ 0 & 1 & 0 & -1 & 1 \\ 0 & 0 & 1 & 0 & 1 \end{bmatrix}$$

This procedure is illustrated in Figure 2.6.

2.3.7.2 *Obtaining B from Q* $Q_c = -B_t'$ or $B_t = -Q_c'$, hence given A,

$$B = [B_t U] = [-Q_c' U] = [-(A_t^{-1} A_c)' U]$$

For the example, the B matrix can be derived directly from the Q matrix given previously:

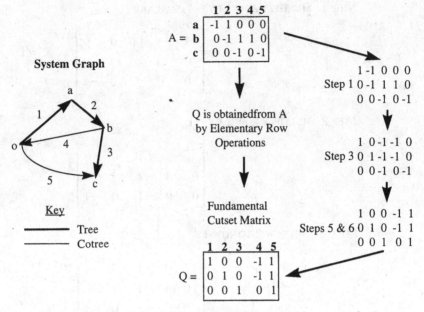

Figure 2.6   Obtaining f-cutset matrix Q from incidence matrix A.

$$B = [- \; Q'_c \; U]$$

$$= \begin{bmatrix} 1 & 1 & 0 & 1 & 0 \\ -1 & -1 & -1 & 0 & 1 \end{bmatrix}$$

This procedure is illustrated in Figure 2.7.

2.4   APPLICATION: FUNCTIONAL OBJECTS

Graph theory facilitates OO software development in four ways. First, one common, rigorous language is used to model all relationships. Second, most of the graph-theoretic algorithms can be automated. Third, it provides a means for defining and checking the interrelationships between objects. This is particularly important for network relationships, which can be hard to trace in OO programs. Fourth, it provides a means for translating the systems models into OO programs.

Figure 2.8 shows how a systems model is represented in graph-theoretic terms. The simplest system is that with one input and one output (two terminal nodes). Such a system can be represented by a tree with one edge. Note that the links (arrows) between boxes in systems models correspond to the nodes in graph theory. More complex systems may have two or more input and

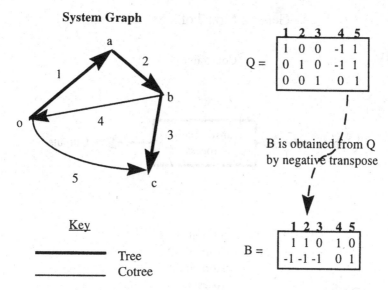

Figure 2.7   Obtaining f-circuit matrix B from f-cutset matrix Q.

output nodes. The trees for these systems will contain a number of edges.[24] More complex systems still may contain more than one tree; that is, they are represented as a forest.

The constraints for a system can be internal or external. In graph theory, an internal system constraint is represented as:

· A relationship between the through and across variables.
· A restricted range of parameter values for the through or across variables.
· The selection of a particular tree if a system is represented as a forest.

External system constraints are represented as node constraints (constraints on inputs and outputs, i.e., on across variables) or as circuit constraints (constraints on through variables) imposed by connecting systems together.

Hierarchical levels of system description can be easily obtained. Lower-level detailed trees can be reduced to a minimum number of edges, that depend solely on the number of terminal nodes (inputs and outputs), to produce simpler higher level trees.

Our functional objects are OO objects that represent real physical things and the relationships between them. The objects may be a single physical entity, such as a component in an assembly; an assemblage of components (an assembly that is not a system); or a system.

A functional object will, in general, contain a minimum of two trees: a functional tree and a structural tree. These trees, in turn, can be of two

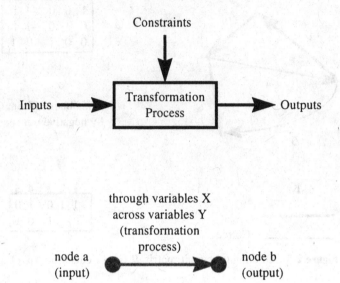

General Model of a System

Figure 2.8   Relationship between system and graph-theoretic models.

types: hierarchical and network (Fig. 2.9). The hierarchical trees correspond to hierarchical connectivity (e.g., parent–child class relationships in OO programming). The network trees correspond to nonhierarchical connectivity that may produce linear or nonlinear overall outputs. It can be seen that the functional objects are capable of representing both conceptual and structural design models.

2.4.1   Illustrative Example

A simple illustration, using a plastic injection mold, will help clarify the concept of functional design as outlined so far. A partial hierarchical functional tree is shown in Figure 2.10. The major functions of a mould are to melt plastic granules, feed the molten plastic into the mold cavity, cool the molten plastic, and eject the plastic part. The ejection function itself can be subdivided into three functions as shown. The "eject" subfunction can be carried out using a "pin ejection system," which is a subassembly of the complete mold assembly. The complete mold assembly has the function "to mold a part"

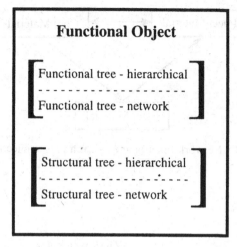

Figure 2.9   Structure of a functional object.

and can perform all the functions shown in the functional tree. The pin ejection system has the function "to eject a part." It does not have the function "to mold a part."

The major functions interact as shown in Figure 2.11; this is a functional network. There is a two-way interaction between "inject" and "cool." The interaction is very complex and cannot be analyzed unless some of the structural characteristics of the mold have been specified. In practice, this interaction is analyzed using mold analysis software. The "eject" function follows in sequence after cooling and must return the mold assembly to a closed position before the next cycle can begin.

A structural hierarchical tree for a mould assembly is shown in Figure 2.12. This tree shows how the mold is assembled. A structural network tree is shown in Figure 2.13. This network is used to calculate the length of the ejector pin automatically.

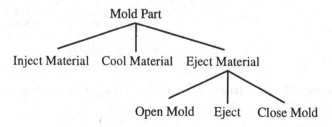

Figure 2.10   Hierarchical functional tree for a plastic injection mold.

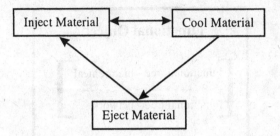

Figure 2.11   Network functional tree for a plastic injection mold.

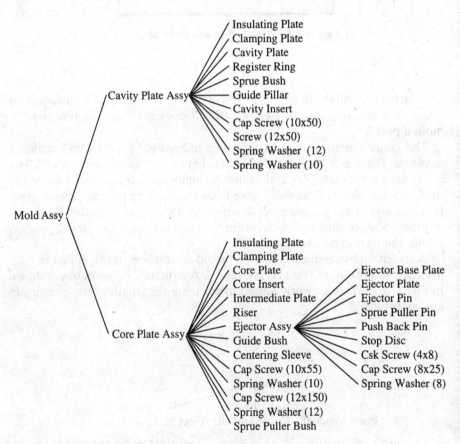

Figure 2.12   Hierarchical structural tree for a plastic injection mold.

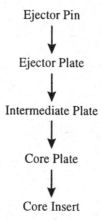

Figure 2.13 Network structural tree for an ejector pin.

The illustrative trees are for combinations of mold components. Each component and subassembly shown in Figure 2.12 will be a functional object and thus contain information about the trees it belongs to and pointers to connect it to other components (objects) in the relevant trees. In addition, the subassemblies contain trees within themselves.

2.5 IMPLEMENTATION: MENTOR

The actual implementation of the functional design in a particular domain requires discipline during software development. This section outlines the methodology we are using.

Object Oriented Pte., Ltd. has developed a full life-cycle, object-oriented, software engineering process suitable for commercial software development. It is called Mentor. Mentor was chosen to implement our functional design approach because it provides both a software engineering process (a method) and a software engineering process architecture (a method for constructing methods); see Figure 2.14.

A software engineering process is a time-sequenced set of process units that is used to transform a user's requirements into a software system. Mentor's software engineering process covers the development of object-oriented software systems, from initiation to deployment. It encompasses all aspects of the software engineering process, including requirements gathering, system design, implementation, testing, project management, and quality. Mentor also provides the key elements required for ISO 9000 quality accreditation and has been used by a number of organizations as part of their overall quality system.

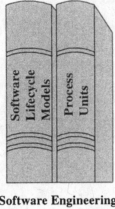
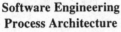

**Software Engineering
Process Architecture**

**Software Engineering
Process**

Figure 2.14   Mentor: a disciplined approach to object-oriented software engineering.

Figure 2.15 shows the underlying principles and constructs used within Mentor, arranged as a set of building blocks. The foundation of Mentor is the set of fundamental object-oriented concepts; for example, class, object, inheritance, and so on. The next level is the infrastructure and software development languages that enable implementation of the concepts.

The third level, also built on the fundamental concepts, is a set of techniques that detail the tasks and guidelines for constructing models. Mentor incorporates a number of techniques derived from other OO methods including Booch,[10] the object modeling technique,[35] object-oriented software engineering,[20] responsibility-driven design,[41] and MOSES.[19] Mentor's techniques are notation independent. This pragmatic approach makes it possible to use a number of different graphical notations within Mentor, as long as the underlying concepts supported by the notation are defined by Mentor.

The techniques need to be packaged as process units before they are useful for commercial development projects. Process units make the techniques robust and usable in industrial settings by providing such things as deliverables templates, resources requirements, and review guidelines. They are the reusable building blocks of industrial-strength approaches. In short, process units provide the activities, guidelines, and tasks that are applied within a software engineering process that is used to execute a software project.

The process units are executed within the framework of a life cycle (next level in Fig. 2.15) and the life cycle is defined by a software engineering process, which is defined within an architecture (top level).

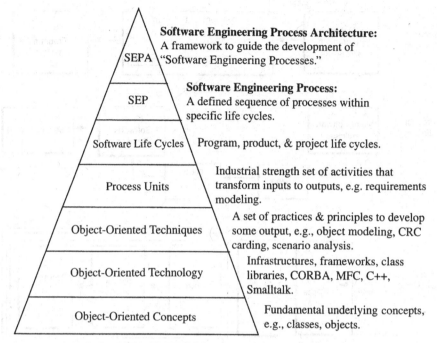

Figure 2.15 Fundamental principles underlying Mentor.

2.5.1 Mentor's Software Engineering Process

One software engineering process does not fit all project types. Software engineering processes can be modeled and constructed just like any other system. The component of Mentor that defines the rules for developing software engineering processes is termed the *software engineering process architecture* (top level in Fig. 2.15). The software engineering process architecture provides the foundation and framework for defining software engineering processes. It is a "method for developing methods."

Currently, one software engineering process, known as M-SEP1, is defined within Mentor. It consists of three related software life cycles:

· Program life cycle
· Product life cycle
· Project life cycle

The relationship between these software life cycles is shown in Figure 2.16. Essentially, the program life cycle consists of a set of projects, each of which undergoes a project life cycle. The software life cycles are briefly described in the following discussion.

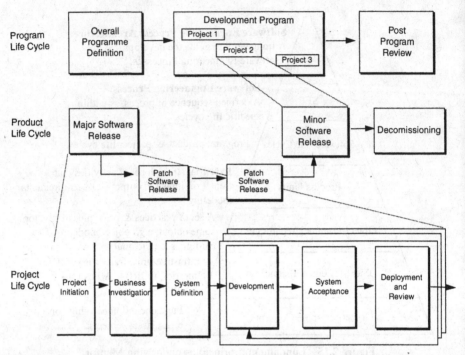

Figure 2.16 Software life cycles of Mentor's software engineering process (M-SEP1).

In large projects there are often many individual projects that are in some way centrally coordinated or managed. A coherent and related set of projects is known as a program of work. A Program life cycle is a software lifecycle that defines a series of phases for such a program of work. A program life cycle therefore encompasses a set of projects and may last for many years. It consists of the following phases:

· Overall program definition
· Development programme
· Postprogram review

A software system usually has a lifespan longer than the individual project that created it. Some software systems may have a very long lifespan during which they are modified, enhanced, and patched. The complete lifespan of a software system is modeled within M-SEP1 as the product life cycle. The product life cycle describes the entire lifespan of a software system from initial development to decommissioning. It consists of:

- Major software releases
- Minor software releases
- Patch software releases
- Decommissioning

Each project undertaken on a software system follows a software life cycle called the project life cycle. The project life cycle defines the phases and software processes for developing a software system from initiation to deployment. It consists of the following phases:

- Project initiation
- Business investigation
- System definition
- Development
- System acceptance
- Deployment and review

2.6 CONCLUSIONS

This chapter has described and explained functional design. Functional design is an approach for designing CAD software that incorporates the representation of functional information, as well as structural information. The objective of functional design is to provide computer-aided tools that can integrate the functional and structural representations developed during a design process.

The authors argue that systems theory provides a philosophical and methodological foundation for functional design. Systems theory can represent the design process and design models. The system models, in turn, can be expressed rigorously using linear graph-theoretic methods. These methods provide the link between the system models and the functional objects. Functional objects are object-oriented objects that represent physical things and the relationships between them. The objects contain functional and structural information and hence can be "viewed" from either perspective. They can represent geometric and nongeometric data and all feasible relationships.

The functional design approach can be implemented using Mentor. Mentor is a commercial, object-oriented, software methodology. It has two main features: (1) process units and (2) a software engineering architecture. These two features can be combined to define a specific software engineering process for any software project. Mentor also provides a software engineering process, M-SEP1, that can be used directly for software development.

The authors have developed functional object prototypes to represent structural relationships (hierarchical and network) for plastic injection mold design. A "QuickMould Project" is currently underway to develop the prototypes for commercial use. Mentor is being used as the software methodology for the project. The project started in August 1996 and will be completed by early 1998.

Further research is continuing to incorporate functional properties into our objects and to incorporate graph-theoretic work on tolerancing for assemblies and manufacturing.

ACKNOWLEDGMENTS

The authors gratefully acknowledge the support of Nanyang Technological University, University of Waterloo, Object Oriented Pte. Ltd., Gintic Institute of Manufacturing Technology, and the National Computer Board.

REFERENCES

1. Ackoff, R. L., 1971, Toward a System of Systems Concepts, *Management Sci.*, Vol. 17, pp. 661–671.
2. Ackoff, R. L., and F. E. Emery, 1972, *On Purposeful Systems,* Tavistock, London.
3. Andreasen, M. M., and J. Olesen, 1990, The Concept of Dispositions, *J. Eng. Des.*, Vol. 1, pp. 17–36.
4. Andrews, G. C., and H. K. Kesavan, 1975, The Vector-Network Model: A New Approach to Vector Dynamics, *J. Mech. Mach. Theory,* Vol. 10, pp. 57–75.
5. Ashby, R., 1964, An Introduction to Cybernetics, Chapman and Hall, London.
6. Asimow, M., 1962, *Introduction to Design,* Prentice-Hall, Englewood Cliffs, NJ.
7. Beer, S., 1996, *Decision and Control,* Wiley, London.
8. Beer, S., 1972, *Brain of the Firm,* Herder and Herder, New York.
9. Black, I., 1990, Embodiment Design: Facilitating a Simultaneous Approach to CAD, *Comput.-Aid. Eng. J.,* Vol. 7, pp. 49–53.
10. Booch, G., 1994, *Object Oriented Design with Applications,* Benjamin-Cummings, Redwood City, CA.
11. Britton, G. A., and H. McCallion, 1994, An Overview of the Singer/Churchman/Ackoff School of Thought, *Syst. Pract.,* Vol. 7, pp. 487–521.
12. Chandrashekar, M., P. H. Roe, and G. J. Savage, 1992, Graph-Theoretic Models—A Unifying Modelling Approach, *Proceedings of the 23rd Annual Pittsburgh Conference on System Modelling and Simulation,* W. G. Vogt and M. H. Mickle (eds.), University of Pittsburgh, pp. 313–321.
13. Chinneck, J. W., and M. Chandrashekar, 1983, Models of Large-Scale Industrial Energy Systems. I. Simulation, *Energy,* Vol. 9, pp. 21–34.
14. Crutchfield, P., and H. Shah, 1997, Moving Design Constraints Through the Process, *Circuits Assembly Asia,* March/April, pp. 40–45.
15. Finger, S., and J. R. Dixon, 1989, A Review of Research in Mechanical Engineering Design. I. Descriptive, Prescriptive and Computer-Based Models of Design Process, *Res. Eng. Des.,* Vol. 1, pp. 51–67.

16. French, M., 1991, Towards a New Science of Functional Design, *ASME*, Vol. DE-31, pp. 241–244.

17. Gui, J. K., and M. Mäntylä, 1994, Functional Understanding of Assembly Modelling, *Comput.-Aid. Des.*, Vol. 26, pp. 435–451.

18. Gupta, S., and M. Chandrashekar, 1995, A Unified Approach to Modelling Photovoltaic Powered Systems, *Solar Energy*, Vol. 55, pp. 267–285.

19. Henderson-Sellers, B., and J. Edwards, 1994, *Book Two of Object-Oriented Knowledge: The Working Object*, Prentice-Hall Australia, Sydney, Australia.

20. Jacobson, I., M., Christerson, P. Jonsson, and G. Övergaard, 1992, *Object-Oriented Software Engineering: A Use Case-Driven Approach*, ACM Press, New York.

21. Jakobsen, I., 1990, The Interrelation Between Product Shape, Material and Production Methods, *Proceedings of International Conference on Engineering Design*, IMechE, London, pp. 775–784.

22. Johnson, A. L., 1990, Functional Modelling: A New Development in Computer-Aided Design, *Proceedings of IFIP TC 5/WG5.2 Second Workshop*, H. Yoshikawa and T. Holden (eds.), Elsevier, Science Publishers B. V., Amsterdam, pp. 203–212.

23. Kesavan, H. K., and M. Chandrashekar, 1972, Graph-Theoretic Models for Pipe Network Analysis, *J. Hyd. Div. ASCE*, Vol. 98, pp. 345–364.

24. Kesavan, H. K., and J. Dueckman, 1982, Multi-terminal Representations and Diakoptics, *J. Franklin Inst.*, Vol. 313, pp. 337–352.

25. Kesavan, H. K., M. A. Pai, and M. V. Bhat, 1972, Piecewise Solution of the Load Flow Problem, *IEEE Trans. PAS*, Vol. 92, pp. 1382–1386.

26. Koenig, H. E., Y. Tokad, and H. K. Kesavan, 1967, *Analysis of Discrete Physical Systems*, McGraw-Hill, New York.

27. Kota, S., and C. L. Lee, 1993, General Framework for Configuration Design. I. Methodology, *J. Eng. Des.*, Vol. 4, pp. 277–293.

28. Kumara, S. R. T., and S. V. Kamarthi, 1991, Function-to-Structure Transformation in Conceptual Design: An Associative Memory-Based Paradigm, *J. Intell. Manuf.*, Vol. 2, pp. 218–292.

29. Kusiak, A., and N. Larson, 1995, Decomposition and Representation Methods in Mechanical Design, *Trans. ASME*, Vol. 117, pp. 17–24.

30. Mäntylä, M., 1989, Directions for Research in Product Modelling, in *Software for Manufacturing*, D. Kockan and G. Olling, (eds.), Elsevier, Amsterdam.

31. Matousek, R., 1963, *Engineering Design: A Systematic Approach*, Blackie, London.

32. McPhee, J., 1996, On the Use of Linear Graph Theory in Multibody System Dynamics, *J. Nonlinear Dynamics*, Vol. 7, pp. 73–90.

33. Mikulecky, D. C., 1993, *Application of Network Thermodynamics to Problems in Biomedical Engineering*, New York University Press, pp. 89–107, 183–195.

34. Pai, M. A., 1979, *Computer Techniques in Power Systems*, McGraw-Hill and Tata, New Delhi.

35. Rumbaugh, J., M. Blaha, W. Premerlani, F. Eddy, and W. Lorensen, 1991, *Object-Oriented Modelling and Design*, Prentice-Hall, Englewood, Cliffs, NJ.

36. Sata, T. (ed.), 1989, *Organization of Engineering Knowledge for Product Modelling in Computer Integrated Manufacturing*, Elsevier, Amsterdam.

37. Shah, J. J., and M. Mäntylä, 1995, *Parametric and Feature-based CAD/CAM*, Wiley, New York.

38. Thulasiraman, K., and M. N. S. Swamy, 1992, *Graphs: Theory and Applications,* Wiley, New York.

39. Tomiyama, T., Y. Umeda, and H. Yoshikawa, 1993, A CAD for Functional Design, *Ann. CIRP,* Vol. 42, pp. 143–146.

40. Welch, R. V., and J. R. Dixon, 1992, Representing Function, Behavior and Structure During Conceptual Design, *ASME,* Vol. DE-42, pp. 11–18.

41. Wirfs-Brock, R., B. Wilkerson, and L. Wiener, 1990, *Designing Object-Oriented Software,* Prentice-Hall, Englewood Cliffs, NJ.

3

TIME-DRIVEN
PRODUCT DEVELOPMENT

MAHENDRA S. HUNDAL
The University of Vermont

3.1 INTRODUCTION

Manufacturing has become more globally competitive in recent years. Smith and Reinertsen[14] emphasize that the key ingredients in a manufacturing company's economic viability are (1) product innovation and (2) quickly developing and bringing products to market. Theproduct development cycle time—from the moment that the companyrealizes that a product should be developed to the time it is in the customer's hands—can be reduced by applying certain techniques of *time-driven development.* This is a procedure used by management to focus on compressing the time taken to develop a product. Other terms used for this technique are *rapid product development* and *accelerated product development.* Companies that have applied the technique of time-driven product development have found time reductions of 50 percent or more in the planning–design–manufacturing cycle. Examples of such companies (products in parentheses) are shown in Figure 3.1.[14] The highly competitive nature of the marketplace requires that manufacturers be able to deliver to the customer products, with the following attributes: (1) high quality, (2) short delivery time, and (3) low cost.

This chapter describes both the management and the engineering aspects of time-driven product development. Steps in the product realization process and the benefits of rapid product development are discussed. Engineering

Integrated Product and Process Development, Edited by John Usher, Utpal Roy, and Hamid Parsaei
ISBN 0-471-15597-7   © 1998 John Wiley & Sons, Inc.

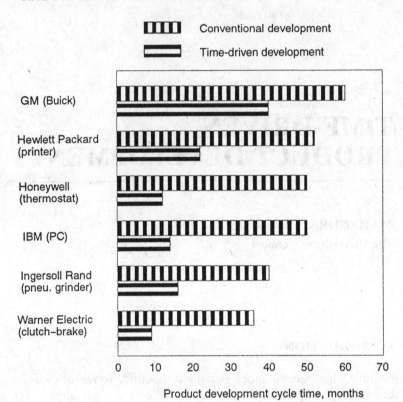

Figure 3.1 Examples of accelerated product development.

aspects include planning, design, and manufacturing. The product specifications list and product structure are the two important determinants of a product's characteristics set during the planning phase. Management considerations include a trade-off study to determine the economic feasibility of rapid product development; product innovation by stages; and the overlapping of activities in planning, design, and manufacture. The chapter includes examples of the aforementioned processes.

3.2 PRODUCT REALIZATION PROCESS

The realization of a product consists essentially of two steps: design and manufacturing. However, in order to understand the process, we must look at the total life of a product. The life of a product begins with its planning and ends with its disposal (i.e., scrapping and/or recycling). Figure 3.2 shows the different life phases of a product. These will be described in brief.

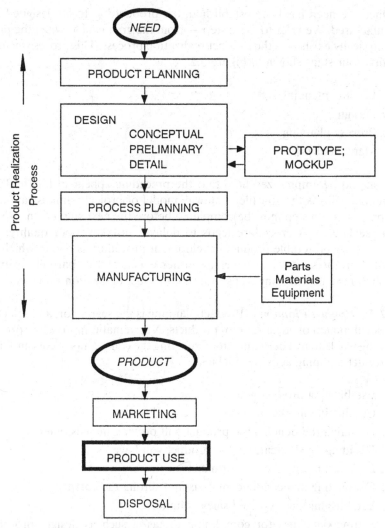

Figure 3.2   Life phases of a product.

3.2.1.   Steps in the Product Realization Process

The need for a product, whether real or imagined, must exist. This may come from external or internal sources. The external causes for a new product might include a direct order from a customer, the obsolescence of an existing product, the availability of new technologies, or a change in market demands.

Internal to the company, new product ideas may come from either new discoveries and developments within the company or the need for a product identified by the marketing department.

Once the need has been established, the product has to be designed and manufactured. We refer to these steps—from the go-ahead to when the product physically exists—as the product realization process. This process includes the first four steps shown in Figure 3.2:

1. Product planning
2. Design
3. Process planning
4. Manufacturing

It should be emphasized here that the preceding appear in Figure 3.2 as sequential. This is the so-called gateway model of project administration. In this process, each step must be completed before the next step begins. As we shall see later, this procedure leads to delays, mistakes, poor quality, and high costs. A preferable product development procedure is one in which the activities in the successive stages are either concurrent or partially overlap. We shall next describe the steps in in the product realization process.

3.2.1.1 Product Planning Product planning is the search for, and selection and development of, ideas for new products. A systematic approach to product planning will lead to a better means of meeting the constraints of cost and time.
Product planning activities include:

1. Establishing product goals
2. Conducting market analyses
3. Detailing the benefits the product will provide the customer
4. Deciding on the features the product will have
5. Establishing product performance
6. Conducting an economic analysis and setting the cost target
7. Establishing the expected sales volume
8. Setting deadlines for completion of tasks, such as design, prototype building, and setting up the manufacturing line

The two most important entities involved in making the decision to develop a product are the company and the market. There are also secondary factors, such as laws, economic policies, and the current state of technology. This interaction is shown in Figure 3.3. Specifically, the company needs to define its objectives and examine its capabilities. The strengths that a company has are its personnel, its facilities, and its financial situation.

The personnel and facilities are distributed among various types of activities or departments (e.g., design, production, and marketing) and among different buildings, such as those for design, test, and production equipment and those for distribution systems. An evaluation of resources and objectives will help

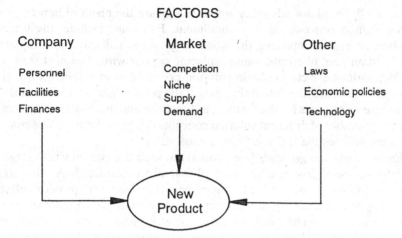

Figure 3.3   Factors influencing new product development.

focus the company on the type of product it should develop. Beitz[1] has provided a methodology for product planning that is strongly based on the VDI Guideline 2221.[18]

3.2.1.2 Design   The first major step in the design of a product is the preparation of the requirements or specifications list. This list includes the overall function of the device and any subfunctions that may be foreseen by the designer. The specific requirements are classified according to (a) life phases of the product (e.g., design and development, manufacturing, marketing, product use, and disposal) and (b) types of requirements (e.g., technical, economic, ergonomic, and legal). The most important are the technical requirements for the product use.

Conceptual design is the most important phase of design; it has the single largest influence on costs. A large portion of the product costs have been committed after only a small portion of the development resources have been expended in this phase. The concept determines the arrangement of elements through the function structure and thus the flows of energy, material, and signals in the system. The starting point in developing the concept is the determination of the functions that must be fulfilled. The function structure depicts the main functions that the product must fulfill, often in a sequential form. Each function can be realized by one or more solutions. Thus, a functional analysis of the requirements is needed, which may lead to a breakdown of complex functions into simpler subfunctions. Functional analysis is also the basis of value analysis, an accepted technique for product improvement. In the next step, physical effects are considered, which determine how the functions are realized. The solution principles used determine how complex the product will be. Major advances in technology are the result of new concepts rather than improvements in embodiment. Examples include: the internal combustion engine replacing the steam engine; the ball point pen replacing the fountain pen; fiberoptic cable replacing copper wire. The next step is to use the various effects to obtain solution possibilities in a sketch form (i.e., without considering the actual shapes that the parts might have). By considering different solutions for the various subfunctions and using systematic combination, a number of different solution concepts can be generated. The concept that best satisfies the specifications is chosen.

Embodiment design leads the process through a more concrete stage as the shapes, materials, and motions (flows) are determined. At this stage, it is no longer economical to change the functions and physical effects. Part shapes and sizes, fastening methods, materials, and processes are in their final forms. The final or *detail design* phase leads to production drawings. The final decisions on dimensions, arrangement, shapes of individual components, and materials are made. The design proceeds from a more abstract level at task clarification and takes on a more concrete form as it approaches this phase.[11]

3.2.1.3 Process Planning   Also called production or manufacturing planning, process planning involves decisions on how the product is to be manufactured, for example, the steps required to manufacture the product, the manufacturing processes, machines, tools required, how the parts are to be assembled, and so forth. Steps in process planning are producibility analysis, initial process design, vendor/sourcing selection, and tooling design.

3.2.1.4 Manufacturing   Under manufacturing, we include materials handling, production of parts, assembly, quality control, and related activities. Many of the decisions regarding manufacturing have already been made during the embodiment design stage, knowingly or unknowingly. At the design stage,[4, 11] the following items have the largest influence on manufacturing:

1. *Overall design arrangement* relates to the manufacturing process due to the nature and number of subassemblies and components.
2. *Form design* of components determines the methods used and their quality.
3. *Materials* selected for components determine the production methods, storage and handling, and quality.
4. Number and type of *standard* and *purchased parts* relate to production capacity and storage.

We look next at the main topic of this chapter, namely, the importance of time in the product development process.

3.3 ACCELERATED PRODUCT DEVELOPMENT

A shorter time to market does not necessarily mean higher costs. A properly managed program can yield high-quality products at low cost and under time constraints. These three properties depend not only on the physical features of the product, (i.e., geometry, materials, and tolerances) but also on how the development process is carried out. This will become apparent from the following discussion.

3.3.1 Cost and Price

From the time the development of a product begins, its cost starts to grow because of the resources used for its realization—personnel, facilities, and equipment. The price of a competitor's product in a free-market economy, however, drops with time. There are several reasons for the latter, for example, rationalization of the manufacturing process, cost-driven improvements in design, and increased knowledge about the product—the "learning process." These two variables—price and cost—are shown in a simplified form in Figure 3.4. The reasons for accelerating the development of products are obvious from this figure.

3.3.2 Benefits of Rapid Product Development

Time-driven product development, as opposed to the conventional development process, benefits a company in many ways:

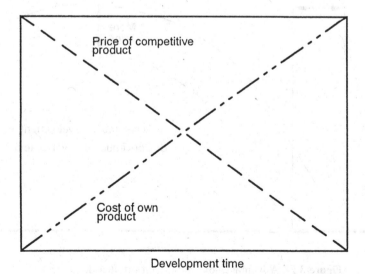

Development time

Figure 3.4 Cost and price change with time.

1. It extends the product's sales life.
2. By early introduction, the product has a marketplace advantage by gaining early customers who lock on to it, develop loyalty, and are less likely to switch.
3. The company gains a pricing advantage and gets on the learning curve ahead of the competition.

These points are illustrated in a simplified form in Figure 3.5. The sales curves of two products, A and B, are shown. Product A is developed faster than B and therefore launched earlier than B. Product A therefore has that much longer a sales life due to its earlier introduction into the market. It also shows higher sales than B due to its earlier introduction, because the market for it begins to build up earlier

A further advantage[16] is that:

4. A company that applies rapid development methods and starts on a product later than a traditional company will use more up-to-date technology in the product.

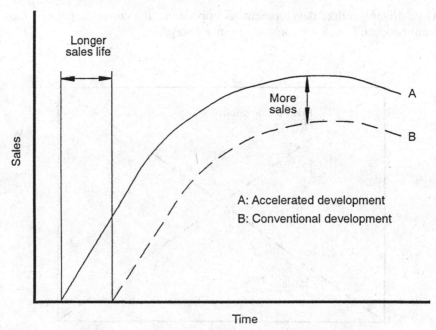

Figure 3.5 Advantages of time-driven product development.

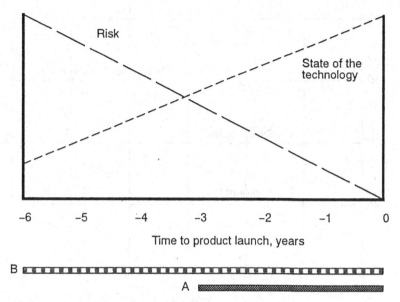

Time to product launch, years

Company B: Conventional development   Company A: Accelerated development

Figure 3.6   State of technology, starting time, and risk.

This point is illustrated in Figure 3.6. The line labeled "State of the technology" indicates how the technology is advancing with time. Company B, carrying out the traditional development process, foresees a six-year development cycle. When it starts its product development, it locks on to the technology at that point in time. The state of the technology in most fields, however, is constantly advancing. Company A applies rapid development methods and foresees a three-year development cycle. Both products reach the market at the same time (time 0). Company A thus starts with a higher state of technology and an advantage over company B. An example of the latter is the Hubble Space Telescope, which uses the technology of the 1970s—when its development began.

Figure 3.6 also illustrates another point. The line labeled "Risk" shows the manner in which the risk that a company takes increases the earlier it starts product development, in relation to the expected launch date. The market is a moving target. The length of time it takes for product development is very critical. The longer the time to product introduction, the more uncertain will be the market forecast and therefore the greater the risk. Unexpected changes may take place in the marketplace. Because company A begins its product development only three years before the launch date, it is taking much less

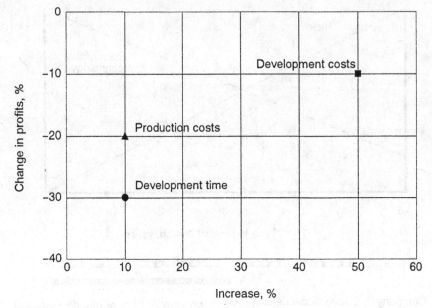

Figure 3.7   Effect on profits.

risk than company B. The latter must look ahead six years when it begins its product development.

The effect on profits[19] due to deviations in development time, production costs, and development costs is shown by the three discrete points in Figure 3.7. The figure shows, for example, that an increase of 50 percent in development costs can decrease profits by 10 percent, but an increase of 10 percent in development time can reduce profits by up to 30 percent. Therefore, it benefits a company to shorten the development time, even at the expense of some increased development costs.

3.4   MANAGEMENT FOR RAPID PRODUCT DEVELOPMENT

It is a well-established fact[12] that the primary initiatives in achieving rapid product development must come from the top management of the company. The decisions that must be made and the procedures to be implemented are several. These include economic decisions, realizing the importance of time in the early stages of product planning, decisions regarding product innovations, rapid product development team setup and dynamics, project management, and allocating resources, to name a few.

3.4.1 Economics of Rapid Product Development

The decision whether to proceed with a product development project must be based on facts rather than intuition. It may appear at first that it is difficult to quantify the costs and benefits of the various development goals. Nevertheless, even the use of gross approximations is better and more easily justifiable than an instinctive decision.

Figure 3.8 shows the four primary factors to be considered in the product development decision. These are the following:

1. *Product Features.* The features of the product determine its performance which is an important determinant of its market success. Product performance is determined by the customer and the marketplace, not merely by satisfying what is in the specifications.
2. *Product Cost.* This is the total product cost over its life cycle, including the purchase price and the operating, maintenance, and disposal costs.
3. *Development Speed.* Time to market is obviously the most critical factor. This is the total time from the moment when someone thinks about developing the product to the time it is in the customer's hands. The speed of product development determines the time to market. As stated

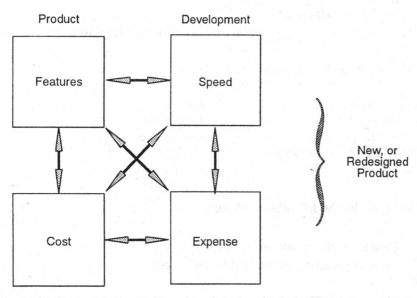

Figure 3.8 The trade-off parameters. Used with permission from *Developing Products in Half the Time: New Rules, New Tools,* Preston G. Smith and Donald G. Reinertsen, Van Nostrand Reinhold, 1988.[14]

elsewhere, it can be crucial to the success of the product. It involves all of the company's departments, not just design engineering.

4. *Development Expense.* These are the one-time costs associated with the development of the product including the one-time costs associated with the product and extra expenses for items such as overtime, facilities, and consultants. Although seldom an overriding concern, this expense must be justified to the upper management. Rapid product development will show extra costs on the balance sheet and must be justified by savings in time that would otherwise be spent.

Each of these factors needs to be weighed against the other three on the basis of costs and benefits before the decision to proceed can be made. In order to make these trade-offs, it is necessary to develop an economic model of the project. At the beginning stages of product development, when not enough information is available, a simple model will suffice. This model leads to better decisions and is usually preferable to more sophisticated and complex approaches and is certainly an improvement over a decision based on intuition only.

As an example of computing two of the six trade-offs (see Fig. 3.8), we consider the following case:

Product: Field trimmers
Expected sales: 2500 units a year
Unit price: $1500 each
Total expected sales: $3,750,000 over a period of five years

The profits and tax figures are as follows:

Gross profit margin: 32%
Profit before tax: 10%
Incremental profit: 15% of sales
(extra profit for each extra $ of
    sales)

We consider the following trade-offs:

1. Product performance versus product cost
2. Product performance vs. development cost

The option to be considered is whether to add a new feature (i.e., a self-starter). The feature is expected to increase sales by 15 percent, product cost by 20 precent, and will cost $15,000 to add. The benefits are increased profits, as follows:

BENEFITS:

Increase sales by 15%:   = $3,750,000 × 15% = $562,000
Increase profits at 15%   = $ 562,000 × 15% = $ 84,375
    profit margin:

The costs will now be computed for the two cases:

COSTS:

Product Performance versus Product Cost	Product Performance versus Development Cost
Product cost: = $3,750,000 × (1 − 0.32) = $2,550,000	Development cost: $15,000
Increase in cost: = $2,550,000 × 20% = $510,000	
Decision: Do not add the feature	Decision: Add the feature

Thus, although the development cost is well below the expected increase in profits, the large increase in product cost is a clear argument against adding the feature.

3.4.2 Early Stages of a Project

The lack of management of a product in its earliest stages can account for the greatest loss of time. A project can become dormant without people realizing it, as there is no one responsible to keep track of it; there may be several reasons for this. There might be a perceived need for the product but not enough importance is assigned to it; the company is not sure it has the technology for all parts of the product; expenses on the product increase only after its importance is recognized, while the market window of opportunity is closing. A management scheme that can shorten product development is to carry out product planning and design concurrently rather than sequentially. The planning and design are broken up into small steps and meshed together. Indeed, manufacturing process planning and certain early steps in manufacturing can also be carried out concurrently with design.

3.4.3 Product Innovation by Stages

All new projects, and new parts thereof, entail learning on the part of the personnel. The greater the degree of novelty, the more are the unknowns and therefore more risk is attached to the product's development. Rather than innovate on the whole product, it pays to use existing technologies as far as possible and to improve only on parts of the product at a time.

An example of such incremental innovation is the development of a heat pump by Mitsubishi as described by Stalk and Hout.[15] Over a period of several years, the product was improved in a step-by-step fashion as shown in Figure 3.9. At each step, Mitsubishi used existing technology rather than dramatic breakthroughs. At the end of a nine-year period, the company had an entirely new unit, but at no time did it expose itself to excessive risk. Other well-known examples of innovating in small steps are[14]:

IBM PC (and Compatibles)

· Started with an 8088 central processing unit (CPU). Upgraded in steps to 286, 386, 486, and Pentium.
· Put in larger memory chips as CPU was upgraded.

Sony Walkman

· Introduced in 1979, it was an unattractive design.
· Many small changes were made sequentially.
· Manufacturing improvements were made that reduced the cost and improved the performance.

1979 (Original design)

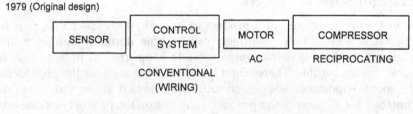

YEAR	FEATURE INTRODUCED	BENEFITS ACHIEVED
1980	Integrated circuits	Improved EER; reliability
1981	Microprocessor	Easier to install and service; increased market share
	"Quick-connect" Freon lines	Sold through Hardware stores
1982	Rotary compressor	Higher efficiency
	Louvered fins; inner-fin tubes	Better heat transfer
	New electronics	Improved Energy Efficiency Ratio (EER)
1983	New sensor	Better control of cycle
	More computing power	
1984	Inverter	Better speed control; higher EER
1985	Shape-memory alloys	Control of air louvers
1986	Fiberoptic sensor	Cycle adjustment; higher efficiency
1987	Remote control	Personalized control
1988	Learning Circuitry	Optimum cycle; timely defrosting

Figure 3.9 Incremental improvement of the Mitsubishi heat pump.

- Mechanical components were gradually replaced with electrical components.

Chrysler Minivan

- Planning was started in 1978, when gasoline prices were high.
- Introduced in 1984, with a four-cylinder engine (gasoline prices had dropped).
- V-6 engine introduced in 1987.

3.5 ENGINEERING ASPECTS OF RAPID PRODUCT DEVELOPMENT

The most important aspects in which engineering can contribute in rapid product development are in developing the specifications, in determining the structure (both the function structure and the subsequent solution structures[11]), and in integrating the product and process design. These will be discussed next.

3.5.1 Product Specifications List

The specifications or the requirements list is a most crucial document, prepared at the start of the development cycle. It should be prepared by design and manufacturing (at the minimum) working together. It should define for the yet nonexistent product, its functions and the constraints on it—not how the functions will be realized or the product's physical features. The specifications should be dictated by the customer's needs, expectations, and problems, benefits to the customer, and expected improvements over existing products. The specifications list is not per se an engineering document or a manufacturing plan. It should be as abstract as possible, as far as its physical features are concerned and not rule out any options by containing false constraints.[11] Examples of false constraints might include, for example, when a company assumes what the customer's needs might be or when a designer thinks what the best manufacturing process might be when in reality such is not the case. The specifications list should pay particular attention to the environment in which it will operate. The following checklist is helpful in developing the specifications list:

Environment	Physical environment	*Ergonomics*	Use patterns
	Other products to interface		User fatigue
	Power and other facilities		User abilities
			Ease of use

Manufacture	Production equipment available Sourcing for parts Assembly methods	*Quality*	Quality control Reliability
Safety	Protection systems Liability	*Marketing*	Transport Distribution
		Legal	Standards Patents

In developing the specifications list, the technique of quality function deployment (QFD) can be helpful but should not supplant the preceding checklist. Quality function deployment is a quantitative means of relating customer requirements, product properties, and design variables to manufacturing requirements.[5] However, an undue reliance on QFD can also lengthen the development cycle.[14]

3.5.2 Product Structure

After the specifications, it is the function structure of the product (see Section 3.2.1.2) that determines, to the largest extent, its complexity and thus its reliability. The solution structure, which is developed from the function structure, determines its modularity, and the decisions made there set the length of the development cycle. It has been estimated that 10 to 20 percent of the total time spent in the conceptual design phase determines 80 to 90 percent of the product cost. Figure 3.10 shows the trend in the degree to which product properties (e.g., functions fulfilled, shapes and materials used, cost, etc.) are determined as we proceed through the realization process. As the process moves through the planning, design, and manufacturing phases, the product becomes increasingly more concrete, and thus the product development team has less and less control on the product properties.

Each element of the solution structure is embodied by a module and the modules are interconnected by interfaces—typically electrical and/or mechanical. The decisions made at the embodiment stage determine which and how many functions are realized in each module and therefore how many modules and interfaces are present in the product. The greater the number of modules, the simpler each can be made because each module incorporates fewer functions. In this way there is also a greater potential for using existing technologies in more of the modules.

Increasing the number of modules increases the number of parts but provides more flexibility in design. However, it also increases the number of interfaces, which, in turn, can worsen reliability problems. On the other hand, integrating more functions into fewer modules reduces the number of parts

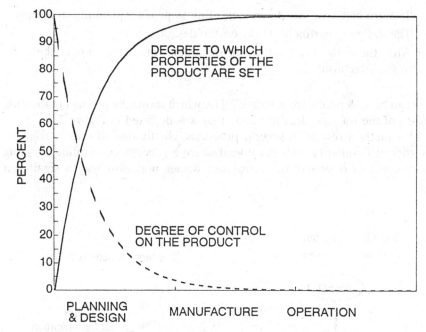

Figure 3.10   Product costs set at different stages of its life cycle.[10]

and generally results in a cheaper product. Kittsteiner[8] discusses the optimum selection of a common mechanical interface, namely, the shaft–hub connection. He presents design rules that lead to the lowest-cost solution, while exploiting the advantages of concurrent design and production planning.

Modular products have the advantage that they can be upgraded incrementally—one module at a time. Functions whose embodiments are likely to see rapid improvement in technology and/or reduction in price should be in separate modules. Further, modular products make it easier to produce design variants to satisfy a range of specifications.

3.5.3   Concurrent Product and Process Design

Present-day technical products tend to be complex in order to fulfill the functional and other requirements imposed on them. Product complexity has led to the emergence of complex organizations, with individual departments addressing the different functions (e.g., design and manufacturing). In most companies there are "walls" between these different departments; that is, the departments are physically separated and/or there is poor communication among the departments. As a result of such separation:

- The product idea goes from marketing to product planning to design.
- The design department works on the design.
- After the design is complete, the documents are "thrown over the wall" to manufacturing.

This can be understood from Figure 3.11a, which shows the product realization phases of the total product life cycle that was depicted in Figure 3.2.

This method results in several problems, chiefly due to most designers' insufficient familiarity with the manufacturing process. A misunderstanding on the part of designers of a customer wishes may also play a significant

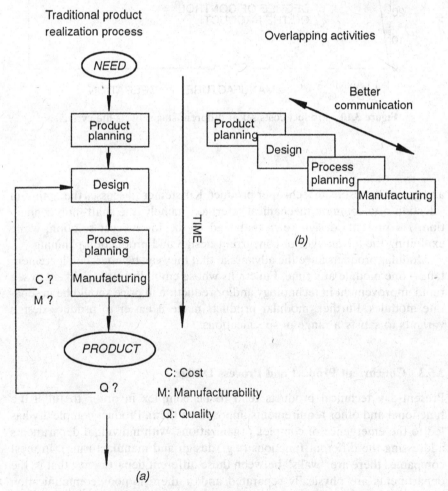

Figure 3.11 Evolution of the product realization process.

role. Often last-minute changes must be made at the manufacturing stage; a consequence is poor product quality as well as higher life-cycle costs. This is indicated in Figure 3.11a. The design process, rather than being isolated, must obtain inputs from all parties involved in the product's life cycle: planning, marketing, manufacturing, and the customer. The parties involved and their number will vary, depending on the product and the company.

There are other reasons for delays in the product development process. The perspectives of the different departments on the product are different. The following list shows how marketing, design, and manufacturing view the process:

Department	Views on the Product
Marketing	As good or better than the competitors'; more features
Design	Uses the latest technology; exciting to design and test
Manufacturing	Has a stable production line; processes easy to set up

It is apparent that these views are in conflict. For example, a stable production line requires that there be no or few changes in the product. New designs may require new production processes to be used. It is incumbent upon the management to develop a sense of common purpose among the various groups.

In recent years, some companies have begun to coordinate design and manufacturing process planning. They have found that by doing so they could address not only the quality and cost issues, but also reduce the time to market. The earlier in the cycle the product is, the more control one has on its properties (shown by the dotted line in Fig. 3.10) and changes are cheaper to implement. Designs do not have to be modified because of manufacturing problems. This evolution of the product realization process is shown in Figure 3.11b. Coupling design and manufacturing shortens the product development time by overlapping the different activities. As stated earlier (see Fig. 3.1), this can amount to 50 percent or more. There is a two-way flow of information between the successive groups that also leads to fewer misunderstandings and errors.[3] We shall look at overlapping activities again later in this chapter.

3.5.4. Concurrent Engineering

As stated previously, the "concurrent" involvement and overlapping of activities and better communication between the departments have proven to reduce development time, reduce costs, and increase quality. The modern product realization paradigm is "the process must be designed at the same time as the product." We shall discuss here how concurrent engineering and other management initiatives may be applied in lowering costs. The benefits obtained by integrating design and process planning can be extended by overlapping more of the activities in the development cycle, particularly the product planning.[2, 17] Concurrent engineering may also include an interplay with marketing, sales, field service, and even suppliers.

3.5.5   Example: Concurrent Engineering in the "Skunk Works"

Lack of communication between departments is a problem in large organizations. This problem can be addressed by the formation of cross-functional teams that operate with a high degree of autonomy. One of the most famous examples of such a unit within a large company is the "Skunk Works" of Lockheed, formed during World War II under Kelly Johnson. This unit was responsible for the quick development and manufacture of many superb airplanes such as the P-80 Shooting Star, the F-104 Starfighter, the U-2, the SR-71 Blackbird, and the F-117 stealth fighter, among others. Kelly Johnson had rules for operating the "Skunk Works,"[13] including:

· "There will be only one object: to get a good airplane built on time."
· "Engineers shall always work within a stone's throw of the airplane being built."
· "Special parts or materials will be avoided whenever possible. Parts from stock shall be used even at the expense of added weight."
· "Everything possible will be done to save time."

The program manager on the development of F-117 stealth aircraft pointed out[13] how concurrent engineering was implemented on this project: "Our engineers were expected on the shop floor the moment their blueprints were approved. Designers lived with their designs through fabrication, assembly and testing. Engineers couldn't just throw their drawings at the shop people on a take-it-or-leave-it basis. Our designers spent at least a third of their day right on the shop floor; at the same time, there were usually two or three shop workers up in the design room conferring on a particular problem." It may be argued that perhaps these people were motivated by national security concerns, but they were also proud to be building excellent products under severe time constraints.

Munro[9] suggests implementing the product designers spend a day on the production line implementing and assembling their designs.

3.6   OVERLAPPING ACTIVITIES

Fully, concurrent engineering is an ideal that is seldom achieved. A frequent compromise is a partial overlapping of each phase of the product realization process with the next phase. The advantage of overlapping activities[6] over that of sequential activities is shown with the aid of Figure 3.12. Let us say that activity i is designing and $i + 1$ is manufacturing. The results of activity i are delivered upon its completion to $i + 1$. The group pursuing activity $i + 1$ needs some time to "digest" the information (designs and documents) given to it. It then begins with its task and proceeds to complete it in a certain length of time, as shown in Figure 3.12a.

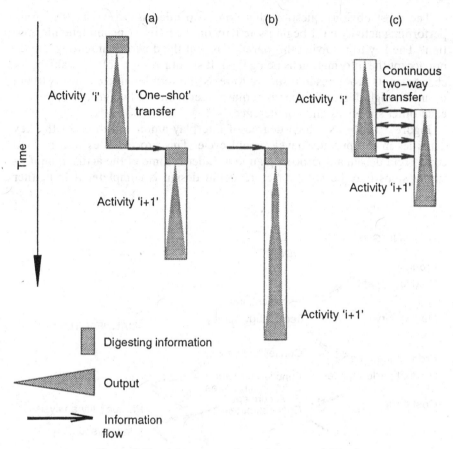

Figure 3.12 Advantages of overlapping activities.[7]

However, such an ideal, trouble-free process does not often take place in real life. There are glitches such as parts that cannot be made economically as designed or an assembly of parts that runs into problems. Thus, quick fixes and changes must be made that lead to mistakes, requiring more fixes. The time taken for activity $i + 1$ gets longer, as shown in Figure 3.12b. At the same time, due to the unplanned changes and hurried decisions, product costs increase and quality suffers.

Figure 3.12c shows an improved mode of operation. There is a continual sharing of information between the groups engaged in the consecutive activities, as well as an early resolution of conflicts. Between product planning and designing such information might concern the state of technology that affects the feasibility of a design. Sharing information between process planning and the design groups can uncover manufacturability concerns earlier than would otherwise happen.

The most obvious question that comes to mind is: How can the group performing activity $i + 1$ begin its activity on the basis of incomplete information? The key to resolving this paradox is that the downstream group should not accept the information as being final. It should also be able to make quick changes in its processes in response to revised information from the upstream group. A high degree of trust is required between the two groups regarding each other's motives and competence.

Figure 3.13 shows a high degree of interplay among all three of the key activities in the product development cycle. This process goes well beyond concurrent design and process planning. Indeed, some of the actual manufacturing steps may be started before detail design is complete. It is neither

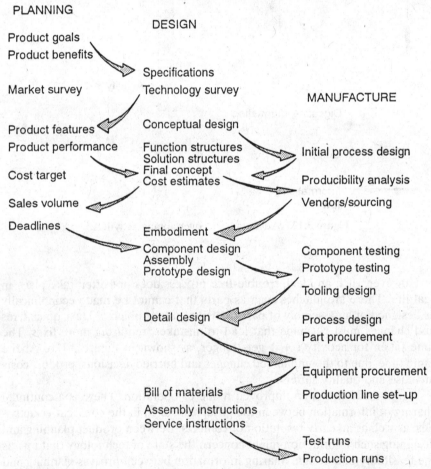

Figure 3.13   Overlapping of activities in planning, design, and manufacture.[7]

necessary for planning to be complete before design starts nor for the latter to be done for manufacturing to begin. There can be continuous two-way flow of information between the activities. Sufficient information is available at appropriate stages of each activity to enable starting some operation in the next. For a large project this interplay can be fostered by dividing it into smaller subprojects.

The sequence of activities is as follows:

1. The design group can begin preparing the requirements list and performing the technology survey as soon as planning has identified the market position of the product, its price range, new features, and its relation to other company products.

2. When specific desired features and performance have been identified and a cost target set, design can use this input to work on the concept.

3. The last stages of planning, namely, sales forecasting and setting deadlines for subsequent activities, are carried out concurrently with, or follow, the concept development.

4. During conceptual design, as the product structure is finalized, manufacturing begins to look at the production process design, analyzes producibility, and begins selecting vendors.

5. At this time, embodiment has started, leading to specific shapes of components and assembly and then to the initial prototype design.

6. Test component production or procurement is succeeded by the building and testing of the prototype.

7. Following development of the prototype, the details of the design are finalized, while final tooling and process design are implemented.

8. The final activities are the procurement of parts and any special production equipment and setup of the production line.

9. Design prepares the final documents on the bill of materials and assembly and service instructions, while manufacturing sets up the test run, followed by production.

3.6.1 Example: Time to Market with Traditional and Concurrent Engineering

Figure 3.14 shows the differences in the time required for bringing a product to market by using the traditional methods and a time-driven approach. The example is that of a company making specialty mechanical power transmissions, described by Smith and Reinertsen.[14] The company had branches in the United States and Japan. The Japanese branch, using cross-functional teams and the overlapping of the various product realization activities, was able to finish the product in less than half the time.

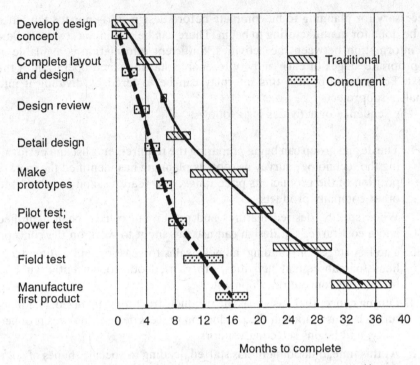

Figure 3.14   Time to market for a mechanical transmission.[14]

3.7 SUMMARY

The chapter began with a discussion of the product realization process, which consists of product planning, design, process planning, and manufacturing. We then showed why development time is such an important factor. Accelerated development of the product results in a longer sales life, a marketplace advantage by gaining early customers, a pricing advantage for the company, and the ability to use more up-to-date technology in the product. An increase in development time has a much greater effect on profits than, say, an increase in production costs or development costs. The decision regarding accelerated product development must consider the trade-off among the four parameters of interest: product features, product cost, development speed, and development expense.

Often, the development of a product is delayed because, in the early stages, no one in the company realizes its importance. Another factor that can cause delays is when a company tries to develop an entirely new product, often using new technology. There is less risk if a product is improved in stages, using only tried and tested technology.

The specifications list is a crucial document in the product realization process. It should be dictated by the customer's needs, pay particular attention to the environment in which it will operate, be as abstract as possible, and not rule out any options by containing false constraints. Conceptual design begins with the development of a function structure, followed by a solution structure. The next step is to embody each element of the solution structure as a module. Modular products provide many advantages: They allow flexibility in design and procurement; they can be upgraded incrementally; and more design variants can be produced.

In closing, the chief factors that facilitate faster development of products are better communication among departments (marketing, planning, design, and manufacturing). Better communication also leads to lower cost and higher-quality products. The company increases its market share and enjoys higher profits by the early introduction of a product.

REFERENCES

1. Beitz, W., 1991, Methodik der Produktplannung, *Proceedings of the International Conference on Engineering Design (ICED '91)*, Heurista, Zurich, Vol. 1, pp. 254–261.

2. Clark, K., and F. Takahiro, 1989, Reducing Time-to-Market: Case of World Auto Industry, *Design Management J.*, Vol. 1, No. 1, pp. 49–57.

3. Dean, J. W., and G. I. Susman, 1989, Organizing for Manufacturable Design, *Harvard Business Rev.*, Vol. 67, No. 1, pp. 28–36.

4. Ehrlenspiel, K., 1985, *Kostenguenstig Konstruieren (Designing for Cost)*, Springer-Verlag, Berlin.

5. Hauser, J. R., and D. Clausing, 1988, House of quality, *Harvard Business Rev.*, Vol. 66, No. 3, pp. 63–73.

6. Hayes, R. H., S. C. Wheelwright, and K. B. Clark, 1988, *Dynamic Manufacturing*, Free Press, New York.

7. Hundal, M. S., 1995, Time and Cost-Driven, *Design for Manufacturability*, ASME, New York, Vol. DE-81, pp. 9–20.

8. Kittsteiner, H. J., 1990, *Die Auswahl und Gestaltung von kostenguenstigen Welle-Nabe-Verbindungen (The Selection and Designing to Cost of Shaft-Hub Connections)*, C. Hanser Verlag, Munich.

9. Munro, A. S., 1995, Let's Roast Engineering's Sacred Cows, *Machine Design*, February 9, pp. 41–46.

10. Nevins, J. L., and D. L. Whitney, 1989, *Concurrent Design of Products and Processes*, McGraw-Hill, New York.

11. Pahl, G., and W. Beitz, 1988, *Engineering Design—A Systematic Approach*, Springer-Verlag, Berlin/New York.

12. Quinn, J. B., 1985, Managing Innovation, *Harvard Business Rev.* Vol. 63, No. 3, pp. 73–84.

13. Rich, B. R., and L. Janos, 1994, *Skunk Works*, Little, Brown, Boston.

14. Smith, P. G., and D. G. Reinertsen, 1991, *Developing Products in Half the Time,* Van Nostrand Reinhold, New York.

15. Stalk, G., and T. Hout, 1990, *Competing Against Time: How Time-Based Competition Is Reshaping Global Markets,* Free Press, New York.

16. Suri, R. N., 1995, Common Misconceptions in Implementing Quick Response Manufacturing, *J. Appl. Manufacturing Systems,* Vol. 7, No. 2, pp. 9–20.

17. Takeuchi, H., and Ikujiro, N., 1986, The New Product Development Game, Harvard Business Rev., Vol. 64, No. 1, pp. 137–146.

18. VDI Guideline 2221, 1987, *Systematic Approach to the Design of Technical Systems and Products,* VDI Verlag, Duesseldorf.

19. Wolfram, M., and K. Ehrlenspiel, 1993, Design Concurrent Calculation in a CAD-System Environment, *Design For Manufacturability,* ASME, New York, Vol. DE-52, pp. 63–67.

4

AN INTEGRATED DATA MODEL OF FUNCTION, BEHAVIOR, AND STRUCTURE FOR COMPUTER-AIDED CONCEPTUAL DESIGN OF MECHANISMS

W. J. ZHANG and D. ZHANG
City University of Hong Kong

K. VAN DER WERFF
Technical University of Delft

4.1 INTRODUCTION

4.1.1 Background

One of the common characteristics in the current literature for intelligent computer-aided conceptual design for general products, as well as more specific products such as mechanical systems, is the interest in the mechanism for matching functions to physical effects. The matching process attempts to generate a solution (physical effect, form, structure, or artifact) given a required function. It is these physical effects that represent the aspects of the design that achieve

Integrated Product and Process Development, Edited by John Usher, Utpal Roy, and Hamid Parsaei
ISBN 0-471-15597-7 © 1998 John Wiley & Sons, Inc.

the desired functions. For example, a rack and pinion is a physical effect that carries out the function of transferring rotation to translation.

It is a common belief that there is a need to build a large knowledge/data base that includes physical effects at various levels of granularity. The term *granularity* means the level of semantics represented in a declarative manner. For example, we may only need to represent the structure of a four-bar mechanism in the knowledge/data base, but not the structure of a six-bar mechanism, as the six-bar mechanism may be composed of two four-bar mechanisms in series. The level of granularity depends on the particular application. Therefore, the generality of an integrated data model should demonstrate the possibility of representing various levels of granularity. Furthermore, such representations must be *integrated* and *neutral*. The sense of an integrated representation is that the function, behavior, and structure of a physical effect are explicitly represented and their "connecting points" are explicitly specified in the representation. The sense of a neutral representation is that (1) the representation of application semantics must be designer oriented and (2) the representation method must be less dependent on any particular programming language.

In the domain of designing production machines, mechanisms or motion-generation devices represent the major physical effects. This chapter examines data representations or models for mechanisms, which can be used to develop a knowledge/data base for computer-aided conceptual design of production machines.

4.1.2 Related Work

In CAD for general product design, the representation of function requirements has been the topic of research in many studies.[2, 13, 20, 22] Function requirements identified from many practical design cases have been classified.[13, 20] The semantics encoded in these representations were expressed by the syntax (e.g., verb + noun, noun + verb + noun, etc.), so a simple data structure is needed for the computer storage and retrieval of them.[13] These past studies have represented function requirements in the computer, but not the structure of physical effects. As a consequence, the computer "knows" nothing about structures, and the capability of automating the design of these systems is thus limited. Many studies were directed at representing different aspects of a product (e.g., from a structural viewpoint, functional viewpoint, etc.).[8, 12] In these studies, one could argue that the connection between functions and structures has been represented in the computer; however, we have found that the representation of product structures in such systems is only available in the form of a bill of material (BOM) for an overall structure. Note that the data representation of the connectivity of a structure is different from that of an overall structure as in a BOM.[44] Henson et al.[12] also point to the need to represent the connectivity of components, which is related to the kinematic behaviors of a machine product.

Data representation of product structures alone has long been of interest to academia.[2, 10, 11, 15, 17, 18, 21, 24, 37, 38, 40] At the Technical University of Delft in the 1980s, research was directed at applying database modeling technology to the description of mechanisms. The main purpose of the research was to develop a more intelligent user-interface system for a general-purpose package for kinematic and dynamic analysis based on a finite-element formulation. A mechanism data model was developed which is semantically based on the finite-element view of mechanisms and methodologically based on the relational modeling approach.[17, 21] It was subsequently recognized by Zhang et al.[38] that it might be possible to create a mechanism data model that is semantically based on an object-oriented description of mechanisms.[38] The main idea was to define a set of mechanism atom bodies and a set of mechanism atom joints and thus any complex mechanism structure can be described by making an analogy to the well-known principle: atom → molecule → matter. Based on this idea and applying the entity–relationship data modeling method,[3] the model was initially formulated,[38] and was later modified by using a semantic data modeling method.[40] The study to be presented in this chapter is a continuation of this work. In comparison with the Delft work, Hardell[10, 11] used a relational data modeling approach, which was considered to have a limited modeling capacity and which may lead to a slow data access speed.[17] The work reported by Otter et al.[24] used a proprietary object-oriented data modeling approach for modeling multibody structures. Both of these studies[11, 24] were limited to spatial multibody systems.

An international working group has been working on formulating a draft version of ISO 10303, Part 105, *Kinematics*[15]—kinematic mechanism information model based on EXPRESS[16] (hereafter referred to as the ISO model). The ISO model is based on the use of *link*[14] as the basic mechanism element, which is a major difference from our previously developed approach.[38, 40] In the ISO model, meaningful machine elements like bars, gears, and cams are not explicitly represented. As a consequence, the specific kinematic behavior of these elements must be specified in terms of their connections with other links. A more user-oriented system based on the ISO model remains to be investigated.

Data models for integrating various design and analysis tools for mechanical systems proposed by Wu et al.[37] makes use of a two-level (global and local) model. The information in the global data model is sharable by more than one application. "Links" between the global data model and local data models are based on the same attribute name being used in the two models. Such sharable attributes are further specified as (1) imported, (2) locally interpreted, and (3) locally determined, which delimits the applicable scope and flow of the data. For example, for an attribute (in a local data model) defined as imported, the contents of this attribute will be derived from the global data model. The focus of their study was on structural analysis. Their work represents one of the earlier attempts to create an integrated representation of structures and behaviors. The evidence is that structural attributes are associated with the results, calculated from analysis tools, which are behavioral in nature. However, their work did not include the representation of a type of behavior, which is described as an input and output relation. That type of behavior is an important concern in mechanism design.

Several other studies along this direction include Baxter et al.[1] and Henson et al.[12] In these studies, when relationships between functions and structures were described, the functions were limited to those that result in something being done, for example, to obstruct the flow of a fluid or isolate a food substance from contaminants.[1] The functions related to input and output relationships in mechanisms were also not covered.

4.1.3   General Ideas for an Integrated Data Model of Mechanisms

First of all, we shall clarify functions, behaviors, and structures, both in concept and in representation. In concept, we follow the widely accepted definitions from the literature.[3, 12] The *behavior* of a system is a description of the system in terms of its allowable states, the system's variables, and how those variables are related. The *function* of a system is the purpose of the behavior to the human user. The *structure* of a system is how the behavior is realized.

It is important to differentiate between them in representation, as one mechanism could have different types of behaviors of interest to different applications, and one behavior could be used for different purposes, that is, functions. This implies that it is unlikely that one can develop a knowledge/ data base about mechanisms to "embody" all situations. A "plug-and-play" strategy is better taken; that is, the system should allow users to plug their defined behaviours and functions into the knowledge base for their particular mechanism system design. Another point in representation is how to specify the "connecting points" between structures, behaviors, and functions. We conclude that there are three methods, which are briefly indicated in Figure 4.1. In Figure 4.1a, an external entity will be defined to model the connections between behaviors and structures (a diagram representation associated with the entity–relationship approach[3] is used here); in Figure 4.1b and c, such

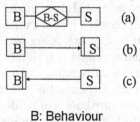

B: Behaviour
S: Structure

Figure 4.1   Plug-and-play strategy for the architecture of data models.

connections are specified in the information block of either the structure or the behavior, accordingly. It is obvious that Figure 4.1a shows a more promising method toward implementation of the plug-and-play strategy, which will be taken in building our model.

We shall develop a *generic* model in the sense that (1) it has been devised as a template to generate instances of individual mechanisms, as well as their behaviors and functions, and (2) it has been devised such that future extensions can be included without changing the existing model, which means, for example, that new types of machine components and joints can be added. Note that adding or removing behaviors and functions has been facilitated with the plug-and-play strategy. Because rich semantics have to be captured in the model, a hybrid semantic data modeling approach proposed by the authors will be used.[40, 44, 45] In this approach, the major model-building blocks are data abstractions such as generalization, aggregation, and grouping, and several basic concepts of data modeling such as type, class, and attribute. Because the data abstractions and these basic data modeling concepts have been essential to most data modeling systems today, a more neutral data representation can thus be envisaged.

4.1.4   Organization of the Chapter

The remainder of this chapter will be developed as follows. Section 4.2 introduces mechanism design as the application background. This will include a new description of the mechanism structure and behavior proposed by the authors. Section 4.3 outlines a hybrid semantic data modeling approach proposed by the authors as a tool to develop an integrated data model for mechanism design. Section 4.4 presents a data representation for the behavior and function of mechanisms, while Section 4.5 presents a data representation for the structure of mechanisms. Section 4.6 discusses how the relationships between structures, behaviors, and functions are represented. Section 4.7 describes the implementation and some uses of the data representations or models within the framework of an implemented computer-aided conceptual design for mechanism and machine (CACDMM) system being developed at Technical University of Delft and City University of Hong Kong. Finally, some conclusions are presented in Section 4.8.

4.2   MECHANISMS AND MECHANISM DESIGN

The mechanical unit of a machine performs motion and force transfer and is composed of machine components or bodies such as *bars* and *gears* and so on. This unit is usually called a mechanism.[14] Figure 4.2 shows the sketch of an example mechanism composed of bar and gear bodies. Some bodies and points (on bodies) are intended to be nonmovable or static. The set of static bodies and points is called the (global) *frame* of the mechanism. In Figure

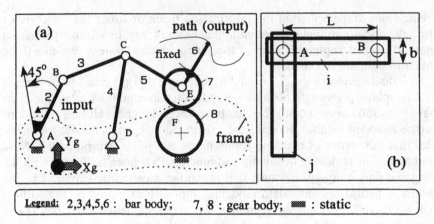

Legend: 2,3,4,5,6 : bar body;   7, 8 : gear body;   ▓▓▓ : static

Figure 4.2   Conceptual mechanism.

4.2*a*, points *A, D,* and *F* and gear body 8 belong to the frame. The static bodies and points (on bodies) may also be called *frame elements.* A coordinate system ($Xg–Yg$) will be attached to the global frame. In Figure 4.2*a*, body 2 performs input motion and a point on body 6 produces a path or trajectory as output.

Machine or mechanism system design is divided into several phases.[20, 29] The body geometry at the conceptual design phase is determined by geometric design parameters, which are called the *kinematic parameters.* The kinematic parameters describe the geometric properties of the mechanism as far as they are important to the kinematic motion. In Figure 4,2*b*, the distance between points *A* and *B* belongs to the category of kinematic parameters; the dimension *b* does not, however, belong to the kinematic parameters because it does not determine the kinematic motion. A data model for mechanism structures must cover the following semantics: (1) various types of machine bodies (their structure and geometry); (2) various types of joints among bodies; (3) various configurations of connected bodies, which form a mechanism; and (4) various configurations of the frame of a mechanism.

In the following discussion, a general description of the behaviors of a mechanism is given in order to lay a semantic foundation for the data representation to be discussed later. It is important to have a general description as the data model is intended to cover all mechanism design problems.

Behaviors in mechanism design can usually be expressed in terms of an *input–output relationship*—a mathematical function generally expressed as $y = f(x)$. Therefore, in the following discussion, we shall interchangeably use the terms behavior, behavior function, and function; otherwise, an explicit indication will be given. Considering a mechanism with one degree of freedom, the behaviors of mechanisms are usually classified into three well-known types as shown in Figure 4.3*a*. For each type shown, further classification is possible. For example, a path generator design problem can further be classified into two types, according to whether or not the points on a path are required to correspond to the prescribed input motion variables (Fig. 4.3*b*); if so, the description of the behavior will be a triple $\langle \alpha, x, y \rangle$ (see Fig. 4.3*b*). Note that the behaviors of a mechanism also include the types of input and/or output motion (e.g., full rotation, reciprocating motion, etc.).

A mathematical function is determined by its function expression $f(\ )$, which can be represented in either a discrete form (i.e., $\langle x_i, y_i \rangle$, $i = 1, 2, \ldots,$ n, or a continuous form. In mechanism system design, both forms could be possible in a design problem description. For example, in the design problem for a path generator, the discrete form is usually taken. In a total problem definition, a mixed expression of $f(\ )$ could be possible (Fig. 4.3*c* and *d*); each subfunction f_i, $i = 1, 2, \ldots, s$, can be expressed as either discrete or continuous. In particular, Figure 4.3*d* further implies that over one region of an active variable, both discrete and continuous forms of function representation may coexist.

Behaviors may also relate to the first-order (or higher) derivative of a mathematical function. For example, a dwell motion at a specified region ($x1$, $x2$) can be represented as $y'|_{x=x1, x2} = 0$, where x is an input motion variable and y an output motion variable. We define a *motion quantity* for y and its derivatives. Therefore, both y and y' are motion quantities. Any mechanism design problem (kinematic in nature) can thus be represented by a set of expressions of motion quantities. (These expressions are called *motion conditions.*) For instance, one expression might be $y'|_{x=30} = 3.5$. It is also possible that an expression for a motion condition could also be composed of more than one motion quantity, for example, the expression $y_{max} - y_{min} = 30$, where both y_{max} and y_{min} are motion quantities. Note that this is a well-known problem called a *stroke generator.*

Although behaviors can be expressed as described previously, a CAD/ CAM system would be much more user friendly if the required behaviors

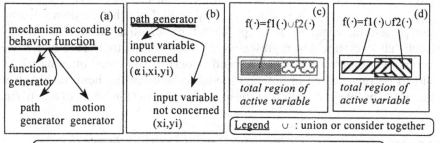

Note: α=input motion variable; (xi,yi) is coordinate of a point on a path

Figure 4.3 Behavior types in mechanism design.

Function features   Feature graphical representation   mathematical expression

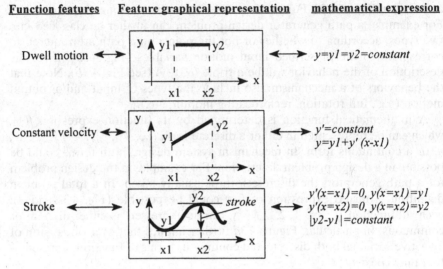

Dwell motion ⟷   ⟷ $y=y1=y2=constant$

Constant velocity ⟷   ⟷ $y'=constant$
                                                                  $y=y1+y'\ (x-x1)$

Stroke ⟷   ⟷ $y'(x=x1)=0,\ y(x=x1)=y1$
                                                                  $y'(x=x2)=0,\ y(x=x2)=y2$
                                                                  $|y2-y1|=constant$

Figure 4.4   Behavior features classification.

could be expressed by the end user (the product developer) using his or her domain-specific terminology.[36, 40] The underlying philosophy of this idea is analogous to that behind feature-based solid modeling, that is,

Behavior Modeling		*Solid Modeling*
Level 1 *Behavior features*	⟷	*Form features*
Level 2 *Motion quantity models*	⟷	*Solid models*

Figure 4.4 lists several well-known behavior features used in mechanism design.

For the *stroke* design problem, at level 1 (the user front end) the problem is simply stated as finding a mechanism with output stroke = 30 (mm); at level 2 the problem is expressed as the motion quantity model: $y'|_{x1} = 0$, $y'|_{x2} = 0$, $y1 = y|_{x1}$, $y_2 = y|_{x2}$, and $|y_2 - y_1| = 30$. Note that we do not intend to represent the complete design problem represented at the motion quantity level; that level should be represented in those synthesis tools. However, some necessary information that can be used to derive the motion quantity model should be represented in the data model to be discussed here, such as (for the stroke problem) the information of x_1, x_2, y_1, y_2, and the length of the stroke.

4.3   HYBRID SEMANTIC DATA MODELING APPROACH

The task to develop an application data model includes a thorough understanding or interpretation of application semantics and the selection of a suitable tool to represent the captured semantics in an appropriate manner. This section introduces our selected tool. Furthermore, we regard the development of an application data model as strong relevance to database technology. Thus, this section first clarifies some commonly used terms in data modeling.

A *database* represents the interesting semantics of an application as completely and accurately as possible.[5] The *data model* defines a framework of concepts to be used to define a database. *Data modeling* is the process of defining a database. The *semantic data modeling approach* focuses on the representation of the semantics of an application more than the object-oriented data modeling approach.[25, 33] Despite the rich expressiveness that semantic data modeling approaches can provide, there is no one agreed set of modeling concepts. Therefore, we have concluded, to meet the needs for modeling mechanism design, a hybrid semantic data modeling approach[40] has been chosen, which consists of a set of building blocks (type, class, and attribute) and several data abstractions (generalization, aggregation, and grouping). In the following, these basic building blocks and data abstractions are briefly introduced; details can be found throughout the literature.[6, 9, 28, 33, 40]

A *type* stands for a structure and a domain of instances of that structure, and a *class* is a named collection of instances of a type. Usually the name of a class is the same as that of its type. A type is divided into base type and nonbase type. A base type is identified as either Integer, Real, and so on. A nonbase type is defined by a list of attributes that can take on other types; for example, a type called "person" is defined by attributes such as "name" and "age." Figure 4.5 gives an example to clarify these concepts. *Aggregation* transforms a relationship between different types of objects into a higher-level type of object; this relationship is sometimes called a *has-a* link. *Generalization* transforms common properties or attributes of types of objects into a higher-level type of object (i.e., generic type); the relationship between a generic type and its specialized types is sometimes called an *is-a* link. A *grouping* transforms the instance objects (the underlying objects) that satisfy a specified condition (*grouping expression*) into a higher-level object or grouping object; the relationship between a grouping object and its underlying objects is called *member of.* Figure 4.6a further clarifies the grouping data abstraction and its representation diagram, where it is shown that the grouping class G is formed from two underlying classes U_1 and U_2. Furthermore, the example instance

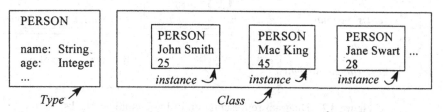

Figure 4.5   Type, class, and instance.

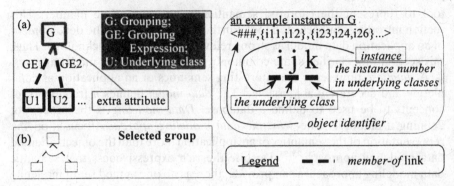

Figure 4.6 Definition of grouping data abstraction and its diagram representation.

of the class *G* is also shown in Figure 4.6. Figure 4.6*b* shows the selected grouping data abstraction with which instances from the two underlying classes are gathered into one file of a grouping object. Note that a powerful data modeling tool EXPRESS[16] supports selected aggregation; however, EX-PRESS does not support the grouping data abstraction. Figure 4.7 shows the representation diagrams for the basic building blocks that will be employed in later discussion and figures.

4.4 CONCEPTUAL DATA MODEL OF BEHAVIOR AND FUNCTION

4.4.1 Data Representation of Behavior

The following discussion focuses on the data representation of the behavior for a design problem classified as a function generator (see Fig. 4.3*a*). Several important features of the model are first elaborated. First, a model is a collec-

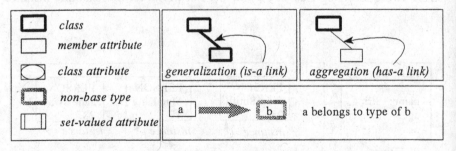

Figure 4.7 Diagram representation of basic building blocks.

tion of object types and it is like a template, whereas a particular design problem corresponds to specific instances of the model (the template). Second, the description of the behaviors consists of two parts: a description of the behavior variables [i.e., x and y in $y = f(x)$] and a description of the relationship between the variables [i.e., $f(\ )$]. Finally, the concepts called *atomic function* and *molecular function* are introduced. When in a defined region of a domain of active variables there is more than one type of behavior defined, each subbehavior is atomic and the combination of them is therefore molecular. For example, in Figure 4.3c and d, f_i , $i = 1, 2, . . . , n$, is atomic. The underlying philosophy of the idea is analogous to the idea of describing a product, that is,

Behavior		*Product*
Atomic behavior	$\leftrightarrow$	Part
Molecular behavior	$\leftrightarrow$	Assembly
Overall behavior	$\leftrightarrow$	Product

The following discussion elaborates on the definitions of the (object) types of classes. Occurrences or instances may be given for a better understanding. Employing the notation introduced in Figure 4.7, Figure 4.8 shows the definition of a type called "VARIABLE." From this definition, some remarks can be made: (1) When the value of the attribute "motion-order" is "1", it means that only the first-order motion quantity is recorded. This is intended for modeling such functions as constant velocity (Fig. 4.4), where possibly the zero-order motion quantities $y1$ and $y2$ are not of interest to the design problem. In general, it is intended for the representation of the problem given directly by the first-order (or higher) motion quantity. (2) When the value of the attribute "time-concerned" is "yes," the first-order (or higher) motion

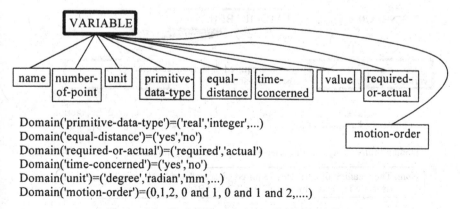

Domain('primitive-data-type')=('real','integer',...)
Domain('equal-distance')=('yes','no')
Domain('required-or-actual')=('required','actual')
Domain('time-concerned')=('yes','no')
Domain('unit')=('degree','radian','mm',...)
Domain('motion-order')=(0,1,2, 0 and 1, 0 and 1 and 2,....)

Figure 4.8 Definition of class "VARIABLE."

quantity is derived with respect to the time; for example, the first-order motion quantity is velocity.

Likewise, Figure 4.9 shows the definition of a type called "ATOMIC-BEHAVIOR." Remarks for this type include: (1) The value of the attribute "motion-order" in the class "ATOMIC-BEHAVIOR" should be consistent with that in the class "dependent-variable" which is of type "VARIABLE." It should be noted that "dependent-variable" serves as an attribute in the type "ATOMIC-BEHAVIOR," while, at the same time, for a class of the type "VARIABLE." Therefore, it makes sense that the attribute "dependent-variable" has the attribute "motion-order." This dependency is not possible to represent by our modeling tool. Therefore, we need to develop (extra) external constraint (EC) rules to represent those data relationships or dependencies beyond a particular data modeling method. For the current problem, an EC rule is then developed as follows.

EC Rule 1: (A cross-reference between the types "VARIABLE" and "ATOMIC-BEHAVIOR"):

atomic-behavior.motion-order = dependent-variable.motion-order

(2) Suppose we have the behavior represented by (x_i, y_i), $i = 1, 2, \ldots$, n. Note that this curve may be like the one described in Figure 4.30, Window 2, where x_i corresponds to the rotation angle of an input motion ($x_1 = 0°$ and $x_n = 30°$) and y_i the angular displacement shown on the vertical axis. Physically, the mechanism could be the one shown in Figure 4.2, where x corresponds to the input of link 2 and y to the rotation of link 7. The following examples present some instances of the corresponding classes for a better understanding. (The symbol w appearing in some of the fields indicates that this value is not

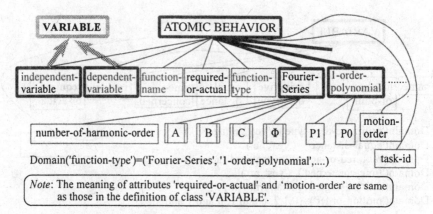

Figure 4.9 Definition of class "ATOMIC-BEHAVIOR."

of interest to our explanation and thus ignored. This notation will be applied hereafter throughout the chapter.)

In the class "VARIABLE":

$\langle$'x__1#',w,w,'REAL',w,w,{0,10,20,30},'REQUIRED',0,$\rangle$
$\langle$'y__1#',w,w,'REAL',w,w,{0,4,12,20},'REQUIRED',0,$\rangle$

In the class "ATOMIC-BEHAVIOR":

$\langle$At__1#, x__1#,y__1#,w,'required','discrete',0,Task__1#$\rangle$

In these examples, x__1#, y__1#, At__1#, and Task__1# are object identifiers, which may be automatically assigned by the database system. (3) The defined schema (Fig. 4.9) also implies that the discrete expression form is common to all continuous expression forms; that is, various continuous representation forms such as Fourier series and polynomials are considered as specializations of the discrete form. In fact, a required behavior in mechanism design is, in most cases, first given in a discrete form $\langle xi, yi \rangle$ and then may be converted into the representation of some continuous forms, for example, Fourier series in the system called TADSOL.[26] In cam mechanism design, the polynomials are very useful; see Figure 4.30 where POL001, POL002, and so forth, are different types of polynomials. The discrete representation serves somewhat as "control" points for those continuous representations. (4) The defined schema can be used for both the required and the actual behaviors.

Figure 4.10 shows the definition of a type called "BEHAVIOR," particularly for the function generator design problem stated earlier. It has been shown that the type "BEHAVIOR" is defined through the grouping data abstraction. The grouping expression is informally stated as grouping all the instances in the class "VARIABLE" on the same value of the attribute "task-id" of the type "ATOMIC-BEHAVIOR." The diagram symbols used in this figure for the grouping expresses the semantics that a behavior itself can be part of another behavior. Consider again the example behavior in Figure 4.30. In addition to the behavior corresponding to the input angle from 0° to 30°, we would also like to represent the behavior called "dwell motion" (i.e., the angle from 30° to 60°) with the model. The following instances show some additional instances of the "VARIABLE" and "ATOMIC-BEHAVIOR" classes.

In the class "VARIABLE":

$\langle$'x__2#',w,w,'REAL',w,w,{},'REQUIRED',0,$\rangle$
$\langle$'y__2#',w,w,'REAL',w,w,{},'REQUIRED',0,$\rangle$

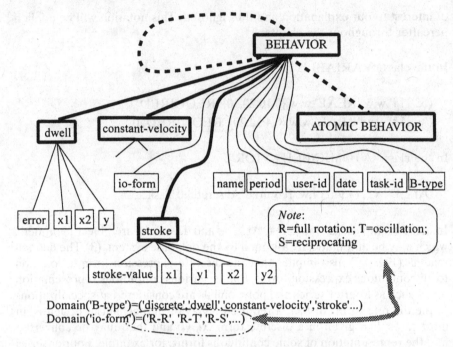

Figure 4.10   Definition of class "BEHAVIOR."

In the class "ATOMIC-BEHAVIOR":

$$\langle At\_2\#,\ x\_2\#, y\text{—}2\#, w, 'required', 'dwell', 0, Task\_1\# \rangle$$

In the class "BEHAVIOUR":

$$\langle B\_1\#, 'R\text{-}R', w, w, w, 05/05/97, Task\text{—}1\#, 'mix', \{At\_1\#, At\_2\#\} \rangle$$

These instances show that the two behaviors, At_1# and At_2#, are part of the overall behavior, B_1#. Furthermore, for the attribute "B-type," the value "mix" is used to indicate that the overall behavior is composed of more than one behavior.

4.4.2   Data Representation of Function

To represent functions (i.e., purposes of the designer), two generic syntactic patterns are considered: (1) noun + verb + noun and (2) verb + noun. The first pattern refers to those functions based on behaviors that have two operands (input and output) and the relationship between input and output can be used to realize a particular function. This type of behavior and its potential

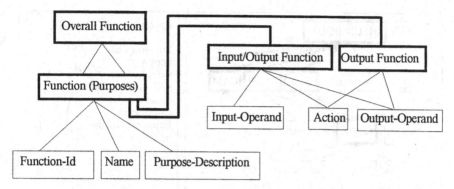

Figure 4.11 Definition of function (purpose).

use is of particular interest to mechanism design. The second pattern refers to those behaviors that result in something being done.

Figure 4.11 shows a conceptual data representation of functions. An overall function structure (i.e., a function is divided into a set of subfunctions) is represented in a bill-of-material manner similar to the method for describing product assemblies.[38]

4.5   CONCEPTUAL DATA MODEL OF STRUCTURE

4.5.1   Modeling of the Atomic Body of a Mechanism

Figure 4.12 shows the model of an atomic body of a mechanism (referred to as "mechanism atomic-body" in the diagrams). The common properties of all types of bodies (e.g., mass, motion status, etc.) are defined as a generic type called "MECH-BODY." As mentioned earlier, at the conceptual design level, the geometric characteristics of mechanism atomic bodies are those features that will only influence the kinematic motion; for example, for a bar body, the two end nodes belong to these features. The kinematic parameters are determined by these features. The number of these features is not predictable in the sense that for an individual body these features are dynamically produced in the design process or, depending on the designer's decision, determined based on how many other bodies are connected with it. For example, Figure 4.13*a* shows a bar with two nodes, whereas Figure 4.13*b* shows a bar with four nodes. It is then difficult to explicitly define these features as attributes of an object type for bodies because this will lead to an unpredictable number of attributes. Therefore, the *inherent feature* and *noninherent feature* concepts are introduced next.

The *inherent features* of a body are the necessary characteristic elements that generically distinguish the underlying type of a body from others,[37] whereas the

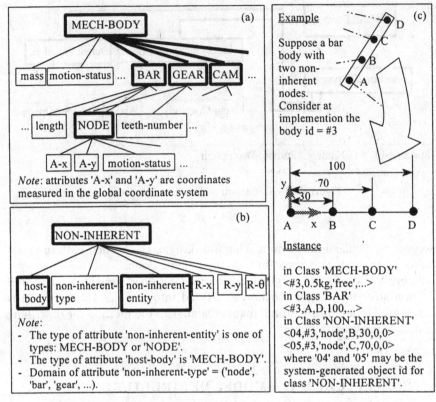

Figure 4.12 Data representation for an atomic body of a mechanism.

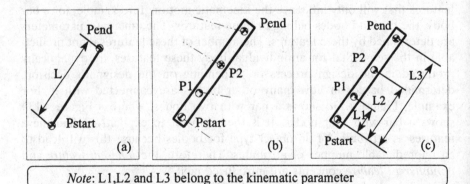

Note: L1,L2 and L3 belong to the kinematic parameter

Figure 4.13 Inherent features versus noninherent features.

remainder of the features are *noninherent.* For example, a bar body has the inherent feature: two end nodes (Fig. 4.13*a*). Node elements P_1 and P_2 in Figure 4.13*b* are noninherent. The inherent features of a body are independent of connections of the body with others and thus the number of inherent features of a type of body is predictable. Inherent features with relevant kinematic parameters are explicitly defined as attributes of an object type for bodies (see Fig. 4.12*a*), whereas noninherent features are represented together in an object type called "NON-INHERENT" (Fig. 4.12*b*). In this class, kinematic parameters are represented by the attributes Rx, Ry, and $R\theta$, which describe the relative position of noninherent features with respect to the reference system on each body. (The reference system on each body will be illustrated in Section 4.5.3.) An example is shown in Figure 4.12*c* where the bar body has two noninherent features. This figure also shows the instances in the classes "MECH-BODY" and "NON-INHERENT," illustrating how the kinematic parameters relevant to these noninherent features are represented. There is an alternative scheme for defining noninherent features by means of the *set-valued attribute* notion (i.e., defining an attribute "noninherent-node" as a set-valued attribute; see Fig. 4.7) for each type of body (Fig. 4.14*a*). Figure 4.14*b* shows the instances filled-in, illustrating how the example bar in Figure 4.12*c* is represented in our data model. However, this scheme has difficulty representing kinematic parameters that are related to those noninherent features explicitly. In other words, the length, the distance between *A* and *C*, cannot be represented in this scheme.

Manipulation of a group of bodies as a whole may be needed in a design step. This group of bodies, called a *composite body,* corresponds to the *composite object* concept in data modeling.[40, 45] Consider an example where, at the conceptual design stage, one gear and one bar, as shown in Figure 4.15*a*, are separately processed. Considering the later design stage (e.g., the embodiment design stage), these two bodies might be considered as a whole (see Fig. 4.15*a*, alternative II). The fact must be specified explicitly that the body depicted in alternative II of Figure 4.15*a* is referenced to two bodies at the conceptual design stage. This leads to the definition of an object type called "COMPOSITE-BODY" (Fig. 4.15*b*). The following remarks concerning this definition will help you clarify the model: (1) A composite body is assigned

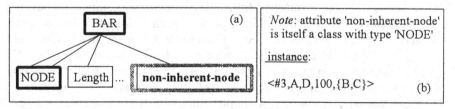

Figure 4.14   Alternative data representation of noninherent feature nodes.

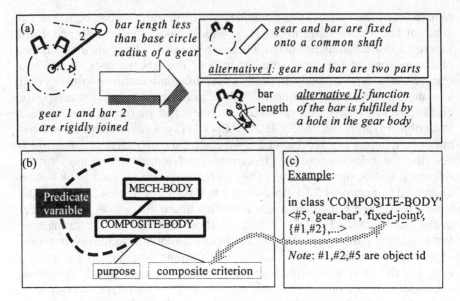

Figure 4.15   Data representation for composite body.

a unique object identification (id) and can have its own attributes such as "purpose." An object id will serve as a reference for a composite body to relate to its component bodies; see the example shown in Figure 4.15c. (2) A composite body is also a subclass of the class "MECH-BODY." Therefore, it can inherit the attributes of "MECH-BODY." (3) The value of the attribute "composite-criterion" can be either predefined at the database schema definition level (e.g., the criterion that any rigidly joined body is viewed as a composite body) or given by the designer through a user interface. The value of this attribute serves as a grouping expression of the grouping data abstraction. Eventually, both the selection criteria (for composition) and the bodies to be grouped with the selected criterion can be end user controlled, which captures the more dynamic nature of the mechanism design.

In the preceding discussion, nodes have been used to describe the geometry of atomic bodies at the conceptual design level (see Fig. 4.12). An important advantage of doing so is that the system based on our model can better support the conceptual design of mechanisms in a "sketch-like" manner. By saying that, we assume that sketching starts with the specification of nodes. In such a system, a human designer describes a mechanism concept in a way similar to that in a manual system (i.e., by drawing a mechanism sketch on a sheet).[41] Another consideration in using a node as a means of defining the geometry of a mechanism is to support different views (or representations) of the geometry of conceptual mechanisms. A simple example is that the length of a bar can be specified either by the length or by the coordinates of the two end nodes of that bar. In the present study, the first method is referred to as an *explicit* representation, whereas the second method is considered as an *implicit* representation. An external constraint must be specified in the model to maintain this data relationship. For example, for the type, bar body, the external constraint would be:

EC Rule 2. (For class object "BAR"):

$$\text{bar.length} = |\text{bar.node-1} - \text{bar.node-2}|$$

It should be noted that the conversion from the implicit representation of kinematic parameters to the explicit representation is straightforward, whereas the reverse process itself requires solving a set of complex nonlinear equations. In most cases, node coordinates in the model are raw data, whereas the length of a bar is dominant data or refined data. When a designer starts with only a drawing of his or her mechanism concept without explicitly specifying a kinematic parameter such as the length of a bar body, the length might be calculated directly from the relevant nodes. It is needed to specify the way the length of a bar is given; that is, the system might enable the designer to specify the length of a bar as raw data when the length is calculated only by nonaccurate node coordinates. In our model, such a requirement has been realized at the implementation level by setting a flag variable for each value in the database as FALSE (raw data) or TRUE (refined data). This facility is directly provided with a dedicated database management system (DBMS) called the ABC system[39] that we used for our implementation. This approach is applied to all other types of kinematic parameters, as well.

In our model, the frame body is modeled using the *grouping* data abstraction as shown in Figure 4.16. To support different views of the same object (e.g., the mechanism frame could be either two ground points or a ground bar

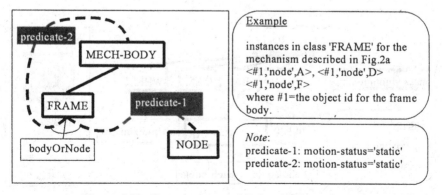

Figure 4.16 Data representation for the frame body.

body), the selected grouping data abstraction is employed on the object types "MECH-BODY" and "NODE." The consistency of the fact that a static bar can be described with two ground nodes is equally maintained by EC Rule 2 described before. Because the grouping data abstraction is particularly useful for describing application semantics, which are dynamic in nature, it is suitable to use the grouping data abstraction to represent the frame semantics. The object type of a frame body may have its own attributes (Fig. 4.16). At later design stages, a complete frame body may be designed for manufacture with an appropriate method such as *casting*.

The model shown in Figure 4.12 is extensible—one of the features of a generic data model. For example, Figure 4.17 shows that adding two new types of bodies (wheel and helical gear) does not affect the existing model at all. The two types of bodies are placed at different levels of the generalization hierarchy, respectively (Fig. 4.17).

4.5.2 Modeling of Connection

Although the nature of the connection of bodies is like a network, it can be viewed as a set of connections between two bodies, *pair connection,* in the interest of data representation. An object type called "CONNECTION" is defined (Fig. 4.18*a*). Several remarks regarding this object type include: (1) Because the instances of connections <body-i, body-j,. . . > and <body-j, body-i,. . .> indicate the same semantics, the system handles the identification and elimination of redundant information using the external constraint:

EC Rule 3. (For the object type "CONNECTION"):

$$\text{connection.body-i} > \text{connection.body-j}$$

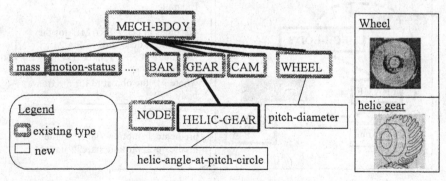

Figure 4.17 Adding new body types.

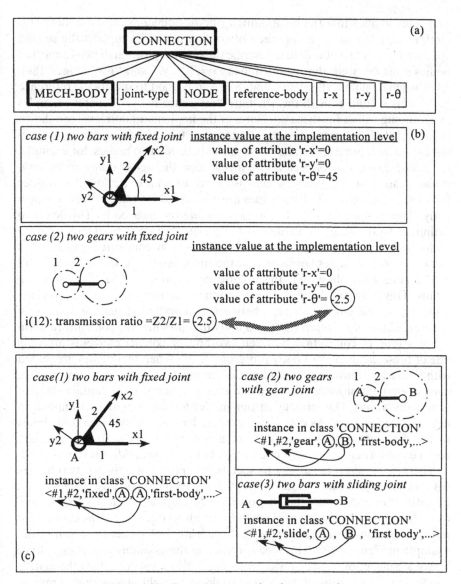

Figure 4.18 Data representation for connections.

(2) The configuration of a mechanism can be determined via two approaches. The first is via the nodes in the global coordinate system attached to the frame body. The second is via the relative positions between two bodies. As mentioned earlier, the coordinates of the nodes are defined in the global coordinate system of a mechanism and the nodes are used in the body descrip-

tion. This implies that the positioning of bodies depends on the nodes. This method supports the first approach of configuration. To support the second approach, a set of attributes that describe the relative position of two connected bodies must be used, that is, attributes "r-x," "r-y," and "r-θ" (Fig. 4.18a). Also, when dealing with relative angles, it is important to specify the reference body. This is fulfilled with the attribute "reference-body." Several examples of specifying values for these attributes at the implementation level are shown in Fig. 4.18b. It is especially noted that in a certain case some attributes may not explicitly represent the relative position between two bodies; for example, for a couple of gears connected with a gear joint the *signed transmission ratio* is used (case 2 of Fig. 4.18b), which also represents inner (ratio > 0) or outer (ratio < 0) engaging. (3) Two bodies are connected via the elements on each body. This is done by selecting a *feature node* on each body. This leads to defining "NODE" as an attribute of the class "CONNECTION." Feature nodes here serve as the *pair element* from each body concerned. Figure 4.18c shows several examples considered at the implementation level. Cases 2 and 3 of Figure 4.18c show that the feature nodes may not be the contacting points. This is an extension to the traditional *pair element* concept[7] in which a kinematic pair implies contacts between two bodies. A feature node is, in general, a kind of representative of a body that provides the total information about a type of connection between two bodies; contact elements are only one of these candidates of body representatives. When two bodies are joined with a relative translation, it does not make much sense to select any specific contact element on each body as a pair element, because the contact element is actually a line. The strategy in our model for this situation is implied in case 3 of Figure 4.18c. As mentioned earlier, the feature node also contributes to the description of the geometry of a conceptual body. In this way, the complete information about a connection between two bodies is represented.

There is an alternative way to specify a connection between two bodies without using the feature node concept. For each body, feature elements can be defined in such a way that the position of the feature elements is known to the local coordinate system of individual bodies (Fig. 4.19a). The connection information can therefore be specified via feature elements as shown in the example in Figure 4.19a. This idea is rooted in the assembly modeling culture where a connection between two geometric entities is specified via the mating condition.[23] The feature element here could be thought of as a kind of mating element. The key point to support the feature element concept is to model the relationship between feature nodes and feature elements. The example of Figure 4.19a implies that the functional correspondence (feature node : feature element) is $(1 : n)$, which means that one feature node corresponds to more than one feature element. Noticing this fact and considering that a feature element is local to an individual body, the object type "FN-FE" is defined (Fig. 4.19b). Another adaptation includes the replacement of the attribute 'NODE" by "feature-element" in the classes "CONNECTION" and "MECH-BODY," respectively.

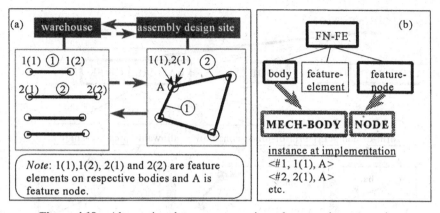

Figure 4.19 Alternative data representation of connection semantics.

Explicit use of nodes seems better than feature elements in supporting the conceptual design in a scenario where a design begins with a sketch of the kinematic mechanism diagram. As such, while a kinematic mechanism diagram is being drawn, corresponding nodes are created in the class "NODE" and written into the class "CONNECTION" as a *marker* for the connection of the two respective bodies. Note that such a marker is also used to describe individual bodies as well (see the discussion in Section 4.5.1). Node coordinates also provide information on a particular *mechanism branch,* which can be stated as follows: Given the same kinematic parameters and degree of freedom, more than one assembly scheme may exist (Fig. 4.20). However, the use of feature elements facilitates the assembly of the mechanism.[23]

The definition of the type "CONNECTION" can easily be extended when needed. For example, to represent the cam mechanism, some extra information about the connection must be included (e.g., the contact form between a cam

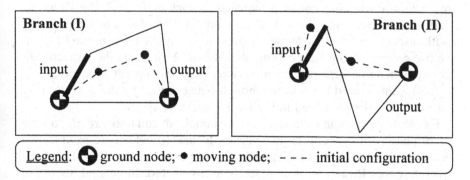

Figure 4.20 Mechanism branch information versus node coordinates.

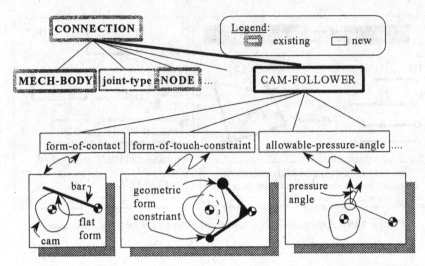

Figure 4.21　Cam–follower connection as an extension to class "CONNECTION."

and its follower bar, pressure angle, etc.). Figure 4.21 shows the inclusion of the extra information, by defining an object type called "CAM-CONNECTION" as a subtype of the type "CONNECTION," without affecting the existing part. The implementation of a system that uses our model as its underlying object model for designing complex cam mechanisms has been successfully implemented (see also Fig. 4.33).[30]

4.5.3 Modeling of Reference Systems to Measure Motion

For mechanisms, motion states are the fundamental elements employed for modeling useful behaviors; other states such as force and kinetic energy depend on the motion states. It is common to describe motion states in such a way that a *local coordinate system* is assigned to each body (e.g., the Denavit–Hartenberg system[4] and Sheth–Uicker system[27]). The relative position of body i with respect to body j is completely determined by the relationship between the two corresponding local coordinate systems. Apart from the local coordinate system, a *user-defined reference system* (a vector, a point, or a coordinate system) is introduced because the motion being viewed by the designer may not necessarily be measured in the local coordinate system.

Figure 4.22 shows an example. In this example an end user expects to view the displacement θ_{12} between two vectors $\mathbf{R}_1$ and $\mathbf{R}_2$, which are fixed to two local coordinate systems X_1–Y_1 and X_2–Y_2, respectively. The definition of the two references $\mathbf{R}_1$ and $\mathbf{R}_2$ is completely user oriented. In general, there are four types of user-oriented references: a point, a free vector, a fixed vector,

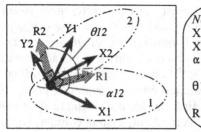
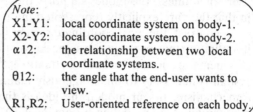

Figure 4.22 User-oriented view versus system-oriented view.

and a coordinate system. Four cases of motion in the context of the user-oriented reference system are considered in our model (Fig. 4.23):

(*a*) On the reference body: a free vector; on the target body: a free vector.
(*b*) On the reference body: a fixed vector; on the target body: a point.
(*c*) On the reference body: a coordinate system; on the target body: a point.
(*d*) On the reference body: a point; on the target body: a point.

In the following discussion, such references are referred to as the *user-oriented reference system*. To raise the system's intelligence with respect to

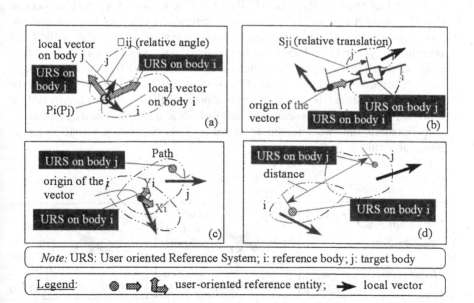

Figure 4.23 User-oriented view of motion states.

user friendliness, the model supports these user-oriented reference systems in that they are explicitly represented in our model (see later discussions).

In contrast, the local coordinate system is much more *system oriented* and its definition may be subject to some conventions to facilitate the establishment of kinematic motion constraint equations.[42] However, the local coordinate system is necessary to unambiguously maintain constraints among bodies of a mechanism. In our model, the idea is that the local coordinate system is maintained at the modeling system level. [A modeling system contains (i) the model, (ii) a set of processes to manipulate the model, and (iii) a user interface to allow the designer to manipulate the model.] At present, the system only considers planar mechanisms, so the local coordinate system is simply represented by *a local (fixed) vector*. The local coordinate system can be created based on this local vector (by taking the vector as the *x* axis and rotating 90° from the *x* axis in a counterclockwise direction to get the *y* axis). Figure 4.24 shows several conventions for defining local vectors with respect to the various body types.

In order to support this approach, an object type called "U-REFERENCE" is defined (in Fig. 4.25a) to represent user-oriented reference systems, as well as their relationships with the *system-oriented reference system,* that is, the local coordinate system. Several remarks concerning this definition are needed: (1) The common properties of each type of user-oriented reference system define a generic class "U-REFERENCE" (U = User). These properties include the name that the end user may want to assign to each reference system and the mechanism body to which the reference system is related to. (2) The values of the attributes "local-x," "local-y," and "local-angle" (defined as individual subtypes of the type "U-REFERENCE" level) are with respect to the local coordinate system of the body. From a user's viewpoint, there are two ways to define the user-oriented reference system: (a) relative to the local coordinate system (Fig. 4.25c) and (b) relative to the global coordinate system (Fig. 4.25d). However, at the system level, only one definition is expressed (i.e., relative to the local coordinate system). It is therefore necessary to maintain consistency between the two origins of the local coordinate system and the user-oriented reference system (O-LCS and O-URS) and the relationship between the two coordinate systems. This requires a fourth external constraint rule.

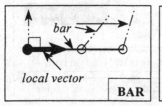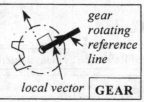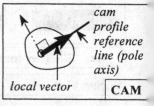

Figure 4.24   Local coordinate systems versus types of body.

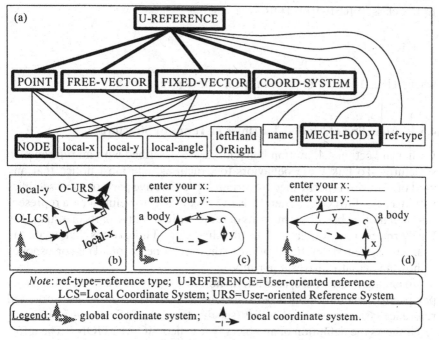

Figure 4.25 Data representation of the user-oriented reference system.

EC Rule 4: (Cross-reference among those subtypes of the type "U-REFERENCE" and the type "MECH-BODY," respectively):

$$O_{URS} - O_{LCS} = [\text{local-x, local-y}]^T \quad \text{where } O_{URS} (O_{LCS}) \text{ is a column matrix}$$

(3) When the end user does not establish his or her own reference system, the model will take the system-oriented reference system as the default. In this case, the value of the attribute "ref-type" will be "system-point" or "system-vector," and so on, and the precise information about the reference can be found in the local coordinate system of the corresponding body.

For a better understanding of how the model works, the following shows some specific instances. Suppose the designer needs to define a motion reference on link 2 in Figure 4.2; in particular, he or she sets up the reference as the one attached on link 2 with an angle of 45° (see Fig. 4.2). The instances to represent this semantics can be written as follows:

In the class "U-REFERENCE":

⟨UR__1#,w,2,'free-vector'⟩

In the class "FREE-VECTOR":

$\langle UR\_1\#,45°\rangle$

where UR_1# is the object id for this motion reference.

4.5.4   Modeling of Motion Variables in Terms of Structures

Earlier in Section 4.4, motion behaviors were described with no reference to structures. To link these behaviors to structures, one must notice that any kind of behavior is performed by certain representative parts or subassemblies within a whole structure. Therefore, our task here is to define a data representation of motion behaviors in terms of these representative parts of a structure. Furthermore, one also has to describe the external operations that excite the change of system states, thus leading to the specific behaviors. This corresponds to how an actuator drives a machine.

To enhance ease of use, user interaction with the system is from a designer's perspective, whereby motion behaviors are defined based on the user-oriented reference systems instead of bodies (see the discussion in Section 4.5.3). Figure 4.26a shows the definition of an object type called "IO-MOTION." The information about external operations such as how an input motion is specified is shown in Figure 4.26b. In Figure 4.26b, the attribute "motion-type" is implied or derived from the combination of two attributes "ref-reference" and 'target-reference" of its supertype "IO-MOTION" and expresses all four cases (e.g., *path, rotation,* etc.) as shown in Figure 4.23. The attribute "motion-order" indicates the required order of motion (e.g., zero-order means the displacement). The attribute "number-of-point" represents the total number of input motion steps for which the output motion is to be calculated. Figure 4.26b also represents for each type of motion the *motion primitive element.* For example, a plane path or trajectory needs to specify two motion quantities (x and y).

Input motion differs from output motion in that the input motion value is predefined (the specification of a detailed prescription of the input motion may be called external operations), whereas the output motion value is calculated by an analysis system. For certain types of input motions, their values can be specified by a mathematical formula with certain parameters given, instead of a set of discrete points. This leads to the definition of an object type called "INPUT" which is a subclass of "IO-MOTION." For the input type "constant-velocity (equal-step)" the formula would be: $S = V_t + S_0$ (where S = displacement and V = velocity). When the end displacement (S_e) is given, the displacement values $S_1, S_2, \ldots, S_e$ (for $t = t_1, t_2, t_3, \ldots, t_e$) are calculated. Consider an example of how the data model works. Suppose that the motion reference defined earlier, that is, the one related to link 2 in Figure 4.2, serves as the input motion with respect to the x axis of the global coordinate

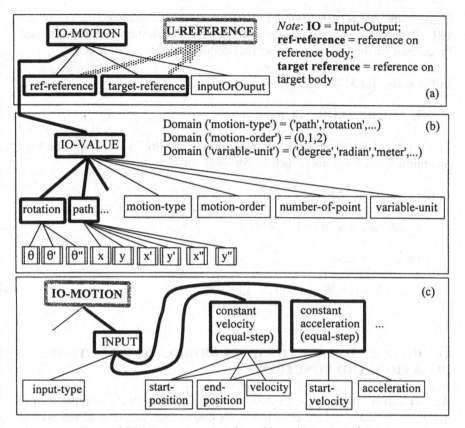

Figure 4.26 Data representation of input/output motions.

system (Fig. 4.2). First, one needs to define this x axis as a reference with the following instances:

In the class "U-REFERENCE":

  ⟨UR__1#,w,2,'free-vector'⟩
  ⟨UR__2#,w,1,'free-vector'⟩

In the class "FREE-VECTOR":

  ⟨UR__1#,45°⟩
  ⟨UR__2#,0°⟩

User reference, UR__2#, has a value of 1 for the "MECH-BODY," which is specifically allocated by the system because this instance represents a global frame body. As well, the defined "FREE-VECTOR" instance for the object UR__2# has a 0° angle representing the x axis. The motion variable can then be defined further based on these two defined references as follows:

In the class "IO-MOTION":

⟨IO__1#,UR__2#,UR__1#,'input'⟩

where IO__1# is the object id. Furthermore, quantities can be specified for this input motion variable, such as constant rotation motion:

In the class "IO-VALUE":

⟨IO__1#,'rotation','0 and 1',w,'degree'⟩

In the class "rotation":

⟨IO__1#, {0,10,20,30,...},{1.0,1.0,...},{0.0,0.0,...}⟩

4.6 INTEGRATION OF THE DATA MODELS OF STRUCTURE, BEHAVIOR, AND FUNCTION

As it is known, the innovative design process starts with the specification of design requirements, including functions and constraints. The functions will further be mapped to the behaviors, which were modeled in the context of mechanism design in our previous discussion (Section 4.4). Use of an integrated data representation means that the representation of the relationships between the required behaviors and their structures should be developed in a declarative manner. For this purpose, an object type called "B-S" is defined as shown in Figure 4.27, where the attribute "required behaviour variable" references

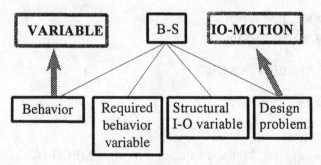

Figure 4.27 Data representation for the behavior–structure (B–S) relationship.

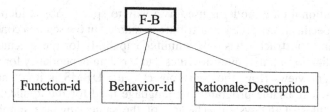

Figure 4.28 Data representation for the behavior–function (B–F) relationship.

the object type "VARIABLE" and the attribute "structural i-o variable" references the object type "IO-MOTION." Note that the representative structural parts are represented in the object type "IO-MOTION" such as (in Fig. 4.2) link 2 with respect to the x axis attached to the global frame body. It may be clear that for a motion problem with one input and one output, there should be two instances of the class "B-S." Given the example in Figure 4.2, assume that the input motion is associated with link 2 and the output motion with link 6, a path. The instances for the class "B-S" can be written as follows:

In the class "B-S":

$\langle$BS_1#,B_1#,x_1#,IO_1#,DesignProblem_1#$\rangle$
$\langle$BS_2#,B_1#,y_1#,IO_2#,DesignProblem_1#$\rangle$

where BS_1# and BS_2# are the object identifiers for the class "B-S." A data model defined as such is very flexible, as the different problems are modeled by filling different instances in the class "B-S" only.

The relationship between a behavior and a function is modeled by the class type "F-B," as shown in Figure 4.28. The attribute "rationale-description" records the purpose for the use of a particular behavior from the human designer's viewpoint.

4.7 IMPLEMENTATION ISSUES AND USES OF THE MODELS

The models described in the previous sections have been implemented at the Technical University of Delft and City University of Hong Kong. To achieve maximal portability and performance, a dedicated database management system was used, which was developed using FORTRAN.[19] The models were specified in terms of the schema constructor of this database management system. The schema constructor of this DBMS mixes some semantic data models with the relational data model. For instance, it is allowed to specify the domain of an attribute as another entity type.[7] The key attribute concept in the relational data model is used as a way to specify object identity, and thus the specification of the key attribute is enabled in the schema constructor. The schema constructor has only a limited capability for the specification of some application-related integrity rules and the data dependency for supporting different views from the same object. This DBMS is implemented in FORTRAN, the same programming language used for the application program modules of the system. The use of the same language simplifies the integration of, and intercommunication between, the system components. A very important advantage of using a dedicated DBMS is its capability to specify raw data in a general way (as we have mentioned in Section 4.5.1) and its efficient data access speed.

There are several applications of this model system. In general, this model is meant to be used as the backbone of a software environment for integrated development of mechanical systems (CACDMM). CACDMM is organized as shown in Figure 4.29a, which distinguishes three different program systems: editors, browsers, and tools. Their functions are also shown. For example, browsers are used to allow the designer to evaluate the actual behaviors of a mechanism, editors permit the human designer to specify his or her design ideas, and tools are needed to analyze the actual behaviors of the designed system and to judge whether the corresponding performances are satisfactory.

The two models (structure and behavior/function) described are mainly used to develop two categories of editor systems (Fig. 4.29b): one category for the mechanism structure and the other for the mechanism behavior. Figure 4.30 shows an editor system that allows the designer to specify various motion design problems. In particular, Window 1 shows that the designer can specify a required path or a required input/output behavior, and Window 2 shows that the designer can specify detailed behaviors outside the region of a dwell motion with options: POL001, POL002, and so forth (types of polynomial functions). The possibility shown in Window 2 (Fig. 4.30) will be particularly useful for designing mechanisms containing cams. Figures 4.31 and Figure

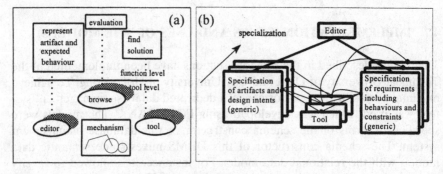

Figure 4.29 CACDMM architecture.

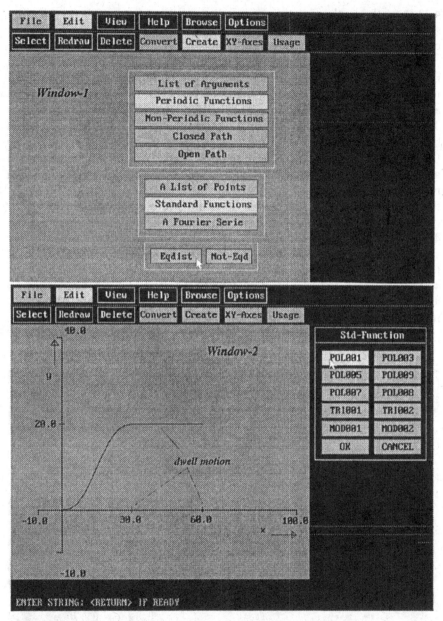

Figure 4.30 Editor for specifying required behaviors.

4.32 show the implementation of an editor system for specifying mechanism structures. In particular, Figure 4.32 also shows the procedure to specify a complex gear–bar mechanism in four steps.

The behavior and function data models are used either as a part of a requirement model or to store actual behaviors and functions. When it is used for the former, after a required behavior is specified by the designer or by another program system (member of the tool family), synthesis tools such as TADSOL[26] will extract the information from the data model for the required behaviors and generate one or more mechanism structures that are stored by the data model. The actual behaviors of these mechanisms will be generated and analyzed by some analysis tools[34, 35, 43] and then stored in a similar data model (which is a kind of template) for the mechanism behavior. Similarly, the data model for the mechanism structure will be used for two scenarios: (i) the specification of a mechanism concept by the designer (Figs. 4.31 and 4.32), and (ii) the storage of generated mechanisms for further processing by synthesis tools. Figure 4.33 shows the implementation of a part of CACDMM for designing mechanisms containing cams. The uppermost portion of figure 4.33 shows a cam–follower function defining the relationship between the cam angular displacement and the follower displacement. In the modeler, this function is referred to as a behavior. The profile of a cam is designed to

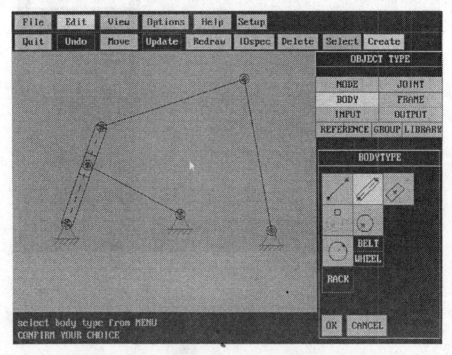

Figure 4.31 Editor for specifying mechanism structures.

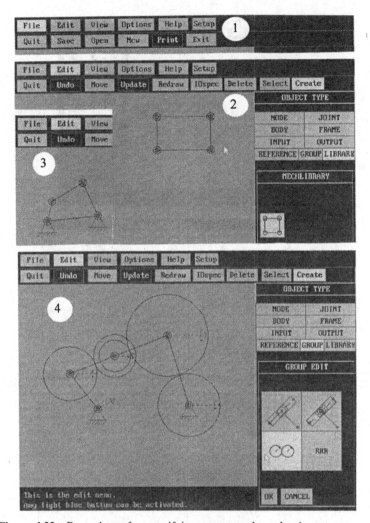

Figure 4.32 Procedures for specifying conceptual mechanism structures.

achieve that behavior. Window 1 in Figure 4.33 shows a configuration of this cam–bar mechanism, which is a possible solution for the required behavior. Through several steps (Windows 2 and 3), the profile of the cam is found (see Window 4 of Fig. 4.33).

This cam mechanism is only one part of the whole system, which contains a gear–bar mechanism as well. Therefore, Figure 4.34 shows how these two mechanisms can be combined either by editing the gear–bar structure upon the cam mechanism or by editing the two mechanisms separately and then using a merge function to glue them. It is noted that the whole mechanism containing the cam–bar and the gear–bar mechanisms will be stored using a

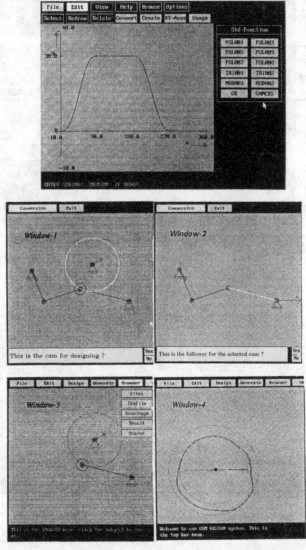

Figure 4.33 CACAMM system for cam–bar mechanism design.

similar data model for the mechanism structure. A dynamic simulation of the whole mechanism can thus be carried out.[34, 35]

It is worth noting that some of the other existing modeling systems only provide a "picture" or a bond graph representation for a generated mechanism[22] (a bond graph representation is for mechanism behavior). This limits further processing activity (with the computer automatically) such as for subsequently editing mechanism structures or modifying any structure parameters

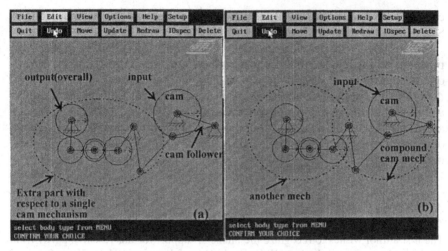

Figure 4.34   Joining two mechanisms.

and interfacing those with general-purpose CAD/CAM systems mainly concerned with product structures (e.g., Unigraphics, etc.).

4.8   CONCLUSIONS

This chapter has described a new strategy called "plug-and-play" for developing data models or data representations of functions, behaviors, and structures for mechanism design. Under this strategy, the data representations for functions, behaviors, and structures are defined separately, and then their relationships are defined in an "external" entity. The chapter has focused on mechanism design. A hybrid semantic modeling approach has been used to develop all the data models. This approach was found to be more expressive than typical object-oriented data modeling methods. For example, the grouping data abstraction has been found very useful for the developed data models, but it is missing in object-oriented data modeling approaches. The data models for mechanism behavior and function have not been reported in the literature. Although there were some reports on a data model for the mechanism structure, the one described in this chapter differs from them both in the views of mechanisms and the representations of those views. For example, in our work, we have taken an object-oriented view of a mechanism; that is, we view the components that constitute a mechanism as identifiable on their own functional and structural right, instead of a link which is identifiable according to the same motion.[14] Development of data models for mechanisms is a complex task, as noted by Eastman et al.[6] Therefore, the work reported in this chapter can certainly contribute to the efforts in such areas as the development of STEP standards for mechanisms.

In the course of developing these models, a user orientation has been pursued. As a result, the integrated design system, CACDMM, that was developed based on these models has shown some unique features of user friendliness. The concepts and data models developed in this study can be useful in an integrated design environment for mechanical systems, as has been shown, and as well be useful to the development of a large knowledge base for product development in a broader sense, where a mechanism is a part of the products being designed.

ACKNOWLEDGEMENTS

The first author wishes to recognize the financial support for this research partly from the Hong Kong Research Grant Council via City University of Hong Kong Strategic Research Grant NR. 7000513.

REFERENCES

1. Baxter, J. E., N. P. Juster, and Alan de Pennington, 1994, Verification of Product Design Specifications Using a Functional Data Model, in *Computer Aided Conceptual Design,* J. Sharpe and V. Oh (eds.), pp. 281–298.

2. Benz, T., 1990, *Function Model in CAD-Systems* (in German), VDI-Verlag, Dusseldorf.

3. Chen, P. P. S., 1976, The Entity-Relationship Model a Basis for the Enterprise View of Data, *ACM TODS,* Vol. 1, No. 1.

4. Denavit, J., and R. S. Hartenberg, 1995, A Kinematic Notation for Lower-Pair Mechanisms Based on Matrices, *J. Appl. Mech.,* Vol. 22, No. 2.

5. Dittrich, K. R., 1986, Object-Oriented Database Systems: The Notion and Issue (Extended Abstract), *1986 IEEE Computer Society Press,* pp. 2–6.

6. Eastman, C. M., A. H. Bond, and S. C. Chase, 1991, A Data Model for Design Databases, *Artificial Intelligence in Design'91,* J. S.Gero (ed.), Butterworth Heinemann, pp. 339–365.

7. Erdman, A. G., and G. N. Sandor, 1984, *Mechanism Design: Analysis and Synthesis,* Prentice-Hall, Englewood Cliffs, NJ, Vol 1.

8. Gui, J. K., and M. Mantyla, 1993, New Concepts for Complete Product Assembly Modeling, *Second ACM Solid Modeling,* Montreal, pp. 397–406.

9. Hammer, M., and D. McLeod, 1981, Database Description with SDM: A Semantic Database Model, *ACM Trans. Database System,* Vol. 6 No. 3, pp. 351–386.

10. Hardell, C., 1996, An Integrated System for Computer Aided Design and Analysis of Multibody Systems, *Engineering with Computer,* Vol. 12, pp. 23–33.

11. Hardell, C., et al., 1993, A Relational Database for General Mechanical Systems, *NATO Advanced Study Institute on Computer Aided Analysis of Rigid and Flexible Mechanical Systems,* Troia, Portugal.

12. Henson, B. W., N. P. Juster, and Alan de Pennington, 1994, Towards an Integrated Representation of Function, Behaviour and Form, in *Computer Aided Conceptual Design*, J. Sharpe and V. Oh (eds.), pp. 95–111.

13. Hundal, M. S., 1990, A Systematic Method for Developing Function Structures, Solutions and Concept Variant, *MMT*, Vol. 25, No. 3, pp. 243–256.

14. IFToMM, 1991, Terminology for the Theory of Machines and Mechanisms, *Mech. Mach. Theory*, Vol. 26, No. 5, pp. 435–539.

15. ISO 10303, Part 105, 1991, *Industrial Automation Systems Exchange of Product Model Data Part 105: Kinematics*, San Diego Version.

16. ISO TC184/SC4/WG5, 1991, *EXPRESS Language Reference Manual*.

17. Kaper, G., 1987, Design of a Data Structure and a Program for the Automatic Generation of the Kinematic Sketch for a Designed Mechanism, M.S. Thesis (in Dutch), Technical University of Delft, The Netherlands.

18. Kimura, F., S. Kawabe, T. Sata, and M. Hosaka, 1984, *A Study on Product Modeling for Integration of CAD/CAM, in Integration of CAD/CAM*, D. Kochan (ed.), Elsevier Science Publishers B.V. North-Holland, Amsterdam, pp. 227–246.

19. Kota, S., and S.-J. Chiou,1992, Conceptual Design of Mechanisms Based on Computational Synthesis and Simulation of Kinematic Building Blocks, *Res. Engineering Design*, No. 4, pp. 75–87.

20. Kuttig, D., 1993, Potential and Limits of Functional Modeling in the CAD Process, *Res. Engineering Design*, Vol. 5, pp. 40–48.

21. Langbroek, A., 1989, Development of a Graphic Interactive System with Intelligent Database for Existing FEM Programs, M.S. Thesis (in Dutch), Laboratory of Engineering Design, Technical University of Delft, The Netherlands.

22. Malmqvist, J., 1993, Towards Computational Design Methods for Conceptual and Parametric Design, Ph.D. Thesis, Chalmers University of Technology, Sweden.

23. Oliver, J. H., and M. J. Harangozo, 1991, Inference of Link Positions for Planar Closed-Loop Mechanisms, *Computer-Aided Design*, Vol. 24, No. 1, pp. 8–26.

24. Otter, M., M. Hocke, A. Daberkow, and G. Leister, 1993, An Object-Oriented Data Model for Multibody System, in *Advanced Multibody System Dynamics*, W. Schiehlen (ed.), pp. 19–48.

25. Peckham, J., and F. Maryanski, 1988, Semantic Data Models, *ACM Computing Surveys*, Vol. 20, No. 3, pp. 153–189.

26. Rankers, H. et al., 1985, Computer Aided Design of Mechanisms: The CADOM Project of the Technical University of Delft, *Proceedings of the Fifth World Congress on Theory of Machines and Mechanisms*, pp. 667–672.

27. Sheth, P. N., and J. J. Uicker, 1971, A Generalized Symbolic Notation for Mechanisms, *J. Engineering for Industry*, Vol. 93, pp. 102–112.

28. Smith, J. M., and D. C. P. Smith, 1978, Principles of Database Conceptual Design, *Proceedings NYU Symposium on Database Design*, pp. 35–49.

29. Tomiyama, T., and P. J. W. Ten Hagen, 1987, The Concept of Intelligent Integrated Interactive CAD System, Report CS-R8717, Center for Mathematics and Computer Science, Amsterdam.

30. Tomiyama, T., et al., 1994, Toward Knowledge Intensive Engineering, in *Computer Aided Conceptual Design*, J. Sharpe and V. Ho (eds), LooseLeaf, United Kingdom, pp. 319–338.

31. Tomiyama, T., Y. Umeda, and H. Yoshikawa, 1993, A CAD for Functional Design, *Ann. CIRP,* Vol. 42, No. 1, pp. 143–146.

32. van Veen, M., 1994, Integrated Design and Manufacture of Cam-Link Mechanisms, M.S. Thesis, Laboratory for Mechanization of Production and Automation, Technical University of Delft, The Netherlands.

33. van der Weerd, P., 1992, Unified Design and Implementation of FPA Control Systems, Ph.D. Thesis, Technical University of Delft, The Netherlands.

34. van der Werff, K., 1977, Kinematic and Dynamic Analysis of Mechanisms: A Finite Element Approach, Ph.D. Thesis, Technical University of Delft, The Netherlands.

35. van der Werff, K., 1983, SPACAR User Manual: A Program System for the Analysis of the Motion of Spatial Mechanisms, Technical University of Delft, The Netherlands.

36. Wingard, L., P. Carleberg, and T. Kjellberg, 1992, Enabling Use of Engineering Terminology in Product Models and User Interfaces of CAD/CAM System, *Ann. CIRP,* Vol. 41, No. 1, pp. 205–208.

37. Wu, J. K., et al., 1991, Data Model for Simulation-Based Design of Mechanical Systems, *Internat. J. SARA,* Vol. 1, pp. 67–87.

38. Zhang, W. J., 1991, An Object-Oriented Definition and Description of Mechanisms with Generic Mechanism Model, Technical Report BM-1.91, Technical University of Delft, The Netherlands.

39. Zhang W. J., 1993, ABC (a Main Memory Database Management System) User Manual (Version 1), Technical Report BM-2.93, Technical University of Delft, The Netherlands.

40. Zhang, W. J., 1994, An Integrated Environment for CAD/CAM of Mechanical Systems, Ph.D. Thesis, Technical University of Delft, The Netherlands.

41. Zhang, W. J., and A. van Dijk, 1994, On Developing an Intelligent Editor System for Machine Conceptual Design: Application of Database Technology, in *Computer-Aided Conceptual Design,* J. Sharpe and V. Ho (eds.), LooseLeaf, pp. 221–240.

42. Zhang, W. J., and A. J. Klein Breteler, 1990, A Computer-Oriented Approach to Kinematic Analysis of Rigid Body Spatial Linkages—Application of Finite Element Method: General Theory and Some Element Models, Technical Report BM-1.90, Laboratory of Production Automation and Mechanization, Delft University of Technology, The Netherlands.

43. Zhang, W. J., and K. van der Werff, 1993, Finite Element Modelling Automation for Mechanism Design and Simulation—Computer System Architecture and Finite Element Mechanism Model, *Proceedings of the Ninth International Conference on CAD/CAM Robotics and Factories of Future.*

44. Zhang, W. J., and K. van der Werff, 1993, Guidelines for Product Data Model Formulation Using Database Technology, *Proceedings of the ICED'93,* pp. 1618–1626.

45. Zhang, W. J., and K. van der Werff, 1994, Critique of Conceptual Data Modelling Notions Relative to the Machine Design Domain, *Proceedings of the ASME Engineering Database Symposium,* pp. 59–66.

5

A WEB-BASED SYSTEM TO ENHANCE IPPD BY AUTOMATING DESIGNER COMMUNICATION AND DATA ACCESS

JAMES E. BAILEY and ROBERT H. RUCKER
Arizona State University

5.1 INTRODUCTION

Integrated product and process development (IPPD) relies heavily on teamwork. However, teaming is much more than simply locating people together and must include information tools to rapidly integrate all aspects of the product and related support systems design. The reasons for this are simple: Responsible individuals may be members of several teams, may reside in distant locations, may not be available when needed, and may change during the design cycle. Design data may reside on a variety of platforms with access protocols that are not widely known. Configuration and project management may not be responsive enough for a rapidly changing environment with individually empowered designers. The resulting overhead and frustration is a limiting factor in the implementation of the IPPD concept and enhanced communication is key to a successful design process.

Many computer-aided design (CAD) vendors are responding to customer requests with more integrated IPPD software systems. Consider several state-

Integrated Product and Process Development, Edited by John Usher, Utpal Roy, and Hamid Parsaei
ISBN 0-471-15597-7 © 1998 John Wiley & Sons, Inc.

ments quoted in Deitz.[5] An IDEAS executive said, "I visit nearly a hundred customers a year, and they're clearly looking at re-engineering their product development process. As a result, our customers are interested in automating and optimizing not only individual tasks but all the process as a whole." An Intergraph Software Solutions executive added, "With features like object linking and embedding (OLE), for instance, an application doesn't have to know anything about the data structure of any object it's using or about how the original program manipulated it. This is the way that applications and data will be integrated in the future." A Unigraphics executive suggested the need for a computerized model of the product with universal access to various data items. He said, "Customers (design organizations) need only a single master assembly model (product breakdown structure). Assembly models can also be set up so that different organizations have control over different sets of data. Design, drafting, and manufacturing engineering can be linked so that each team will be updated with the latest version of a part whenever team members access their files; any work that they've already done is associatively updated. This not only makes it easier to control who has access to assembly models but it makes it more practical for teams of engineers to work concurrently." That these models can be predefined and reused was suggested by a Pro/ENGINEER executive, "Many users of Pro/ENGINEER tell me that around 80 percent of their designs are standard and about 20 percent are customized. We've taken much of the pain out of creating new models on the basis of existing ones and evaluating the changes that will be needed downstream." Clearly, the CAD industry is adopting an object-based predefined model of the design deliverables and process as a facilitator of concurrent engineering.

A computerized design support system is needed to simplify data access and designer communication in a dynamic and complex environment. Consider several characteristics of such a system. The system should support consistency in the application of approved design processes so that the quality of the design output is more robust to variations in the design team. The system should facilitate team communication so that potentially significant decisions are not lost in the haste to completion. The system should facilitate awareness of, and access to, company-accepted design tools so that the utility of those tools is maximized. The system should simplify access to widely scattered design data so that everyone on the team is better informed. Finally, the system should facilitate designer level project management so that task scheduling decisions are made with more accurate real-time data.

The concept presented in this chapter supports the IPPD process by employing templates of standardized design protocols. It enhances communication by establishing point-and-click electronic links to other team members. It makes the design process more consistent by specifying deliverables and giving each team member easy access to standardized tools, data, and models. Finally, it facilitates real-time update and access to project status information. The system has been demonstrated in prototype code but a production level version does not yet exist.

As suggested in Figure 5.1, design team members are linked through a Web-based system that has three major components: a Web server set, a concurrent engineering project management system (CEPM), and a set of product development protocols (PDPs). Each designer has a Web server and the CEPM system loaded into his or her workbench. The CEPM system exists as a series of Web pages with JavaScript-enabled interactive buttons. Instead of accessing different resources through a wide variety of complex commands, the designer merely highlights the object he or she wants and clicks a button. A JavaScript module that resides behind the button would access item specific data from the PDP's database and initiate the request. In this fashion, designers can easily access every deliverable for which they are responsible. They can bring up any tool that might be helpful. They can also access read-only copies of all related deliverables without knowing where they reside. They can send messages to interested people without knowing who they are or where they reside. They can request electronic approval of their decisions. They can examine activity status data for any deliverable in the project. Finally, they can update estimates of percentage completion and expected completion dates every time they update their own deliverables.

The product development protocols are models of the approved design processes. Large product development programs are typically broken into subsystem projects. At some breakdown level, the need for real-time communi-

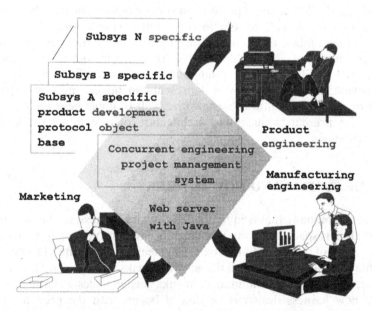

Figure 5.1 The Web based-communication and data-sharing system.

cation and data sharing becomes so great that a well-coordinated IPPD team is appropriate. Most companies perform variations of the same types of projects repeatedly, and for a specific project type, the activities have been shown to be similar every time the project is performed.[1, 17] In a hierarchical sense, each activity can be broken down into specific tasks which vary somewhat between repetitions of the same project. Associated with most tasks are deliverables which, in turn, are related to responsible people, design tools, and communication links. The combination of activities, deliverables, and resources is referred to here as the design culture for a specific project. The PDPs are standardized templates of the culture for all repeated projects.

Consider the advantages that this concept brings to the design process. A preferred design process for given projects is established and applied more consistently. The preferred process will be easier to employ than ad hoc approaches, making adoption simpler. Every design tool that the company has which could be helpful for developing a given deliverable is available instantly to the designer. Designers can access electronic read-only copies of any deliverable in any project without having to know where the item resides or who is developing it. E-mail can be sent to the owner of any deliverable without knowing who that person is. The suggestion is that these new capabilities will greatly reduce the frustration and time consumed in team communication and design data sharing.

The development of the design culture concept and its role in creating PDP models, as well as the generation of a demonstration prototype of the CEPM, were accomplished in a three-year National Science Foundation (NSF) funded effort. The ideas reported here are still in the research phase with the next step being to build and test a production version of the concept. The following presentation should help anyone interested in such an effort.

To further explain this concept and the underlying theories, this chapter is divided into four additional sections. After this introduction, Section 5.2 will describe the CEPM/PDP system and how it works. Section 5.3 presents the relevant literature. Section 5.4 presents an underlying object-oriented model of the design culture that must be built to implement such a system. Section 5.5 presents a suggested architecture of a Web system based on 1997 technology.

5.2 THE SYSTEM AND HOW IT WORKS

Consider a company that frequently designs wire harnesses. This design process at an aerospace company was used to envision functionality during the development of the CEPM prototype. Engineering management could build a reusable template specifying all the activities, tasks, and deliverables, required skills, design tools, and communication links needed to design harnesses. Each time a new harness design is needed, it begins with the predefined PDP template. The design teams would adapt the template to the specific project by adding and deleting tasks/deliverables and assigning specific people with appropriate skills. This is done using a PDP builder, which displays lists of deliverables and resources with their availability. Adding an item to the PDP brings prewritten code modules for the expected actions that might occur when the item is employed. These code modules contain variable fields which are item specific. The PDP builder also instantiates a PDP database with item-specific variables such as drawing numbers, E-mail addresses, and application software call routines. The modified PDP template is then loaded into the CEPM's database and links to each user's workstation are established. The PDP builder module has not been prototyped.

In the demonstration system, the CEPM offers the designer two types of working windows: a project navigator and home pages for each active project. The designer uses the navigator window, as illustrated in Figure 5.2, to search through projects and select the one he or she wishes to work on. Note, from the illustration, that the navigator allows project selection in the context of the program, program phase, or subsystem project, which will be discussed later in the chapter. Note also that each project exists in a neighborhood of related projects. A parent project is that effort that initiated the project and established the design requirements for the project. Sibling projects are precedent and antecedent efforts, which likely affect or are affected by the project in question. If the project's work is to be subdivided and passed on to other teams of designers, then the project has children.

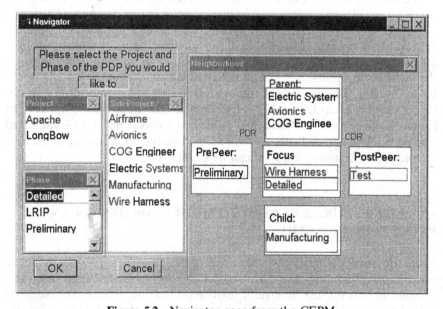

Figure 5.2 Navigator page from the CEPM.

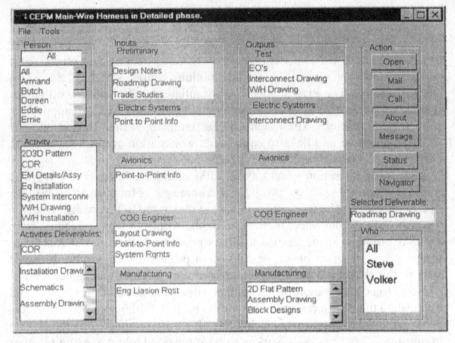

Figure 5.3　Home page of the CEPM.

Selecting a focus project, the designer can call up that project's home page, which is key to the way the system facilitates the IPPD process and is illustrated in Figure 5.3. Recall that design consists of activities intended to generate deliverables. On the left-hand side of the window, the designer can highlight a list of all deliverables associated with a person or activity. Depending on the highlighted option, a list of all deliverables appears in the middle of the page. On the right-hand side of the window is a set of action buttons that can be activated once a deliverable is highlighted. The item can be opened, an E-mail message can be written and mailed to a predefined list of interested parties, a telephone call can be invoked, and the configuration management or completion status of the item can be examined.

5.3.　LITERATURE RELATED TO MODELS OF DESIGN AND PROCESS INTEGRATION

There are many publications in the areas of design process/product modeling and design integration. For an extensive bibliography, see Sullivan et al.[13] Process models appear in several textbooks. For example, Pugh[10] suggests four phases: specification formulation, concept design, detail design, and manufacturing. Ulrich and Eppinger[13] offer a similar five-phase process: concept development, systems level design, detail design, testing and refinement, and production ramp up.

Steward[12] proposes a process model where all tasks can be viewed as a diagonal of precedent-related blocks from the upper left to the lower right with data-sharing requirements represented by the linkages where output and input data are represented by horizontal and vertical lines between blocks. Task nodes are also connected with feedback links representing iteration during design. Eppinger et al.[6] offer a similar model of tasks and information flows. Although these models do not include resources, they do offer a useful view of the design process and its deliverable data requirements.

Johnson and Brockman[9] present a two-part cyclical model of the design process. In the first part, the process is modeled in an object context as a four-level architecture where level 1 represents elemental process classes. Level 2 represents instantiations of process elements for specific designs. Level 3 represents the metadata describing the process, artifact, and task status. Finally, level 4 includes the actual design data viewed as a series of dependent tasks that have been modeled using construction rules,[4] Petri nets,[3] and networks.[2]

The second part of the Johnson–Brockman concept is a problem-solving model that is defined in terms of functional and deliverable objectives. Complex problems can be simplified or decomposed in two ways, as either subobjectives or subfunctions. A third interesting simplification of a problem is the transformation to another technology abstraction. Using these process and problem models, one can integrate design deliverable objectives and process steps with the data that link them.

Erens et al.[7] report on an object-oriented product modeling effort at the University of Leeds. Because products belong to families, they can be modeled with a combination of family characteristics and variant specifications. For example, all automobiles have a chassis, engine, drive train, and so on. However, an individual car has specific versions of each of these subsystems. In addition, each chassis has a frame, suspension, steering mechanism, and so forth. Hierarchically, each suspension has a standard set of components. At some level of abstraction, all members of the automobile product line have an identical hierarchy of family components and the design efforts for these components follow very similar processes. Eren's et al.[7] object model for design management has four levels of data abstraction that are required to navigate any product: product data structure, product family description, project management data, and product variant specifications. Project management data include: who is responsible, when they should start and finish, other product family membership, and where the component is to be manufactured. This object database is implemented in the Leeds Structure Editor, which allows engineers to model and navigate products and design efforts using collections of subsystem families and individual selections of subsystems. Engineers can also identify and access nodes of the same type as a selected node.

These concepts are incorporated into the system reported here through the PDP protocol and CEPM navigator.

Finally, standardized product modeling has been conducted in conjunction with product data management (PDM) efforts; see Teeuw et al.[14] and Gain.[8] Product data management systems manage the flow of data between applications during the product design cycle. The intent is to give many users access to the correct product data in a complex network of application packages and computers. During design, access paths must also be cognizant of the many changes occurring to the product data themselves. This objective can only be met by systematically organizing the data, which requires a good product data model. Unfortunately, generic product models are not reported.

Our efforts to build a unified model based on work done in the aerospace industry can easily be adapted to model the design culture of other industries. The presentation will cover each component separately; a brief presentation of the underlying theory will be followed by suggestions for building an object model useful for the CEPM concept.

5.4 OBJECT MODEL OF DESIGN CULTURE

For this system to work, there must be a model of the design domain that relates individual tasks, deliverables, and design resources. We shall refer to that domain as the design culture. The purpose of the section is to show a theoretical model underpinning the product development protocol needed to integrate the design culture.

We propose, a design culture that has three primary components: product breakdown, design process, and organizational resources. As illustrated in Figure 5.4, the convergence of these three structures are "work packages," each consisting of a task, its resulting deliverables, and the required resources. Predefined combinations of these work packages define a project.

5.4.1 Product Breakdown Structure

The product breakdown structure models the product's deliverable requirements. One such product structure model is the work breakdown structure (WBS) required by the Department of Defense (DoD), as partially illustrated in Figure 5.5. In theory, the product structure begins at level 0 with a strategic mission requirements document and decomposes through systems and subsystems to every deliverable of the design effort.

The WBS can be extended to also include internal design process deliverables such as meeting agendas and analysis reports that support the best design practices of the company. At some level of decomposition, requirements are so highly interrelated that the design should be accomplished by a project team. Thus, the product structure acts to define what a project is in terms of a set of the deliverable requirements that must be configuration managed.

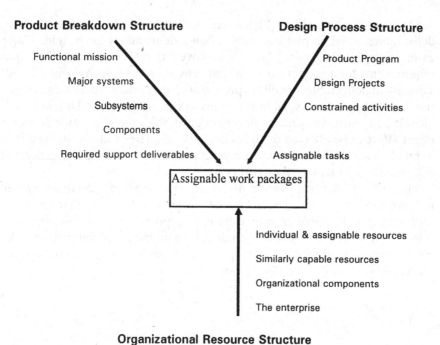

Figure 5.4 Model relating the product, process, and organizational structure within a design protocol.

Level I	Level II	Level III	Level IV
Weapon system	1. Air vehicle	1.1. Airframe	1.1.1. Fuselage
			1.1.2. Lifting surfaces
			1.1.3. Control surfaces
		1.2. Propulsion	1.2.1. Inlet
			1.2.2. Engine
			1.2.3. Exhaust
	2. Logistic support	2.1. Maintenance	2.1.1. Manuals
		2.2. Support	2.2.1. Tools Equip.
			2.2.2. Schedule

Figure 5.5 Partial air weapons systems work breakdown structure.

One can start by building WBS models for often-repeated projects defining deliverables based on past experience. Other deliverables can be added later as management determines how to improve the design process. The input requirements for a project can be included as a project launch document. All subsequent deliverables are either project output or internal documents. Once the deliverable list exists, an interest network needs to be established, which includes, for each deliverable, all other deliverables the generation of which might affect or be affected by the deliverable in question. Thus, the first step in building a PDP is to define the project's scope, all its computer generated deliverables, and its interest network.

Figure 5.6 illustrates a simple object-oriented product breakdown model, including the deliverable types and interest networks. Note that each box represents a deliverable class and the lines represent relationships between classes. An analyst developing a product domain model would interview designers experienced with a specific product to identify, more specifically, the object classes and the specific relationships.

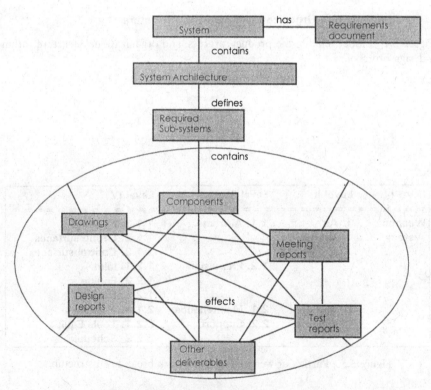

Figure 5.6   Object-oriented product model.

5.4.2   Process Structure

An interesting combination of the product and process models is the integrated product and process development (IPPD) model illustrated in Figure 5.7. This model, based on Steward's[12] diagonal model of processes, is superior to others in that it includes inputs and outputs with their source or destination, as well as a process flow diagram. Many larger companies are developing some version of an IPPD product model. Such models should include input/output (I/O) documents because they are key to ensuring that all requirements are met with a given design.

Note that each box in the IPPD model can be decomposed into a similar lower-level model. Note also that each box at any level represents the solution to some engineering problem and the problem-solving process employs the same systems engineering philosophy depicted in Figure 5.8. That is, at any level, design is a recursive process containing five steps:

1. Receive and analyze higher-level requirements with associated quality metrics.
2. Decompose requirements and specify the functions needed to satisfy them.
3. Group similar functions together and logically specify subcomponents.
4. Design the subcomponents directly or break the effort into logical pieces and allocate subcomponents to children design processes in a recursive fashion.

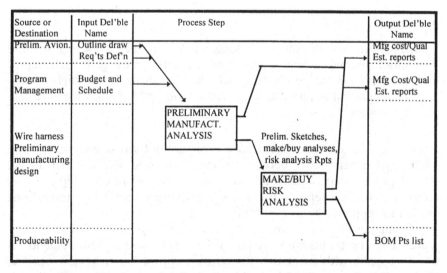

Figure 5.7   Integrated product and process development model.

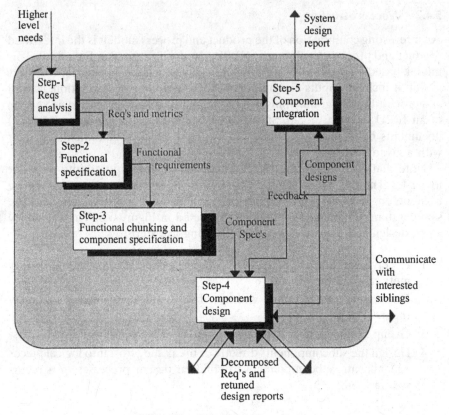

Figure 5.8   Generic design process.

5. Integrate the subcomponent designs to ensure they work together and satisfy the higher-level requirements. If satisfactory, generate a report to document the results and pass that report up to the higher level. If integration is unsatisfactory, return the problematic subcomponent for rework.

The recursive nature of the process is important from an integrating computer support perspective because we can create one object model for the generic process that contains all common objects and relationships, and by referring to it when defining any project's PDP, we inherit the computer code needed to support that specific effort.

For our purposes, the process structure models include: inputs, outputs, precedence constraints, time requirements, and resource requirements. A whole program with its phases, subsystems, critical design reviews, and so forth, is seen as a predefined set of projects. Each project is viewed as a five-step design effort, as shown in Figure 5.8. The "component design" step in the design effort is viewed as a set of work packages managed by the team leader. Work packages themselves are made of primary activities such as: conduct meetings, generate drawings, run tests, and write reports. Each meeting, drawing, test, or report becomes an assignable task with due dates and time budgets. At this level, project management passes from the team leader to the individual designer. In any case, the objects from the process model identify the steps of the process, the precedent requirements, start and finish targets, and resource requirements.

As illustrated in Figure 5.9, a practical process model is not complicated. The design process model is intended to provide links between projects and between tasks and their associated inputs and outputs, tools, skill requirements, and budgets. This model also places tasks on a time line. Finally, the process model relates projects to the parent that launched it, children that it should launch, and siblings that are taking place in parallel.

Each project consists of activities with precedent constraints and predictable start and finish times. Activities can be broken into tasks, which can be assigned

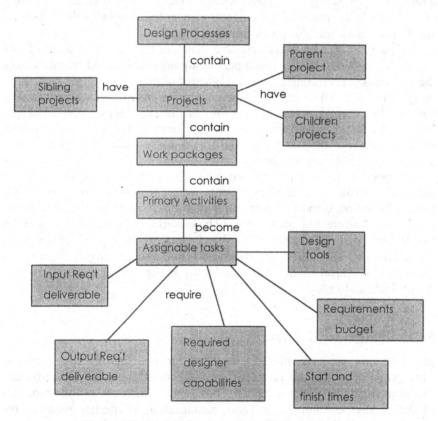

Figure 5.9   Practical object-oriented process model.

to a specific person to generate specific deliverables. Therefore, tasks are the efforts that must be related to certain skills, a resource budget, design tools, input requirement files, and output deliverable destination files. These last objects are referenced in the PDP and accessed by clicking on buttons. To build the process model, a team should interview experienced designers to determine all the input and output deliverables for a given project and then determine the inputs, and so on, for these deliverables.

5.4.3   Resource Structure

The resource structure models the tools and supplies required to accomplish the design. This model establishes the linkages between: people, time, budget, software, and datafiles that need to be assigned to each work package. The computerized component of project management is the allocation and control of these resources. Sonnewald[11] developed a descriptive model of designer communication. Her analysis discovered 11 roles played by designers: champion, consumer, consumer representative, external star, agent, technician, internal star, gatekeeper, manager, creator, and boundary scanner. The utility of this view is the potential to assign generic roles to individuals as parts of an "interest network" for given work packages.

The path to these roles is a resource structure that can be thought of as a tree with a root that leads to each partnering organization. At the next level, the resources of each company are categorized as: time, money, people, equipment, software, and datafiles. These categories are further decomposed to individually assigned project and task budgets. The need is to establish linkages between deliverables and the items used to generate them. Project management becomes the allocation and control of those resources in order to generate the deliverables on schedule.

The point is that every time we plan a project, we start with a set of deliverables with resource requirements. When we link a given deliverable to specific software tools, object components, a responsible designer, and a time budget, we create pathways through networks to the tools, code modules to call the tools, links to destination object components, designer communication addresses, start times, due dates, and time budgets targets. Figure 5.10 illustrates the object model that permits the generation of resource budgets for projects and tasks.

5.4.4   Utility of the Concept

In this section, we have offered a structure for modeling the design culture. Three components of this structure are the product, process, and resources. An object model of the structures has been offered. The utility of this model is that communication and management of design requires the dynamic integration of all three structures. The fact that the structures

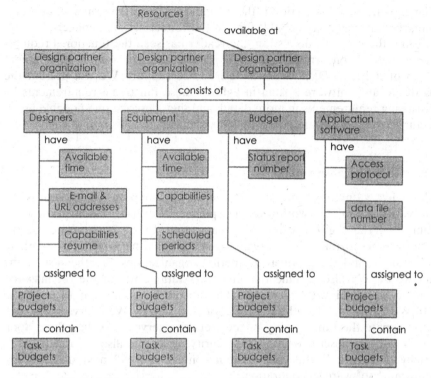

Figure 5.10   Practical object model of the resource structure.

intersect in a task with resources and deliverables is significant in that together they form a manageable work package. All design projects can be defined as a set of these work packages. The packages can be associated with start and finish dates and a time budget. From a computer integration perspective, the work package object is linked to data records containing information about everything associated with the specific design effort. The fact that other models of design omit parts of the culture limits their utility as a PDM integrating tool.

5.5   A SYSTEMS ARCHITECTURE TO DELIVER THE CEPM/PDP FUNCTIONS

To be useful, however, the concept must be implemented. It could be implemented as a commercial software package or in-house development effort. To aid in implementation, we describe an approach based on 1997 technology.

The audience for this section is the computer systems developer, and, therefore, the language may be a bit foreign to product design managers.

From the previous discussion, the reader has seen the functional requirements that the concurrent engineering project manager/product development protocol (CEPM/PDP) system is designed to support. We will present the hardware and software system in light of these functional requirements by discussing requirements in more detail, and then describe one possible architecture for the hardware and software services that will support them.

5.5.1   Baseline Assumptions

The delivery environment we have chosen for our prototype system is the Web because of its worldwide acceptance. We have assumed network connectivity as a given within the organization. In this context, we use the term *intranet* to refer to an organization's internal network and *internet* to refer to the linkage between partnering organizations and the rest of the net world. We also assume that the workstations within the organization will be equipped with browsers with capabilities equivalent to Netscape 4.0. We anticipate that the organization will deploy Web servers that will approximate the functionality of the Netscape server set called Suite Spot, which will be described later. A security system is also assumed and is briefly discussed. Within this general context, we will next describe our hardware/ software configuration.

The modern enterprise is richly interconnected with local-area networks (LANs) and wide-area networks (WANs) and is already or will soon be interconnected to corporate intranets, as well as reaching the outside world via specialized network level interfaces called *firewalls*. The overall architecture of this globally distributed object system will be in compliance with a set of specifications called CORBA (common object request brakes). To facilitate the discussion, we assume that the corporation has already built an intranet or is well on the way to doing so. Additionally, we assume a level of hardware and software technical sophistication that includes knowledge about and experience in browser and server technologies, Java, JavaBeans, JavaScript, HTML, relational databases, and object bases. The following is a capsule description of some of these terms:

1. *Hypertext Markup Language* (*HTML*). This language is being extended to act as a container for all other forms of browser content. It is a subset of SGML (standardized general markup language).
2. *Java.* An object-oriented programming language specifically designed for net software development and deployment. Java applets are downloadable, platform independent, and locally executable Java programs.

3. *JavaBeans.* A platform-independent software component architecture that allows true software reuse.

4. *JavaScript.* A network scripting language that ties together Java applets, JavaBeans, JavaScript code, plugins, and HTML code.

5. *Plugins.* Helper applications that are platform specific and provide extended functionality to Java and JavaScript.

6. *Object Bases.* Persistent data and functionality storage systems based on the object paradigm.

5.5.2   The Functional Requirements at the three Levels

To break up the inherent complexity of the following description, we will categorize the CEPM/PDP functional requirements and hardware/software requirements into three levels:

1. The individual workstation level where local files and applications reside and access to local resources are effected. These resources are specific to the designer's platform.

2. The enterprise or intranet level where organizational files, applications, and resources reside. One new set of items at this level is "Web" specific, while the remaining resources make up the company's standard tool set.

3. The global or Internet level where partner organization files and applications reside as well as generic net resources. These are the network services outside of the organization's control involving external tools and protocols.

5.5.2.1   Workstation Level   For individual workers at the workstation level, the browser is their universal window into the networked world. Using the browser, a designer can access everything needed for his or her computer-related tasks. The functional requirements at this level are the following:

1. Information production, sharing, and the management of related tasks is required. In this context, the user is able to browse and retrieve all the documents and resources on the net that he or she is authorized to obtain. The user needs access capabilities in order to interact with both his or her own personal object base as well as the corporate level repositories. Word documents, CAD drawings, virtual reality models, Excel spreadsheets, or anything else that the browser could possibly interpret must be accessible, including special CEPM functions and platform specific plugins.

2. A communication and "groupware" type of collaboration between engineering designers is required. E-mail is the premier example in this

category. Multiple-team real-time interaction collaboration over the net must be supported. This capability allows secure, threaded multiperson interactions over the net, which permits efficient communication regardless of location. This is the functionality that the CEPM/PDP is designed to facilitate by providing point-and-click access to specific resources.

3. Searching, information retrieval, and indexing of net-worthy information for retrieval by others is also required. Functionality here involves capabilities to find specific engineering resources and personnel, as well as net level queries.

4. Finally, access to application software is required at this level. This is of singular importance for the individual because it defines the capability limits of the designer. The user must be able to invoke downloadable executable content from a server whether that server is local on the intranet or elsewhere on the Internet. This would involve, for example, the capability to download a Java applet served from an internal enterprise server or to download an executable applet from an external server under security restrictions. Finally, the user should be able to invoke local applications either as a consequence of downloading a particular type of file or as a direct invocation of the local executable.

5.5.2.2 Corporate Level   To efficiently serve the individual user, there are network level tasks that need to be supported, as well as the organization's standard operational server tool support. The preferred approach to this requirement is to establish "net level services."[20] Instead of collections of particular operating system–dependent programs specialized for a set of functionality, these net level tasks are implemented by a set of "net level" Web servers, as well as "application servers." Thus, these modern servers are (relatively) platform and operating system independent and are each specialized for particular sets of net level tasks. In general, the net level requirements are:

1. Keeping track of information about resources and personnel. These are directory services that track and manage all of the information needed to access personnel data, server configurations, and application-oriented resources. End users must be able to look up information about people such as E-mail addresses, phone numbers, as well as security keys. In addition, users need to be able to find out about application resources throughout the organization, and if authorized, external to the corporation as well.

2. Keeping track of and enforcing security policies. Security is a major issue in this modern era and will become even more important as intranet/internet access expands to include the global community. Thus, there is a requirement for encryption, authentication, verification, and validation. E-mail and group collaboration communications must both be se-

cure and protected. In addition to individual designer access to information sources, applications themselves are now subject to stringent security restrictions as to their use and authenticity. As an aside, Java has the capability to implement the most stringent of security measures that have guaranteed its endorsement and deployment by the commercial community.

3. Distributing data and functionality across the network. Replication is often needed in order to achieve speed of response or ease of administration in distributing copies of portions of large applications or data sets, Web contents, directories, collaborative group messages, or, in general, any slowly time varying data or processes that can be refreshed asynchronously. In addition, provision must be made for reconciling these remote changes back into the server process repository.

4. Managing the network itself. As the network expands by including more and more resources, the need for management of the net itself becomes a prominent issue. There needs to be ways to manage the net, preferably from an administrative browser using standard network protocols such as the simple network management protocol (SNMP). The managed objects must include: user information, access control parameters, server(s) configuration information, and all of this being available in a secure mode.

5.5.2.3 Global Level   The realization of the importance of customers, partners, suppliers, and competitors, who are all just a "mouse click away," drives the recognition of the importance of the external network linkages of an enterprise. Fortunately, the Internet and much of this infrastructure is already in place or is being built at unprecedented rates. There is, however, still the issue of how to smoothly connect to other parties on the Web given the proliferation of operating systems, hardware/software applications, and general and pervasive incompatibilities at all levels. These interface issues involve all levels of the corporation, but we will consider only the issues of security and protocols and standards. The role of firewall interfacing software and hardware assumes prominence at this point. Going from inside the corporation to the global net poses unique challenges and the security interface is a basic and pervasive one. Commerce at this level already involves enormous numbers of distributed objects. To manage this incredible melee of systems and products requires standards and protocols. The issues here are implementing those standards and protocols. The most important standard in this area is the common object request broker (CORBA), a specification of how all these distributed objects could conceivably intercommunicate. Our software approach to this issue has been to concentrate on products that support the CORBA specifications and, in particular, support interORB (object request broker) commerce, specifically our desire to communicate engineering objects throughout the world.

5.5.3 Systems Architecture, Hardware, and Software

Given the functional requirements for the CEPM/PDP system described previously, we now suggest one possible structure for the delivery mechanisms
that would implement it. From Figure 5.11 we can see some of the software/
hardware that would be required to deliver CEPM/PDP functionality in an
open standards environment.

Figure 5.11 shows that the computing part of the enterprise intranet/internet
must include at least the following software/hardware components:

1. A Web server set, hardware, and software that provide required functionality to the user community
2. Object bases, relational databases, and other types of databases such as
   hierarchical and CODASYL network databases
3. Application servers
4. Workstation browsers or other network computers
5. CORBA-compliant Web servers and clients
6. The Internet itself
7. Other enterprises with their own intranets

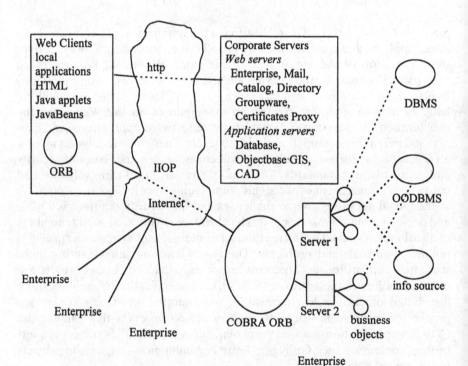

Figure 5.11 Delivery architecture for the CEPM/PDP protocol.

5.5.3.1   Web Client Workstation Browser Level: Hardware and Software   It is necessary to assume that the local level consists of a workstation that is running a net-aware browser (in the future there will be browsers that are stripped-down workstations, JavaStations, for example). The browser or client should also include a CORBA ORB. One good example of a fully functional browser is the Netscape Navigator 4.0. This browser encapsulates all of the browser functionality discussed in this chapter. The browser should be able to support HTML, Java, JavaBeans, JavaScript, and plugins. Given the assumption that the browser is the "universal client," all of the user's computing needs will be met via interaction with the browser screen. The net to both corporate level resources as well as their local applications and local files (assuming the browser machine has disk storage and personal relational/object bases for workstation level persistent storage) links the workstation browser. In addition, the local client "ORB" allows communication across the net to business objects residing on remote CORBA servers.

5.5.3.2   Corporate Level: Hardware and Software   Consider two major sets of "servers." The more recent set is characterized as "Web servers" or "network level service providers." The second set is the more familiar set that encompasses current processors that are general enough to be used by a group of clients rather than only a few. The distinctions between these two sets are not black and white and they will most likely all be simply called "corporate servers." Our interest in these server sets arises because the CEPM/PDP makes critical use of the Web server sets as well as the standard corporate server tool set.

Consider now the hardware/software needed to support the CEPM/PDP functional requirements. The server requirements are roughly those embodied in the Netscape Suite Spot server set or comparable product offerings from IBM or Microsoft.

1. *Enterprise Server.* Manages content that is to be served to users. This is the mechanism that allows deliverables such as word processing documents or CAD drawings to be downloaded over the net. This server also allows Java applets that can run on any browser. The reader is likely familiar with the static displays that the Web delivers when items on the screen are clicked on. Java applets are modules of code that permit dynamic displays over the Web such as opening a CAD application or accessing a list of deliverables from a database.

2. *Mail Server* This provides the enterprise with its standard E-mail protocols and functionality.

3. *Catalog Server:* This processor provides indexing, searching, and browsing throughout the network.

4. *Certificate Server:* Supplies and manages public key certificates and security keys. This allows secure network operations.
5. *Groupware Server:* This allows secure groupware-style interactions among teams or individuals.
6. *Directory Server:* This is the net level directory of services and resources including people characteristics and access control.
7. *Proxy Server.* Replicates and distributes data and functionality netwide.

Finally, consider standard servers that traditionally have supported the corporation. These servers are specialized to perform basic communications and analysis work for clients that is beyond the resources of the individual workstation.

1. *Database Servers.* These systems store simple data that can be specified as belonging to one of a few fixed types. They are useful for situations that are highly constrained as to format and content. Our initial CEPM/PDP repository uses a relational base (MS SQL Server) that captures some essential parts of the complexity of the engineering design system we are modeling.
2. *Object Base Servers.* These are database management systems that hold both data structures and methods on the data. Because these systems hold objects natively without the translations required by the relational models, they are the preferred storage method of choice for complex data structures. For example, a CAD drawing or an image or audio file is more easily stored and manipulated in an object base than a relational base and, therefore, the object base ultimately provides a more efficient mechanism for their manipulation. Our second-generation CEPM/PDP will utilize an object base as its main repository.
3. *CAD and CAE Servers.* These are the systems on which the companies design support systems reside.
4. *GIS Servers.* These are the geographical support systems that are supplied by telecommunications suppliers.
5. *Telecommunications Servers.* These are the systems that permit voice and video exchanges.

Perhaps a short note should be inserted here concerning databases and their associated servers. Database management systems form the backbone of any enterprise and provide much of the source of the enterprise's knowledge, in addition to that of its personnel. In keeping with its importance, database technology has steadily improved and evolved. With the explosive emergence of Web-based solutions, databases have again assumed premier importance, but now as data retrieval and storage sources for Web applications. The most prominent database management system currently used by corporations is the relational model. There is another model, however, that promises even more capability to the corporation because it is based on the object paradigm—the object-based data management system. The CEPM/PDP system is currently using and evaluating both models in concert.

5.5.3.3 *Global Level: Hardware and Software* At the point where the organization interfaces with the outside world, issues of security, standards, protocols, legalities and liabilities, and competition and cooperation all emerge as considerations. Our CEPM/PDP system is designed to be embedded into the local corporate environment and to conform to the corporate policies as established. Our design uses the standard open internet protocols such as HTTP, HTMP, TCP/IP, SNMP, POP3, and so forth, and soon CORBA and IIOP, but again this is within the corporate context with its constraints.

Our templates and programs are designed to be downloadable to a corporation's servers, and it is then a corporate decision as to whether to allow this to happen. This issue of corporate policy and security software must, be addressed by corporate level initiatives.

5.6 SUMMARY

In summary, this chapter presents a concept developed during a National Science Foundation–funded effort to facilitate design data access and communication. A laboratory level prototype was built around a real-world design process employed in an aerospace company. The next phase of development is to build an alpha testable product version of the software system. The concept works by attaching a network-based concurrent engineering project management overlay on each designer's workstation. All active design projects are represented in the CEPM by object-oriented design protocols. These PDPs permit point-and-click access to people, deliverables, and design tools. It is hypothesized that using this system will speed the design process by reducing access overhead and automating communication between interested parties.

At the core of the concept is an IPPD model, which can be broken into a recursive set of work packages with a common systematic structure. Taking advantage of that commonality, it is possible to build a system that provides point-and-click access to a large portion of the people, data, and tools needed during design. To build that system, a model of the design culture is needed, which consists of hierarchies of the product, process, and resource structures that make up the design culture. It has been shown that with the same culture, the way we design is consistent, and design projects are repeated.

ACKNOWLEDGMENTS

The material presented in this chapter was developed under NSF Grant DMI 9414273. The authors want to recognize the contributions of McDonnell Douglas Helicopter

Systems, Dr. Gurshaman Baweja, and Jarrod Rogers, Scott Brandon, and Kevin Bradbury.

REFERENCES

1. Bailey, J., Brandon, S., Bradbury, K., and Baweja, G., 1994, Consistency of Concurrent Engineering Design Projects, Activities, and Tasks, *Proceedings of the First Conference on Concurrent Engineering,* pp. 83–88.

2. Bosch, K., P. Bingley, and P. van der Wolf, 1991, Design Flow Management in the NELSIS CAD Framework, *Proceedings of the 28th ACM/IEEE Design Automation Conference,* Spring, pp. 711–716.

3. Bretschneider, F., C. Kopf, and H. Lagger, 1991, Knowledge Based Design Flow Management, *Proceedings of the IEEE International Conference of Computer Aided Design,* pp. 350–353.

4. Brockman, J., and S. Director, 1991, The Hercules CAD Task Management System, *Proceedings of IEEE International conference of Computer Aided Design,* pp. 254–257

5. Deitz, D., 1996, Next Generation CAD Systems, *Mech. Eng.,* Vol. 11, No. 8, pp. 68–72.

6. Eppinger, S., D. Whitney, R. Smith, and D. Gebala, 1994, A Model Based Method for Organizing Tasks in Product Development, *Res. Eng. Des.,* Vol. 6, No. 1, pp. 1–13.

7. Erens, F., A. McKay, and S. Bloor, 1994, Product Modeling Using Multiple Levels of Abstraction Instances as Types, *Computers in Industry,* Vol. 24, pp. 17–28.

8. Gain, P., 1996, New Generation of PDM Emerges, *Computer-Aided Eng.,* Vol. 15, No. 11, pp. 52–58.

9. Johnson, E., and J. Brockman, 1996, Towards a model for Electronic Design Process Refinement, *Computers in Industry,* Vol. 30, No. 1, pp. 27–36.

10. Pugh, S., 1992, *Total Design, Integrated Methods for Successful Product Engineering,* Addison-Wesley, Reading, MA.

11. Sonnenwald, D., 1992, Communications in Design, *Bull. Amer. Soc. Inform. Sci.* Vol. 8, No. 4, pp. 15–16.

12. Steward, D. U., 1991, The Design Structure System: A Method of Managing the Design of Complex Systems, IEEE Transactions on Engineering Management, Vol. EM-28, No. 3, pp. 71–74.

13. Sullivan, W. G., P. Lee, J. Luxhoj, and R. Thannirpalli, 1994, A Survey of Engineering Design Literature: Methodology, Education, Economics, and Management Aspects. *Engineering Economist,* Vol. 40, pp. 7–40.

14. Teeuw, W., J. Liefting, R. Demkes, and M. Houtsma, 1996, Experience with Product Data Interchange: On Product Models, Integration, and Standardization, *Computers in Industry,* Vol. 11, No. 3, pp. 205–211.

15. Ulrich, K., and S. Eppinger, 1995, *Product Design and Development,* McGraw-Hill, New York.

SECTION II
PROCESS DESIGN

6

RAPID PROTOTYPING AND MANUFACTURING: THE ESSENTIAL LINK BETWEEN DESIGN AND MANUFACTURING*

CHUA CHEE-KAI and LEONG KAH-FAI
Nanyang Technological University

6.1 INTRODUCTION

6.1.1. Historical Development

The declining cost of computers, especially that of personal computers and minicomputers, is changing the way a factory works. Increase in the use of computers has spurred advancements in many computer-related areas such as computer-aided design (CAD), computer-aided manufacturing (CAM), and computer numerical control (CNC) machine tools. In particular, the emergence of rapid prototyping (RP) systems could not have been possible without the presence of CAD. However, when one examines the numerous RP systems in existence today, it is easy to conclude that, besides CAD, many other technologies and advancements in other fields such as manufacturing systems and materials are just as essential. Table 6.1 traces the historical development of various technologies with the estimated date of inception.

* ©John Wiley & Sons (Asia) PTE Ltd., 2 Clementi Loop, #02-0 Jin Xing Distripark, Singapore 129809.

Integrated Product and Process Development, Edited by John Usher, Utpal Roy, and Hamid Parsaei
ISBN 0-471-15597-7 © 1998 John Wiley & Sons, Inc.

TABLE 6.1 Historical Development of Rapid Prototyping and Related Technologies

Year of Inception	Technology
1770	Mechanization[19]
1946	First computer
1952	First numerical control (NC) machine tool
1960	First commercial laser[10]
1961	First commercial robot
1963	First interactive graphics system (early version of computer-aided design)[28]
1988	First commercial rapid prototyping system

6.1.2 Three Phases of Prototyping

Prototyping or model making is an age-old practice. The intention of having a physical prototype is to realize the conceptualization of a design. Thus, a prototype is usually required before the start of the full production of a product. The fabrication of prototypes has been experimented in many forms—material removal, castings, molds—and with many material types—zinc, urethanes. In this context, it is appropriate therefore to define *prototype* as shown in Figure 6.1.

Prototyping development has gone through three phases, the last two having emerged only in the past 20 years. [1, 2] The three phases are described as follows.

6.1.2.1 First Phase: Manual (or Hard) Prototyping The first phase began several centuries ago. In this phase, prototypes are not very sophisticated and fabrication of prototypes takes, on average, about four weeks.[22] The techniques used in making these prototypes are craft based, extremely labor intensive, and usually take a long time to complete.

6.1.2.2 Second Phase: Soft or Virtual Prototyping As application of CAD/CAE/CAM became more widespread, the early 1980s saw the evolution of the second phase of prototyping—*soft or virtual prototyping*. Soft prototyping takes on a new meaning—computer models can be stressed, tested, analyzed,

> *A prototype is the first or original example of something that has been or will be copied or developed; It is a model or preliminary version; E.g.: A prototype supersonic aircraft.*

Figure 6.1 Definition of a prototype.[4]

and modified as if they were physical prototypes. For example, analysis of stress and strain can be accurately predicted because of the ability to specify exact material attributes and properties.

Meanwhile, prototypes tend to become more complex—about twice the complexity as before.[22] Correspondingly, the average time required to make the physical model increases tremendously to about 16 weeks as the building of physical prototypes is still dependent on craft-based methods, although the introduction of higher-precision machines such as CNC machines has helped to reduce this.

6.1.2.3 Third Phase: Rapid Prototyping
Rapid prototyping of physical parts, otherwise known as solid freeform fabrication, desktop manufacturing, or layer manufacturing technology, represents the third phase in the evolution of prototyping. The inventions of these RP methodologies are described as "watershed events"[16] because of the tremendous time savings, especially for complicated models. Although parts are three times as complex as before, the time needed to make such a part averages only three weeks.[22] Since 1988, more than 20 different systems have emerged.

Even with the advent of rapid prototyping, there is still strong support for soft prototyping. Lee[20] argued that there are still unavoidable limitations with rapid prototyping. These include material limitations (either because of expense or through the use of materials dissimilar to that of the intended part), the inability to perform endless what-if scenarios, and the likelihood that little or no reliable data can be gathered from the rapid prototype to perform finite-element analysis (FEA).

6.1.3 Fundamentals of Rapid Prototyping

Fundamentally, the development of RP can be discussed in terms of four primary areas: input, method, material, and applications. These are illustrated in the rapid prototyping wheel in Figure 6.2.

6.1.3.1 Input
Input refers to the electronic information required to describe the physical object with three-dimensional data. There are two possible starting points—a computer model or a physical model. The computer model created by a CAD system can be either a surface model or a solid model. For the physical model, it requires data acquisition through reverse engineering to obtain information required by the RP systems. In reverse engineering, a wide range of equipment can be used, including a coordinate measuring machine (CMM) or a laser digitizer.

6.1.3.2 Method
Although there are currently more than 20 vendors for RP systems, the method used by each vendor can be simply classified into several categories: photocuring, cutting and gluing/joining, melting and solidi-

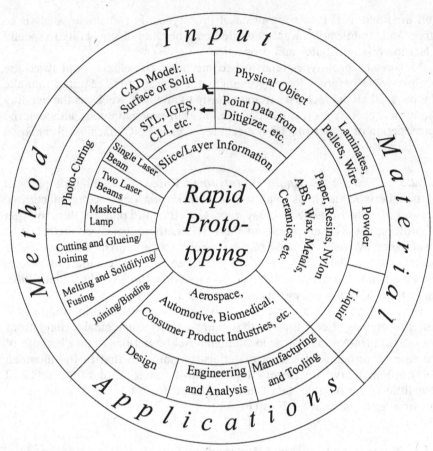

Figure 6.2 Rapid prototyping wheel depicting the four major aspects of RP.

fying/fusing, and joining/binding. Photocuring can be further divided into single or double laser beam and masked lamp.

6.1.3.3 Material   The initial state of the material can be either solid, liquid, or powder. In a solid state, it can come in various forms such as pellets, wire, or laminates. The current range of materials include paper, nylon, wax, resins, metals, and ceramics.

6.1.3.4 Applications   Most of the RP parts are finished or touched up before they are used in specific areas of applications. Applications can be grouped into (1) design; (2) engineering, analysis, and planning; and (3) tooling and manufacturing. A wide range of industries can benefit from RP and these include, but are not limited to, aerospace, automotive, biomedical, consumer, electrical, and electronics products.

6.1.4 Advantages of Rapid Prototyping

Today's RP systems can directly produce functional parts in limited production quantities. Parts produced in this way have an accuracy and surface finish inferior to those made by machining. Neither are the material and mechanical properties of these parts identical, or very close, to those of the metals and plastics used commonly today. However, the time taken to produce any part— once the design data are available—will be a matter of hours.

The full production of any product encompasses a wide spectrum of activities. Kochan and Chua[18] describe the impact of RP technologies on the entire spectrum of product development and realization. In Figure 6.3, the activities for full production in a conventional model and with an RP model are comparatively depicted. Depending on the size of production, savings on time and cost could range from 50 to 90 percent!

6.1.5 Classification of Rapid Prototyping Systems

Although there are many ways in which one can classify the numerous RP systems in the market, one of the better ways is to classify an RP system broadly by the initial form of its material. In this manner, all RP systems can

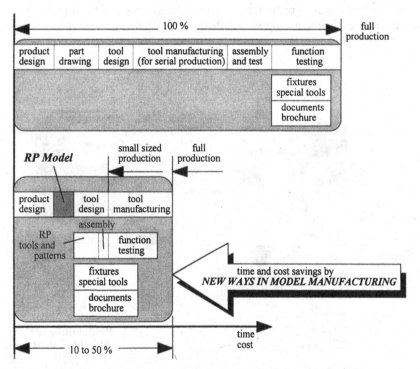

Figure 6.3 Results of the integration of RP technologies.

be easily categorized into (1) liquid-based, (2) solid-based, and (3) powder-based systems.

6.2   RAPID PROTOTYPING PROCESS CHAIN

All RP systems generally have a similar sort of process chain. Such a generalized process chain is shown in Figure 6.4. There are a total of five steps in the chain: three-dimensional modeling, data conversion and transmission, checking and preparing, building, and postprocessing. Depending on the quality of the model and part during checking and postprocessing, the process may be iterated until a satisfactory model and part is achieved.

Like other fabrication processes, process planning is an important step before RP itself commences. In process planning, the steps of the RP process chain are executed. The first step is three-dimensional geometric modeling. In this instance, the requirement would be a workstation and a CAD modeling system. The factors and parameters that influence the performance of each operation are examined and decided upon. For example, if stereolithographic apparatus (SLA) is used to build the part, the orientation of the part is an important factor that could, among other things, influence the quality of the part and the speed of the process. Needless to say, an operation sheet used in this manner requires proper documentation and guidelines. Good documentation, such as a process logbook, allows future examination and evaluation,

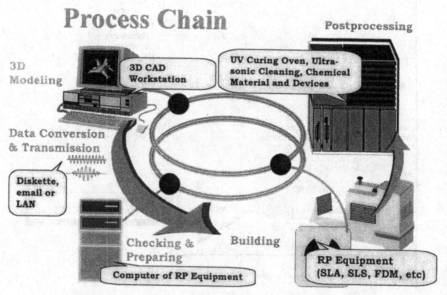

Figure 6.4   Process chain of the rapid prototyping process.[17]

and subsequent improvements can be implemented to process planning. The five steps in the process chain are discussed in the following sections.

6.2.1 Three-Dimensional Modeling

Advanced three-dimensional CAD modeling is a general prerequisite in RP processes and is usually the most time-consuming part of the entire process chain. It is most important that such three-dimensional geometric models can be shared by the entire design team for many different purposes, such as interference studies, stress analyses, finite element method (FEM) and analysis, detail design and drafting, planning of manufacturing, including numerical control (NC) programming. Such a model, which is an archetype of the physical part to be built, must be represented as closed surfaces that unambiguously define an enclosed volume. Thus, the data must specify the inside, outside, and boundary of the model. An enclosed volume ensures that all horizontal cross sections are closed curves to create the solid object.

There are commonly two misconceptions among new users of RP. First, unlike NC programming, RP requires a closed volume of the model, whether the basic elements are surfaces or solids. This confusion arises because new users are usually acquainted with the use of NC programming where a single surface or even a line element can be an NC element. Second, new users also assume *what you see is what you get.* These two misconceptions often lead to underspecifying parameters to the RP systems, resulting in poor performance and nonoptimal utilization of the system. Examples of considerations that have to be taken into account include orientation of the part; need for supports; difficult-to-build part structures such as thin walls, small slots, or holes; and overhanging elements. Thus, RP users have to learn and gain experience from working on the system. The problem is usually more complex than one can imagine as different RP machines have different requirements and capabilities. Whereas SLA requires supports, solid ground curing (SGC) does not, and SGC works most economically if many parts are nested together and processed simultaneously.

6.2.2 Data Conversion and Transmission

The solid or surface model to be built is next converted into a format dubbed the .STL file format because it originates from 3D Systems, which pioneered the stereolithography system. The .STL file format approximates the surfaces of the model using tiny triangles. Highly curved surfaces must employ many triangles, which means that .STL files for curved parts can be very large. Almost, if not all, major CAD/CAM vendors supply the CAD–STL interface. Since 1990, a large number of CAD/CAM vendors have developed and integrated this interface into their systems.

This conversion step is probably the simplest and shortest of the entire process chain. However, for a highly complex model coupled with an extremely low performance workstation or personal computer, the conversion can take several hours. Otherwise, the conversion to an .STL file should take only several minutes. Where necessary, supports are also converted to a separate .STL file. Supports can alternatively be created or modified in the next step by third-party software, which allows verification and modification of models and supports.

The transmission step is also fairly straightforward. The purpose of this step is to transfer the .STL files that reside in the workstation to the RP system's computer. It is usual that the workstation and the RP system are situated in different locations because of different facilities and environmental requirements of the equipment. The workstation as a design tool is usually located in a design office, whereas the RP system is usually located on the shop floor. Data transmission via agreed data formats such as .STL or IGES may be carried out through a diskette, E-mail, or LAN (local-area network). No validation of the quality of the .STL files is carried out at this stage.

6.2.3 Checking and Preparing

The computer term, *garbage in garbage out,* is applicable to RP. Many first-time users are frustrated at this step to discover that their .STL files are faulty. However, more often than not, this is due to both the errors of CAD models and the nonrobustness of the CAD–STL interface. Unfortunately, today's CAD models, whose quality is dependent on CAD systems, human operators, and postprocessing, are still afflicted with a wide spectrum of problems, including the generation of unwanted shell punctures (i.e., holes, gaps, cracks, etc.). These problems, if not rectified, will result in the frequent failure of applications downstream.

At present, CAD model errors are corrected by human operators assisted by specialized software such as MAGICS, a software developed by Materialise, N.V., Belgium.[21] This process of manual repair is very tedious and time consuming, especially if one considers the large number of geometric entities (e.g., triangular facets) encountered in a CAD model.

Once the .STL files are verified to be error free, the RP system's computer analyzes the .STL files that define the model and slices the model into cross sections. The cross sections are systematically recreated through the solidification of either liquids or powders or the fusing of solids to form a three-dimensional model.

In SLA, for example, each output file is sliced into cross sections, between 0.076 mm (minimum) to 0.5 mm (maximum) in thickness. Generally, the model is sliced into the thinnest layer as it has to be very accurate. The supports are created using coarser settings. An internal cross-hatch structure is generated between the inner and the outer surface boundaries of the part. This serves to hold up the walls and entrap liquid that is later solidified in the presence of ultraviolet (UV) light.

Preparing building parameters for positioning and stepwise manufacturing in light of the many available possibilities can be difficult if there is no proper documentation. These possibilities include determination of build orientation, spatial assortments, arrangement with other parts, necessary support structures and slice parameters. They also include determining technological parameters such as cure depth, laser power, and other physical parameters as in the case of SLA. Thus, user-friendly software, user support in terms of user manuals, dialogue mode, and on-line graphical aids will be very helpful to users of the RP system.

6.2.4 Building

From the prepared "slice," cross sections are recreated layer by layer through either the solidification of liquids or the binding of powders to form a three-dimensional model. Alternatively, the cross sections are already thin, solid laminations that can be glued together. Other similar methods may also be used to build the model.

For most RP systems, this step is fully automated. Thus, it is usual for operators to leave the machine on to build a part overnight. The building process may take several hours, depending on the size and number of parts required. The number of identical parts that can be built is subject to the overall build size constrained by the builder of the RP system.

6.2.5 Postprocessing

The final task in the process chain is postprocessing and generally some manual operations are necessary. As a result, the danger of damaging a part is particularly high. Thus, the postprocessing operator has a high responsibility to ensure successful process realization. The necessary postprocessing tasks for some major RP systems are shown in Table 6.2.

TABLE 6.2 Essential Postprocessing Tasks for Different RP Processes ($\checkmark$ = Is Required, × = Not Required)

	Rapid Prototyping Technologies			
Postprocessing Tasks	Selective Laser Sintering (SLS)	Stereolithography (SLA)	Fused Deposition Modeling (FDM)	Laminated Object Manufacturing (LOM)
1. Cleaning	$\checkmark$	$\checkmark$	×	$\checkmark$
2. Postcuring	×	$\checkmark$	×	×
3. Finishing	$\checkmark$	$\checkmark$	$\checkmark$	$\checkmark$

The cleaning task refers to the removal of excess materials that may have remained on the part. Thus, for SLA parts, this refers to the removal of excess resin residing in an entrapped portion of the part such as a blind hole, as well as the removal of supports. Similarly, for SLS parts, the excess powder has to be removed. Likewise, for LOM, pieces of excess woodlike blocks of paper that acted as supports have to be removed.

As shown in Table 6.2, the SLA procedures require the highest number of postprocessing tasks. More importantly, for safety reasons, specific recommendations for postprocessing tasks have to be prepared, especially for the cleaning of SLA parts. Furthermore, accuracy is related to the post-treatment process.[23] Specifically, this refers to the swelling of SLA-built parts with the use of cleaning solvents. Parts are typically cleaned with solvent to remove unreacted photosensitive resin. Depending on the "build style" and the extent of crosslinking in the resin, the part can be distorted during the cleaning process. This effect was particularly pronounced with more open "build styles" and aggressive solvents. With "build styles" approaching a solid fill and more solvent-resistant materials, damage from a cleaning solvent can be minimized. With newer and better cleaning solvents, part damage due to the cleaning solvent can be reduced or even eliminated.[23]

The SLA parts are built with pockets of liquid embedded within them. Therefore, postcuring in an ultraviolet (UV) oven is required. All other nonliquid RP methods do not undergo this task. Finishing refers to the secondary processes such as sanding and painting used primarily to improve the surface finish or aesthetic appearance of the part. It also includes any additional machining processes such as drilling, tapping, and milling to add necessary features to the parts.

6.3 LIQUID-BASED RAPID PROTOTYPING SYSTEMS

Most liquid-based rapid prototyping systems build parts in a vat of photocurable liquid resin, an organic resin that cures or solidifies under the effect of exposure to light, usually in the UV range. The light cures the resin near the surface, forming a hardened layer. When a layer of the part is formed, it is lowered by an elevation control system to allow the next layer of resin to be similarly formed over it. This continues until the entire part is completed. The vat can then be drained and the part removed for further processing, if necessary.

6.3.1 3D Systems' Stereolithography Apparatus

3D Systems was founded in 1986. Among all commercial RP systems, SLA was the pioneer, being first marketed in 1988. It has been awarded more than 40 U.S. patents and 20 international patents, with additional patents filed or pending internationally.

3D Systems produces a wide range of machines to cater to various part sizes and throughputs. There are several models available, including those in the series of SLA-190, SLA-250, SLA-350, and SLA-500. The SLA-190 is an entry-level machine that uses an He–Cd laser. The SLA-250 is the most widely installed solid-imaging system in the world as its work volume is suitable for many applications. The SLA-350 is the new generation of SLA machine, which has a better solid-state Nd : YVO_4 laser and is up to 53 percent faster than the SLA-250 in part building. The SLA-500 series is the top-of-the-line machine manufactured by 3D Systems.

All these machines use one-component, photocurable liquid resins as the material for building. There are several grades of resins available and usage is dependent on the laser on the machine and the mechanical requirement of the part. Specific details of the correct type of resin to be used are available from the manufacturer. The other main consumable used on these machines is the cleaning solvent required to clean the part of any residual resin after the building of the part is completed on the machine.

3D Systems' stereolithography process creates three-dimensional plastic objects directly from CAD data. The process begins with a vat filled with the photocurable liquid resin and an elevator table set just below the surface of the liquid resin. The operator loads a three-dimensional CAD solid-model file into the system. If necessary, supports are designed to stabilize the part during building and postcuring. The translator converts the CAD data into an .STL file. The control unit slices the model and support into a series of cross sections ranging from 0.076 to 0.5 mm thick. The computer-controlled optical scanning system then directs and focuses the laser beam so that it solidifies a two-dimensional cross section corresponding to the slice on the surface of the photocurable liquid resin. The elevator table then lowers enough to cover the solid polymer with another layer of the liquid resin. A leveling wiper moves across the surfaces to recoat the next layer of resin on the surface. The laser then draws the next layer. This process continues, building the part from bottom up, until the system completes the part. The part is then raised out of the vat and cleaned of excess polymer.

The main components of the SLA system are a control computer, a control panel, a laser, an optical system, and a process chamber. The workstation software used by the SLA system, known as Maestro,[29] includes the software modules to: verify and perform minor repairs on .STL files; view the .STL files graphically for visual inspection and orientation; merge two or more .STL files; automatically generate support structures for the part files; prepare a part for building; slice the part into horizontal crosssections; and create the final build files to be used by the SLA.

The SLA process is based fundamentally on the following principles[11] (see Fig. 6.5):

1. Parts are built from a photocurable liquid resin that cures when exposed to a laser beam (i.e., undergoing the photopolymerization process), which

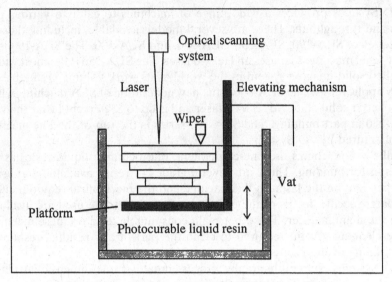

Figure 6.5 Working principles of the SLA process.

scans across the surface of the resin. Photopolymerization is polymerization initiated by a photochemical process whereby the starting point is usually the induction of energy from a radiation source.[11]

2. The building is done layer by layer, each layer being scanned by an optical scanning laser and controlled by an elevation mechanism that descends at the completion of each layer.

The SLA technology provides manufacturers cost-justifiable methods for reducing time to market, lowering product development costs, gaining greater control of their design process, and improving product design. The range of applications include: models for conceptualization, packaging, and presentation; prototypes for design, analysis, verification, and functional testing; parts for production tooling and low-volume production; patterns for investment casting, sand casting, and molding; and tools for fixture and tooling design and prototype tooling.

The software developed to support these applications includes QuickCast, a software tool that enables highly accurate resin patterns for investment casting to be built rapidly to produce high-quality metal castings.

3D Systems' research focus is on improving the process and developing new materials and applications, especially rapid tooling. In 1996, 3D Systems introduced product enhancements to its popular SLA-250 with a 40-mW laser and the new Zephyr recoating system.[26] The Zephyr system eliminates the need for the traditional "deep dip" in which a part is dunked into the resin vat after each layer and then raised to within one layer's depth of the top of the vat followed by a wiper blade sweep across the surface of the vat to remove excess resin. The Zephyr system has a vacuum blade that picks up resin from the side of the vat and applies a thin layer of resin as it sweeps across the part. This speeds up the build process by reducing the time required between layers and greatly reduces problems involved when building parts with trapped volumes.

In addition, 3D Systems and Ciba-Geigy have formed a joint research and development program to continually work on new resins that have better mechanical and processing properties, are faster and easier to process, and are able to withstand high temperatures.[25]

One other important area of research is in rapid tooling, that is, the realization of prototype molds and ultimately production tooling inserts.[12] 3D Systems is involved in 15 cooperative rapid tooling partnerships with various industrial, university, and government agencies. The methods being studied, tested, or evaluated include those used for soft tooling and hard tooling.

6.3.2 Cubital's Solid Ground Curing

The solid ground curing (SGC) system is produced by Cubital Ltd., which began its operations in 1987 and commercial sales in 1991. Cubital's products include the Solider 4600 and Solider 5600. The Solider 4600 is Cubital's entry-level three-dimensional model-making system based on SGC. The Solider 5600, Cubital's sophisticated high-end system, provides a wider range and options for varied modeling demands. The systems use two kinds of resins, liquid resin and cured resin, as materials to create parts, water-soluble wax as support material, and ionographic solid toner for creating an erasable image of the cross section on a glass mask.

The Solider system has the following advantages:

1. *Parallel Processing.* The process is based on instant, simultaneous curing of a whole cross-sectional layer area (rather than point-by-point curing). Its throughput is about eight times faster than its competitors and costs can be 25 to 50 percent lower.
2. *Self-Supporting.* It is user friendly, fast, and simple to use. It has a solid modeling environment with unlimited geometry. The solid wax supports the part in all dimensions and therefore support structures are not required.
3. *Fault Tolerance.* It has good fault tolerances. Removable trays allow job changing during a run and layers are erasable.
4. *Unique Part Properties.* The part that the Solider system produces is reliable, accurate, sturdy, and can be mechanically finished.
5. *CAD to RP Software.* Cubital's RP software, Data Front End (DFE), processes three-dimensional CAD files before they are transferred to Cubital's machines. The DFE is an interactive and user-friendly software.

Cubital's SGC process, shown in Figure 6.6, includes three main steps: data preparation, mask generation, and model making.[15]

In the first step of data preparation, the CAD model of the job to be prototyped is prepared and the cross sections are generated digitally and transferred to the mask generator. The software used, Cubital's Solider DFE (Data Front End) software, is a motif-based special-purpose CAD application package that processes solid-model CAD files prior to sending them to the Cubital Solider system. Data Front End can search and correct flaws in the CAD files and render files on screen for visualization purposes. Solider DFE accepts CAD files in the .STL format and other widely used formats exported by most commercial CAD systems.

After data are received, the mask plate is charged through an "image-wise" ionographic process in mask generation. The charged image is then developed with electrostatic toner.

In the model-making step, a thin layer of photopolymer resin is spread on the work surface. The photomask from the mask generator is placed in close proximity above the workpiece, and aligned under a collimated UV lamp.

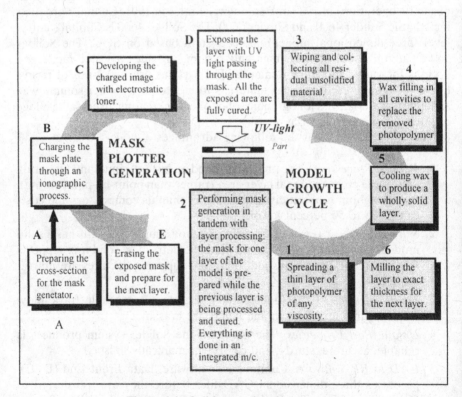

Figure 6.6 Solid ground curing process.

The UV light is turned on for a few seconds. The part of the resin layer exposed to the UV light through the photomask is hardened. Note that the layers laid down for exposure to the lamp are actually thicker than the desired thickness. This is to allow for the final milling process. The unsolidified resin is then collected from the workpiece. This is done by vacuum suction. Following that, melted wax is spread into the cavities created after collecting the liquid resin. Consequently, the wax in the cavities is cooled to produce a wholly solid layer. Finally, the layer is milled to its exact thickness, producing a flat solid surface ready to receive the next layer.

The main components of the Solider system are a Data Front End (DFE) workstation; the model production machine (MPM), which includes a process engine; an operator's console; and a vacuum generator. An optional automatic dewaxing machine is also available.

Cubital's RP technology creates highly physical models directly from computerized three-dimensional datafiles. Parts of any geometric complexity can be produced without tools, dies, or molds. The process is based on the following principles:

1. Parts are built, layer by layer, from a liquid photopolymer resin that solidifies when exposed to UV light. The photopolymerization process is similar to that used in the SLA, except that the irradiation source is a high-power collimated UV lamp and the image of the layer is generated by masked illumination. The mask is created from the CAD data input and "printed" on a transparent substrate (the mask plate) by a nonimpact ionographic printing process, a process similar to the xerography process used in photocopiers.[13] The image is formed by depositing black powder, a toner that adheres to the substrate electrostatically. This is used to mask the uniform illumination of the UV lamp. After exposure, the electrostatic toner is removed from the substrate for reuse and the pattern for the next layer is similarly "printed."

2. Multiple parts may be processed and built in parallel by grouping them into batches (runs) using Cubital's proprietary software.

3. Each layer of a multiple-layer run contains cross-sectional slices of one or many parts. Therefore, all slices in one layer are created simultaneously. Layers are created thicker than desired. This is to allow the layer to be milled precisely to its exact thickness, thus giving overall control of the vertical accuracy. This step also produces a roughened surface of cured photopolymer, assisting adhesion of the next layer to it.

4. The process is self-supporting and does not require the addition of external support structures to emerging parts because continuous structural support for the parts is provided by the use of wax, acting as a solid support material.

The applications of Cubital's system can generally be divided into four areas:

1. *General Applications.* Conceptual design presentation, design proofing, engineering testing, integration and fitting, functional analysis, exhibitions and preproduction sales, market research, and interprofessional communication.
2. *Tooling and Casting Applications.* Investment casting, sand casting, and rapid, tool-free manufacturing of plastic parts.
3. *Mold and Tooling.* Silicon rubber tooling, epoxy tooling, spray metal tooling, acrylic tooling, and plaster mold casting.
4. *Medical Imaging.* Diagnostic, surgical, operation, and reconstruction planning and custom prosthesis design, and so forth.

Cubital is doing research on faster-processing, higher-performance, higher-resolution graphics and more accurate renderings and shadings. Following increasing demand for RP parts with improved mechanical properties and for direct utilization in rapid tooling processes, Cubital is developing an extension to the SGC process that will be upgradable on standard Solider systems. The extended process will enable production of parts made of enhanced thermoset, thermoplastic, and metallic materials, thereby increasing the user's ability to select materials suitable for specific applications or production processes. In particular, parts made of castable wax can be used directly for investment casting, and parts produced of metal-sprayed zinc can be used directly for plastic injection molding.

6.3.3 Other Liquid-Based RP Systems

Other liquid-based RP systems that are similar to the stereolithography system include: Sony's Solid Creation System (SCS), Mitsubishi's Solid Object Ultraviolet-Laser Plotter (SOUP), EOS's Stereos Systems, Teijin Seiki's Soliform, Meiko's Rapid Prototyping System for the Jewelry Industry, Denken Engineering's SLP, Mitsui Zosen's COLAMM, Fockele & Schwarze's LMS, and Light Sculpting's Light Sculpting device.

6.4 SOLID-BASED RAPID PROTOTYPING SYSTEMS

Solid-based rapid prototyping systems are very different from liquid-based photocuring systems. They are also quite different from one another. The basic common feature among these systems is that they all utilize solids (in one form or another) as the initial medium to create the prototype. A special group of solid-based RP systems which uses powder as the prototyping medium will be covered separately in Section 6.5.

6.4.1 Helisys' Laminated Object Manufacturing

Helisys, Inc. was founded in 1985 to manufacture and market laminated object manufacturing (LOM) systems and shipped its first commercial system in

1991. Helisys, Inc. produces two models of LOM machines, the LOM-1015 and the LOM-2030. Both these systems use a CO_2 laser, with the LOM-1015 operating a 25-W laser and the LOM-2030 operating a 50-W laser. The optical system, which delivers a laser beam to the top surface of the work, consists of mirrors that reflect the laser beam and a focal lens that focuses the beam to about 0.25 mm. The control of the laser during cutting is by means of an XY positioning table that is belt driven as opposed to the galvanometer mirror system. The LOM-2030 is a larger machine and produces larger prototypes. The work volume of the LOM-2030 is 810 mm × 550 mm × 500 mm and that of the LOM-1015 is 380 mm × 250 mm × 350 mm.

The patented Laminated Object Manufacturing (LOM)[7–9] process is an automated fabrication method in which a three-dimensional object is constructed from a solid CAD representation by sequentially laminating the part cross sections. The process consists of three phases: preprocessing, building, and postprocessing.

The preprocessing phase includes generating an image from a CAD-derived .STL file of the part to be manufactured, sorting input data, and creating secondary data structures. These are fully automated by LOMSlice, the LOM system software. Orienting and merging the part on the LOM system are performed manually.

In the building phase, thin layers of adhesive-coated material are sequentially bonded to each other and individually cut by a CO_2 laser beam (see Fig. 6.7).

The build cycle involves the following steps:

1. LOMSlice creates a cross section of the three-dimensional model, measuring the exact height of the model and slicing the horizontal plane accordingly. It then images cross hatches, which define the outer perimeter and convert the excess material into a support structure.
2. The computer generates precise calculations, which guide the focused laser beam to cut the cross-sectional outline, the cross hatches, and the model's perimeter. The laser beam power is designed to cut exactly the thickness of one layer of material at a time.
3. The platform with the part descends and a new section of material advances. The platform ascends and the heated roller laminates the material to the stack with a single reciprocal motion, thereby bonding it to the previous layer.
4. A vertical encoder measures the height of the stack and with the information LOMSlice calculates the cross section for the next layer as the laser cuts the model's current layer.

This sequence continues until all the layers are built. The product emerges from the LOM machine as a completely enclosed rectangular block containing the part.

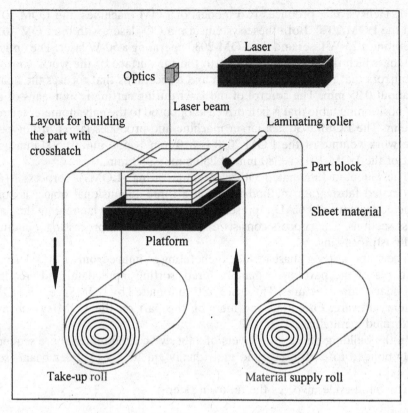

Figure 6.7 LOM process: building the part.

The last phase, postprocessing, includes separating the part from its support material and finishing it. The separation sequence is as follows:

1. The metal platform with the newly created part is removed from the LOM machine.
2. Normally, a hammer and a putty knife are all that is required to separate the LOM block from the platform. However, a live thin wire may also be used to slice through the double-sided foam tapes that hold the LOM stack onto the platform.
3. The surrounding wall frame is lifted off the block to expose the cross-hatched pieces of the excess material. Cross-hatched pieces may then be separated from the part using wood-carving tools.

After the part is extracted from the surrounding cross hatches, the woodlike LOM part can be finished. Traditional model-making finishing techniques, such as sanding, polishing, painting, and so on, can be used. After the part has been separated, it is recommended that it be sealed immediately with urethane, epoxy, or silicon spray to prevent moisture absorption and expansion. If necessary, LOM parts can be machined—by drilling, milling, and turning.

Laminated object manufacturing can be applied across a wide spectrum of industries, including aerospace or automotive products, consumer items, and medical devices. LOM parts are ideal in design applications for visualization purposes, or to test for form, fit, and function, as well as in a manufacturing environment to create prototypes, make production tooling, or even produce a small volume of finished goods.

Helisys reinvests more than 20 percent of its annual revenue into engineering and research and development. The focus of its current research activities is in accuracy improvements through software modifications and laser beam size reduction. Establishing a thorough understanding of residual stress through mathematical modeling and simulation studies will aid in understanding associated issues in the LOM process. Methods to speed up the bonding and laser cutting speeds are also under examination.

In early 1994, Helisys won a three-year ARPA (Advanced Research Program Agency) contract for upwards of $3,500,000 to develop a solid freeform fabrication process, based on the Helisys LOM process, which would manufacture fully functional ceramic matrix composite parts with virtually no human intervention. The process, referred to as LOM/COM, is currently under development.[14] The project team consists of world-renowned research centers, material suppliers, federal agencies, and end users, including Wright Patterson Air Force Base and the University of Dayton Research Institute.

6.4.2 Stratasys' Fused Deposition Modeling

Stratasys Inc., the company that invented and markets fused deposition modeling (FDM) machines, was founded in 1989 and shipped its first commercial machine in 1991.

Stratasys' main product, the FDM 1600 Rapid Prototyping System, consists of proprietary QuickSlice and SupportWork preprocessing software and the FDM machine. Each system is custom configured, depending on the customer's material requirements. Material packages available are investment casting wax, ABS, polyamide, and MABS (methyl methacrylate ABS). A system can include one or several material packages.

The FDM 1650 Rapid Prototyping System is the new system that is set to replace the FDM 1600. The FDM 1650 has many feature improvements over the FDM 1600 and is able to handle additional types of modeling materials including elastomers.

The key strengths of the FDM products and process include the following features:

1. *Ease of Use.* The product and process are easy to use.
2. *Multiple, Nontoxic Modeling Materials.* Up to five different modeling materials are available, including ABS, investment casting wax, polyamide, MABS, and elastomers. These materials are nontoxic and easy to process.
3. *Automatic Support Generation and Breakaway Support System.* The FDMs software, SupportWork, can automatically determine if supports are needed and generate them if they are. They can be easily broken away when the model is complete.
4. *Speed.* The FDM modeling process is simple, accurate, and fast.
5. *Office Environment Operation.* FDM machines can be installed in an office environment without any special facility requirement.

In this patented process,[5] a geometric model is created on CAD software, which uses IGES or .STL formatted files. It can be imported into a workstation and processed with the QuickSlice and SupportWork proprietary software. Within QuickSlice, the CAD file is sliced into horizontal layers after the part is oriented for the optimum build position, and any necessary support structures are automatically detected and generated by the SupportWork software. The slice thickness can be set anywhere between 0.051 and 0.762 mm. The tool paths of the build process are then generated and downloaded to the FDM machine.

Modeling materials come in spools in a form very much like a fishing line (see Fig. 6.8). The filament on the spools is fed into an extrusion head and heated to a semiliquid state. The semiliquid material is extruded through the head and then deposited in ultra-thin layers from the FDM head, one layer at a time. Because the air surrounding the head is maintained at a temperature below the material's melting point, the exiting material quickly solidifies. Moving on the X–Y plane, the head follows the tool path generated by QuickSlicec creating the desired layer. When the layer is completed, the head moves on to create the next layer. The horizontal width of the extruded material can vary between 0.254 and 2.54 mm. This feature, called "road width," can vary from slice to slice. Two modeling materials are dispensed through a dual-tip mechanism in the FDM machine. A primary modeling material is used to produce the model geometry and a secondary material, or release material, is used to produce the support structures. The release material forms a bond with the primary modeling material. When minimal force is applied to these supports, the supports break away easily due to the weak nature of this bond.

FDM models can be used in the following general application areas:

1. *Models for Conceptualization and Presentation.* Models can be marked, sanded, painted, and drilled and thus can be finished to closely resemble the actual product.

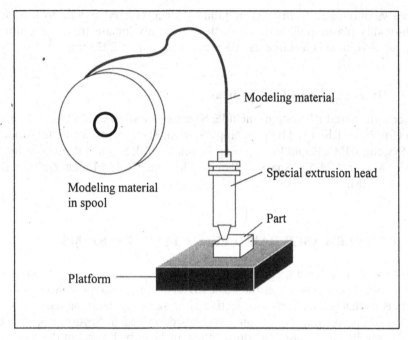

Figure 6.8 Working principle of the FDM system.

2. *Prototypes for Design, Analysis, and Functional Testing.* The system can produce fully functional prototypes in ABS. The resulting ABS parts have 85 percent of the strength of the actual molded part. Thus, actual functional testing can be carried out.
3. *Patterns and Masters for Tooling.* Models can be used as patterns for investment casting, sand casting, and molding.

Stratasys continues to develop new modeling materials to work with the FDM technology. It announced the addition of MABS (methyl methacrylate ABS) to the family of modeling materials available for use with the FDM 1600 in November 1995. This material, designed for medical applications, has been approved by the Food and Drug Administration (FDA), can be sterilized with gamma radiation, and has the chemical resistance required for the medical device industry. Other new materials such as elastomers are also becoming available.

Stratasys is also working on improving the software to reduce the processing time of .STL files, reduce the model build time, and increase the number of automatic functions for easier use. In 1996, Stratasys announced a new system, called Genisys, which functions like a printer but uses a slightly different technology. Based on the IBM three-dimensional printer system whose rights

Stratasys purchased from IBM in January 1995, Genisys is able to produce high-quality plastic polymer models that are ready for use from the printer. The build volume is as large as 203 mm × 203 mm × 203 mm.

6.4.3 Other Solid-Based RP Systems

Other solid-based RP systems include: Kira Corporation's Selective Adhesive and Hot Press, Kinergy-HUST's Zippy System, 3D Systems' Multi-Jet Modeling System, IBM's Rapid Prototyping System (RPS), Sanders Prototype Inc.'s Model Maker MM-6B, Sparx AB's Hot Plot, and Scale Models Unlimited's LaserCAMM.

6.5 POWDER-BASED RAPID PROTOTYPING SYSTEMS

This section describes a special group of solid-based rapid prototyping systems that primarily use powder as the basic medium for prototyping. Some of the systems in this group, such as selective laser sintering, bear similarities with the liquid-based rapid prototyping systems described in Section 6.3, that is, they generally have a laser to "draw" the part layer by layer, but the medium used for building the model is a powder instead of a photocurable resin. Others, such as three-dimensional printing and multiphase jet solidification, have similarities with the solid-based rapid prototyping systems described in Section 6.4. The common feature among these systems is that the material used for building the part or prototype is invariably powder based.

6.5.1 DTM's Selective Laser Sintering

DTM Corporation was founded in 1987 to commercialize selective laser sintering (SLS) technology and it shipped its first commercial machine in 1992. The current DTM product is the Sinterstation 2000. The Sinterstation 2000 system is capable of producing objects measuring 305 mm in diameter and 380 mm in height, accommodating most prototyping applications. The SLS process is the only technology with the capability to directly process a variety of engineering thermoplastic materials, investment casting wax, metallic materials, and thermoplastic composites.

The key strengths of DTM Corporation's Sinterstation 2000 system are:

1. *High Throughput.* The throughput of the system is high due to its ability to utilize powdered materials to quickly create the part in layers inside the part build envelope.
2. *Wide Range of Processing Material.* DTM offers a wide range of materials including nylon, polycarbonates, and metals, providing flexibility and wide scope of applications.

3. *Does not Require Support.* The system does not require CAD-developed support structures in the process and therefore does not need time for support structure removal.

4. *Requires Little Postprocessing.* The finishing of the part is reasonably fine and requires only minimal postprocessing such as sanding.

5. *Requires no Postcuring.* The completed part from the Sinterstation 2000 system is by itself solid enough and does not need postcuring.

The SLS process creates three-dimensional objects, layer by layer, from CAD data using powdered materials with heat generated by a CO_2 laser within the Sinterstation 2000 system. CAD datafiles in the industry-standard .STL file format are first transferred to the Sinterstation 2000 system where they are sliced. From this point, the SLS process starts and operates as follows[6] (see Fig. 6.9):

1. At the start, a thin layer of the heat-fusible powder is deposited onto the part-building cylinder within the process chamber.

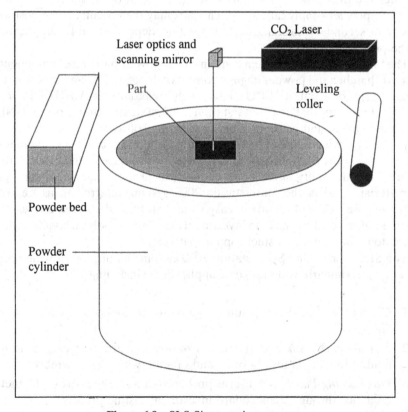

Figure 6.9 SLS Sinterstation process.

2. A cross section of the object under fabrication is selectively "drawn" (or scanned) on the layer of powder by a heat-generating CO_2 laser. The interaction of the laser beam with the powder elevates the temperature to the melting point, fusing the powder particles into a solid mass. The intensity of the laser beam is modulated to melt the powder only in areas defined by the part's geometry. The rest would remain in its powder form.

3. When the cross section is complete, an additional layer of powder is deposited via a roller mechanism on top of the scanned layer. This prepares the next layer for scanning.

4. Steps 2 and 3 are repeated, with each layer fusing to the layer below it. Successive layers of powder are deposited and the process is repeated until the part is complete.

As SLS materials are in powdered form, the powder not melted or fused during processing serves as a customized, built-in support structure. There is no need to create support structures within the CAD design prior to or during processing and thus no support structure to remove when the part is complete.

After the SLS process, the part is removed from the build chamber and the loose powder simply falls away. The part may then require some postprocessing or secondary finishing, such as sanding, depending on the application of the prototype built.

The Sinterstation 2000 system contains the following hardware components: process chamber and powder engine; controls cabinet—CPU and disk storage; atmospheric control unit (ACU); and rough breakout unit (RBO). The software components that the Sinterstation 2000 system includes are a UNIX operating system and its proprietary application software.

In theory, a wide range of thermoplastics, composites, ceramics, and metals can be used in the SLS process. This is the key advantage of the process, that is, providing flexibility in the material used for prototyping. The main types of materials used in the Sinterstation 2000 system, referred to as Laserite materials, are safe and nontoxic, easy to use, store, and recycle, and can be disposed of easily. They include investment casting wax, polycarbonate, nylon, fine nylon, and metal in a steel/copper matrix.

The SLS process and Sinterstation 2000 system can produce a wide range of parts in a similarly wide range of applications, including:

1. *Concept Models.* Visualization models used to review design ideas, form, and style.

2. *Functional Models and Working Prototypes.* Parts that can withstand limited functional testing or fit and operate within an assembly.

3. *Wax Casting Patterns.* Patterns produced in wax, then cast in the metal of choice through the standard investment casting process.

4. *Polycarbonate (RapidCasting) Patterns.* Patterns produced using polycarbonate, then cast in the metal of choice through the standard investment casting process. These build faster than wax patterns and are ideally suited for designs with thin walls and fine features. These patterns are also durable and heat resistant.
5. *Metal Tools (RapidTool).* Direct rapid prototypes of tools of molds for small or short production runs.

DTM has focused its research on developing new and additional advanced materials for new applications, particularly in metal composites. It is also working on improving its process. One example is the development of automatic part placement software that further enhances the productivity of the Sinterstation 2000 system. The software allows for rapid and automatic placement of parts in the build envelope on the Sinterstation.

6.5.2 Soligen's Direct Shell Production Casting

Soligen Inc. was founded in 1991 and it first installed its Direct Shell Production Casting (DSPC) System at three "alpha" sites in 1993. Direct Shell Production Casting (DSPC) creates ceramic molds for metal parts with integral coves directly and automatically from CAD files. Soligen's Direct Shell Production Casting machine, the DSPC 300, includes: a powder holder, which contains the manufacturing material—powder; a powder distributor, which disperses a thin layer of powder; rollers, which compress each layer before binding; a print head, which sprays binder on each layer; and a bin, which holds the mold.

DSPC is the only rapid prototyping process that creates ceramic molds directly for metal casting. As a result, functional metal parts (or metal tooling, such as dies for die casting) could be made directly from CAD data of the part.

The DSPC technology is derived from a process known as three-dimensional printing and was invented and developed at the Massachusetts Institute of Technology (MIT). The process (see Fig. 6.10) contains the following steps[30]:

1. A part is first designed on a computer, using commercial CAD software.
2. The CAD model is then loaded into the shell design unit—the central control unit of the equipment. Preparing the computer model for casting mold requires modifications, such as scaling the dimensions to compensate for shrinkage, adding fillets, and removing characteristics that will be machined later. Then it is necessary to decide the number of mold cavities on each shell and the type of gating system, including the basic sprues, runners, and gates. Once the CAD mold shells are modified to the desired configuration, the shell design unit generates an electronic model of the assembly in slices to the specified thickness. The electronic model is then transferred to the shell production unit.

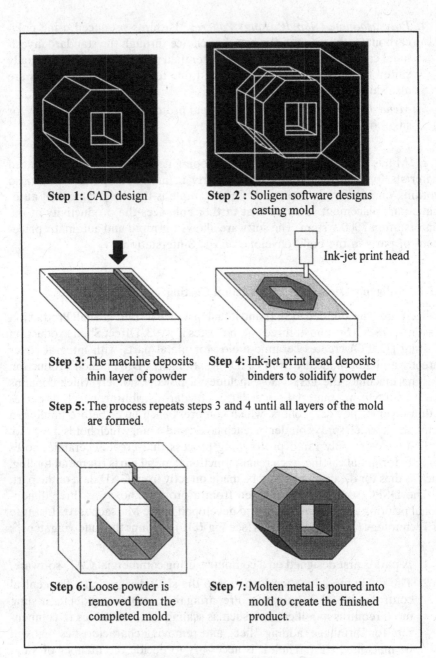

Step 1: CAD design

Step 2 : Soligen software designs casting mold

Ink-jet print head

Step 3: The machine deposits thin layer of powder

Step 4: Ink-jet print head deposits binders to solidify powder

Step 5: The process repeats steps 3 and 4 until all layers of the mold are formed.

Step 6: Loose powder is removed from the completed mold.

Step 7: Molten metal is poured into mold to create the finished product.

Figure 6.10 Soligen DSPC process.

3. The shell production unit begins depositing a thin layer of fine alumina powder over the shell working surface for the first slice of the casting mold. A roller follows the powder, leveling the surface.

4. An ink-jet print head, similar to those in computer printers, moves over the layer, injecting tiny drops of colloidal silica binder onto the powder surface from its 128 ink jets. Passing the pressurized stream of binder through a vibrating piezoelectric ceramic atomizes it as it exits the jet. The droplets pick up electric charge as they pass through an electric field, which helps to align them to the powder. The binder solidifies the powder into ceramic on contact and the unbounded alumina remains as support for the following layer. The work area lowers and another layer of powder is distributed.

5. The process through steps 3 and 4 is repeated until all layers have been formed.

6. After the building process is completed, the casting shell remains buried in a block of loose alumina powder. The unbound excess powder is then separated from the finished shell. The shell can then be removed for postprocessing, which may include firing in a kiln to remove moisture or preheating to the appropriate temperature for casting.

7. Molten metal can then be poured in to fill the casting shell or mold. After cooling, the shell can be broken up to remove the cast, which can then be processed to remove gatings, sprues, and so on, thus completing the casting process.

The hardware of the DSPC system contains a personal computer, a powder holder, a powder distributor, rolls, a print head, and a bin. The software includes a CAD system and Soligen's slicing software.

The DSPC technology is used primarily to create casting shells for the production of parts and prototypes. Soligen Inc. has acquired a foundry in Santa Ana, for the California, commercial production of prototypes and molds. Called the Parts Now Division, it serves as a service center, as well as a place for equipment and process research and development. It aims to be a premier "one-stop shop" for functional cast metal parts produced directly from a CAD file, and with no need for prefabricated tooling to produce the first article. The DSPC system has been used in the automotive and aerospace industries, for computer manufacture, and in medical prostheses.

Soligen Inc. continues to conduct research on part and mold design. Using knowledge and experience gained from application development, the Parts Now Division is developing technology to produce intricate aluminum die casts that could improve throughput by 50 to 250 percent while simultaneously improving accuracy and durability. Improving mold surface finishes with a DSPC-cast electrode used with electric-discharge machining (EDM) and coating technologies are also under development.

6.5.3 Other Powder-Based RP Systems

Other powder-based RP systems include: Fraunhofer's Multiphase Jet Solidification (MJS), EOS's EOSINT Systems, BPM Technology's Ballistic Particle Manufacturing (BPM), and MIT's 3-Dimensional Printing (3DP). These systems are mostly similar in nature to those described earlier, with perhaps the exception of BPM, which uses a particle-jetting technology and proprietary thermoplastic material.

6.6 APPLICATIONS

Areas of application are closely related to the purposes of prototyping and, consequently, the materials used. As such, the closer the RP materials are to traditional prototyping materials in physical and behavioral characteristics, the wider will be the range of applications. Unfortunately, there are marked differences in these areas between current RP materials and traditional materials in manufacturing. The key to increasing the applicability of RP technologies therefore lies in widening the range of materials.

In the early development of RP systems, the emphasis of the tasks at hand was oriented toward the creation of "touch-and-feel" models to support design, that is, the creation of three-dimensional objects with little regard to their function and performance. These were broadly classified as "applications in design," and they were influenced and in many cases limited by the materials available. As the initial costs of these machines were high, vendors began to look for more areas of applications, with the logical search for functional evaluation and testing applications, and eventually tooling. This not only called for improvements in RP technologies in terms of the process to create stronger and more accurate parts, but also in terms of developing an even wider range of materials. Applications of RP prototypes were first extended to "applications in engineering, analysis, and planning" and later extended further to "applications in manufacturing and tooling." These typical application areas are summarized in Figure 6.11.

The major breakthrough of RP technologies in manufacturing has been their ability to enhance and improve the product development process and, at the same time, reduce the costs and time required to take the product from conception to market. This has allowed manufacturers and industrialists to shorten the ever-vital time to market, thereby enhancing their competitiveness in the market. There are many case examples of applications in the aerospace, automotive, biomedical, jewelry, coin, and tableware industries, which are recorded in many journals and other publications.

6.7 FURTHER DEVELOPMENT TRENDS OF RP

As seen in Sections 6.3 to and 6.5, RP vendors are spending much time, effort, and money on R&D so as to develop better and more accurate RP systems

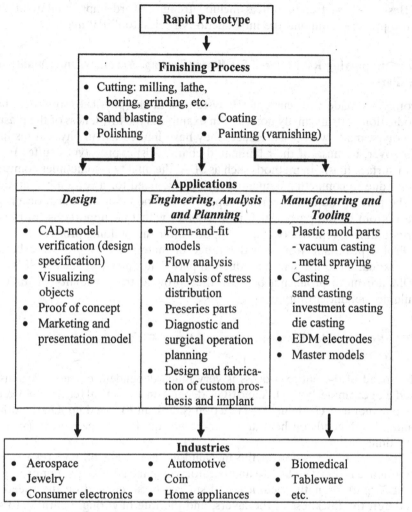

Figure 6.11 Typical application areas of RP parts.

to overcome the current restrictions of RP technologies. Restrictions include the following:

1. *Limited Range of Materials.* This implies limited applications such as engineering testing and analysis due to dissimilar mechanical and material properties.
2. *Accuracy.* This falls short when compared to existing machining technology.
3. *Curling, Warping, and Growing.* These are relevant to several RP processes such as SLA and SGC.

Besides these trends, others include trends toward improvement of the entire process chain and the introduction of "desktop" RP machines.

6.7.1 Improving RP Material for Better Testing, Accuracy, and Quality of RP Parts

Prototypes made with current RP systems lack the physical properties that production parts normally possess. An example is the brittleness of the plastic prototype made with SLA. Hence, they have limited use in physical testing. Moreover, because of their laminated characteristic, prototypes suffer from poor surface finish. In methods such as SLA, the quality is sometimes compromised due to conflicting demands, such as the need for a certain orientation of the part or the use of certain supports. The poor surface finish occurs at the bottom of the part, which is in contact with the supports as burrs are formed during removal of these supports after the building process.

Prototypes made from a wider range of materials may provide a better approximation to the actual production part. Thus, methods such as SLS and LOM are more promising because they provide the possibility of directly building metallic prototypes.

6.7.2 Speed of the Entire Process and Ease of Operation of the Mechanism

The speed of the entire process is very much dependent on the mechanism used. For example, both SLA and SGC depend on a chemical reaction between the polymer and UV light. This reaction is difficult to speed up. On the other hand, SLS depends on heat and, as a result, the build time can be three to four times faster.

The ease of operation of the mechanism should not be underestimated. Experience is required before an operator can make good parts. For example, in SLA, part strength, accuracy, and surface finish depend on the speed of the laser, the thickness of the layers, and the rate of curing. Learning to set these parameters correctly is more an art than a science.

6.7.3 Means of Coupling to Downstream Manufacturing Activities

The ability to make molds with the output of RP systems improves the economics of this technology and will certainly make it more attractive to industrial users. This trend can also be described in terms of the shift in focus in demands on various application areas over the years, as shown in Figure 6.12. In the early years of RP, users focused on the applications of RP parts for design rather than for engineering, analysis, and planning, with virtually little or no emphasis on tooling and manufacturing. However, in the last five years, applications in engineering, analysis, and planning have grown in emphasis, as have applications for manufacturing and tooling. It is clear that the trend will continue at the ex-

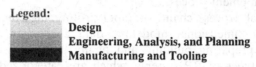

Period	Focus on Applications (in percentage)				
From 1988 - 1991	80%			15%	5%
From 1991 - 1996	40%		40%	20%	
1996 and beyond	20%	40%		40%	

Legend:

Design
Engineering, Analysis, and Planning
Manufacturing and Tooling

Figure 6.12 Shift in focus in demands on application areas.

pense of applications in design, as users are no longer satisfied with only limited applications and seek to maximize their investment in RP systems.

6.7.4 "Desktop," "Office," or "Concept" Modelers

There is a growing trend toward "desktop," "office," or "concept" modeler-type RP systems. These are small RP systems priced well below $100,000 and can be operated in an office environment. The high cost of RP machines is the biggest problem facing the industry, as was reported in a reader survey conducted by the Rapid Prototyping Report.[24] Because they are intended as "concept" modelers, speed and ease of use are of premium, usually at the expense of fine accuracy and part durability.

Currently, there are at least four vendors selling these desktop RP systems. Sanders Prototype's remarkable sales in 1995 is a testimony of the strong niche market for such kinds of RP systems. Recently launched systems such as 3D Systems' Actua (MJM) and Stratasys' Genisys are strategically targeted at the same market segment.

BPM's Personal Modeler, introduced at Autofact 1995, also claims to be a "concept" modeler. However, it is doubtful that these machines are really speedy. Reasonably complicated industrial parts may not be built within 20 hours on these machines, as a benchmark by Ford shows.[27] If parts take a long time to build, then the "concept" modelers do not serve their main purpose, which is to allow a user to verify a conceptual design quickly.

Nevertheless, concept modelers are still much cheaper compared to their more expensive "cousins" and are within the reach of smaller enterprises. They are especially relevant for teaching and educational institutions used in laboratory or practical exercises or projects.

6.8 SUMMARY

Rapid prototyping and manufacturing technologies have not only reduced prototyping time they have also increased the complexities and usages of prototypes. They are used not simply for visualization or technical evaluation but also for preliminary manufacturing and tooling for production. As a result, the time taken from design to manufacturing is significantly reduced, thereby leading to a shorter product development cycle and better control of the product development process.

The general process chain for rapid prototyping consists of a five-step process: three-dimensional modeling, data conversion and transmission, checking and preparing, building, and, finally, postprocessing. There are primarily three types of RP systems, which are categorized according to the processing materials used.[3] They are, namely, liquid-based systems, solid-based systems, and powder-based systems. Applications of RP are wide ranging, including design, engineering, analysis, planning, manufacturing, and tooling. The range of industries for such applications is also very wide, from aerospace to biomedical.[3]

The impact of RP in manufacturing cannot be underestimated. It is the main watershed event of the 1990s for design and manufacturing and will continue to dominate the product development and manufacturing field into the next century.

REFERENCES

1. Chua, C. K., 1987, Solid Modeling—A State-of-the-Art Report. *Manufacturing Equipment News,* Vol. 9, pp. 33–34.

2. Chua, C. K., 1994, Three-Dimensional Rapid Prototyping Technologies and Key Development Areas, *Computing and Control Engineering J.,* Vol. 5, No. 4, pp. 200–206.

3. Chua, C. K., and Leong, K. F., 1997, *Rapid Prototyping: Principles and Applications in Manufacturing,* Wiley, New York.

4. Cowie, A. P., 1989, *Oxford Advanced Learner's Dictionary of Current English,* 4th ed., Oxford University Press.

5. Crump, S., 1992, The Extrusion of Fused Deposition Modeling, *Proceedings of the Third International Conference on Rapid Prototyping,* pp. 91–100.

6. DTM Corp., 1995, *Product Brochure: Sinterstation 2000.*

7. Feygin, M., 1988, Apparatus and Method for Forming an Integral Object from Laminations, U.S. Patent 4,752,352, 6/21/1988.

8. Feygin, M., 1994, Apparatus and Method for Forming an Integral Object from Laminations, European Patent 0,272,305, 3/2/1994.

9. Feygin, M., 1994, Apparatus and Method for Forming an Integral Object from Laminations, U.S. Patent 5,354,414, 10/11/1994.

10. Hecht, J., 1992, *The Laser Guidebook,* 2nd ed., McGraw-Hill, New York.

11. Jacobs, P. F., 1992, *Rapid Prototyping and Manufacturing, Fundamentals of Stereolithography,* Society of Manufacturing Engineers, Vol. 1, pp. 11–18.

12. Jacobs, P. F., 1995, Insight: Moving Toward Rapid Tooling, *The Edge,* Vol. IV, No. 3, pp. 6–7.

13. Johnson, J. L., 1994, *Principles of Computer Automated Fabrication,* Palatino Press, pp. 2, 44.

14. Klosterman, D., R. Chartoff, and S. S. Pak, 1995, *Affordable, Rapid Composite Tooling via Laminated Object Manufacturing.*

15. Kobe, G., 1992, Cubital's Unknown Solider, *Automotive Industries,* Vol. 8, pp. 54–55.

16. Kochan, D., 1992, Solid Freeform Manufacturing—possibilities and Restrictions, *Computers in Industry,* Vol. 20, pp. 133–140.

17. Kochan, D., and C. K. Chua, 1994, Solid Freeform Manufacturing—Assessments and Improvements at the Entire Process Chain, *Proceedings of the International Dedicated Conference on Rapid Prototyping for the Automotive Industries.*

18. Kochan, D., and C. K. Chua, 1995, State-of-the-Art and Future Trends in Advanced Rapid Prototyping and Manufacturing, *Internat. J. Inform. Technol.* Vol. 1, No. 2, pp. 173–184.

19. Koren, Y., 1983, *Computer Control of Manufacturing Systems,* McGraw-Hill, Singapore.

20. Lee, G., 1995, Virtual Prototyping on Personal Computers, *Mechanical Engineering,* Vol. 117, No. 7, pp. 70–73.

21. Materialise, N. V., 1994, *Magics 3.01 Materialise User Manual,* Materalise Software Department, Kapeldreef 60, B-3001 Heverlee, Belgium.

22. Metelnick, J., 1991, How Today's Model/Prototype Shop Helps Designers Use Rapid Prototyping to Full Advantage, Society of Manufacturing Engineers Technical Paper, MS91-475.

23. Peiffer, R. W., 1993, The Laser Stereolithography Process—Photosensitive Materials and Accuracy, *Proceedings of the First User Congress on Solid Freeform Manufacturing.*

24. Rapid Prototyping Report, 1995, BPM Technology Introduces Low-Priced Personal Modeler, CAD/CAM Publishing, Vol. 5, pp. 3–5.

25. Rapid Prototyping Report, 1996, Ciba Introduces Fast Polyurethanes for Part Duplication, CAD/CAM Publishing, Vol. 6, No. 3, p. 6.

26. Rapid Prototyping Report, 1996, 3D Systems Introduces Upgraded SLA-250 with Zephyr Recoating, CAD/CAM Publishing, Vol. 6, No. 4, p. 3.

27. Rapid Prototyping Report, 1996, Conference Highlights, CAD/CAM Publishing, Vol. 6, No. 5, p. 3.

28. Taraman, K., 1982, *CAD/CAM: Meeting Today's Productivity Challenge,* Computer and Automated Systems Association of SME, Michigan.

29. 3D Systems, 1995, *Product Brochure: Maestro.*

30. Uziel, Y., 1995, art to part in 10 days, *Machine Design,* Vol. 8, pp. 56–60.

7

CASE-BASED PROCESS PLANNING FOR THREE-DIMENSIONAL MACHINED COMPONENTS

MICHAEL M. MAREFAT and JOHN M. BRITANIK
University of Arizona

7.1 INTRODUCTION

Process planning maps the information in a part design into work instructions for its manufacture. Manual process planners develop their skills through many years of experience. Automating process planning therefore requires capturing expertise and knowledge in the explicit form of procedures and structures understandable to a computer. These systems must have knowledge about the available processes and the capabilities of these processes, which includes: the shapes that a process can generate; the size limitations, such as the boundaries of the machine tools and the fixtures for each process; the dimensional and geometric tolerances attainable by each process; and the surface finish produced by a process. An automated process planning system must therefore be able to:

1. Interpret the part geometry and topology to find the high-level machining-oriented shape information, such as holes, pockets, and slots in a part description.

Integrated Product and Process Development, Edited by John Usher, Utpal Roy, and Hamid Parsaei
ISBN 0-471-15597-7 © 1998 John Wiley & Sons, Inc.

2. Determine the relationships between the high-level shape features of a part and discover the parameters, such as axis and length, describing each one.

3. Select an appropriate set of processes and tools to produce each shape feature.

4. Determine an appropriate machining sequence.

7.1.1 Approaches to Process Planning

Generative process planning systems automatically synthesize a process plan for a new component. They generate a process plan for a part from scratch based on manufacturing information stored in a database and decision-making logic and algorithms. However, the generative approach does not utilize the experience gained from past solutions. Also, they generally lack the option of generating alternate plans for a given part, and system knowledge cannot be extended or modified without significant reprogramming. A case-based process planner avoids the duplication of solution effort found in generative systems by reusing past experiences to solve new problems. Old solutions are retrieved and adapted to fit the new scenario.

Variant process planning systems group parts into a family. Plans for a family of parts are stored in files. A new part is classified into one of the families. A standard plan for that family is retrieved and manually modified based on the new part's dimensions and features. However, variant approaches are labor intensive to implement. They work only for parts similar to those planned previously. Experienced process planners are still required to modify the standard plan; hence, variant process planners are not completely automated. A case-based system reuses previous experiences to *automatically* generate new solutions.

Our case-based process planner combines the advantages of the generative and variant approaches by automatically reusing old solutions and planning generatively when no appropriate old solution exists.

7.1.2 Overview

A typical case-based process planner must be capable of:

1. Retrieving past experiences from the plan memory.
2. Modifying the old solution fragments for the new part.
3. Abstracting and storing the new generated solutions in the plan memory.

The process planner must use abstract and detailed information about the component and the knowledge of the processes and tools to automatically generate a plan to manufacture the component. In order to successfully achieve the tasks set forth for a case-based approach to process planning, the planner should be able to compare the old plans and decide on the extent of their suitability. It must also be capable of abstracting the detailed information associated with particular solutions and divide the abstract solutions into smaller reusable chunks.

As shown in Figure 7.1, the first step in a case-based system is to retrieve old solutions (cases) that match an abstracted form of the input design description. The best matching candidate case is then chosen using the similarity metric. This best old solution is then adapted to fit the new design description and verified for consistency to form the new solution. If the new solution is sufficiently different from others stored in the case library, then it is abstracted and stored for reference in subsequent problem solving scenarios.

Case-based planning overcomes the shortcomings of previous approaches, which are mostly rule based. There are several clear advantages of casebased systems over rule-based systems, including:

- *Efficiency.* Case-based systems have the ability to become more efficient by abstracting and storing previous solutions and reusing these solutions to solve similar problems in the future. A rule-based system will always generate solutions from scratch, duplicating previous solution effort.
- *Ability to Learn.* Case-based systems have the ability to learn from their mistakes. Once a solution is corrected and stored as a case, the case-based system will not make the same mistake again. It will retrieve the corrected case and use it to develop a solution. A rule-based system will repeat mistakes until its rule base is updated with new rules.

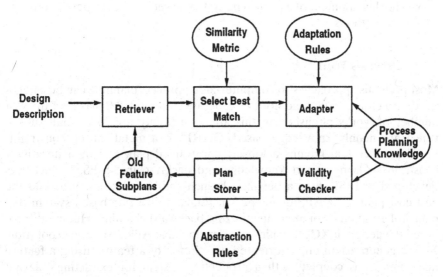

Figure 7.1 Schematic of a case-based process planning system.

· *Maintainability.* It is difficult to maintain and ensure the consistency of the rule base as it gets larger. Case-based systems update their knowledge through the addition of example scenarios (cases). The only check that has to be done in storing a new case is to ensure that it is different from other cases. This is done through the case-indexing mechanism.

· *Knowledge Acquisition.* Knowledge engineers must carefully form consistent rules to encode knowledge from the domain expert. Domain experts typically remember how to perform specific tasks in the form of scenarios or stories rather than specific detailed rules. In a case-based system, this problem is alleviated because the knowledge engineer can encode the experiences of the domain expert directly as cases, rather than developing a set of rules.

Our planner[23] is part of a completely integrated computer-aided design, process planning, and inspection system known as IDP.[24, 25] The contributions of this part of our research can be expressed as follows:

1. The development of a process planning system that utilizes casebased techniques for multilevel process selection (feature process planning) for three-dimensional prismatic parts. This involves retrieving old feature plans generated from past experiences, modifying them to fit the part at hand, and abstracting and storing the new plan for future use.

2. The development of a novel plan merging mechanism that combines the feature plans generated by the casebased planner into a global plan for the part. This merging mechanism considers optimizations such as minimizing the number of fixture and tool changes.

3. The development of a prototype system based on the proposed concepts and methods.

7.1.3 Previous Work

Most previous approaches to computer-aided process planning can be categorized into either variant methods or generative methods.[6] Because of the vast number of works published in this area, we briefly review a small subset, which are mainly knowledge based. GARI[10] is a metal-cutting constraint-based planner that creates a loosely constrained plan and then iteratively constrains it using expert knowledge and backtracking. SIPS,[26] which was developed in LISP, uses a best-first branch-and-bound strategy to find the best-cost plan. EXCAP,[9] developed in England, is a rule-based system that plans for rotational components utilizing backward chaining. Hummel[15] proposed the design of XCUT, which was developed at Allied Signal Corporation. XCUT generates an object-oriented description of a feature using a feature taxonomy. It is coupled with a production system for extracting relevant information from feature volume representation. Other knowledge-based systems include IMACS,[27] FBAPP,[11] IOOPP,[32] and Joshi et al.[17] Algorithmic approaches to process planning can be found in the work of Prabhu et al.,[28] Irani et al.,[16] Korde et al.,[20] Mani and Raman,[22] and Cho et al.[7] All of these systems belong to the category of generative approaches. Group technology is the most popular effort toward the development of CAPP systems based on the variant approach.

Case-based reasoning is a general problem-solving approach based on the reuse of previous solutions.[19, 21] One of the first case-based planners was Hammond's CHEF.[13] The first attempt to develop a case-based approach to process planning was reported by Tsatsoulis and Kashyap.[29–31] They describe a system for rotational parts which considers the machining surfaces one at a time. Other case-based process planners for rotational parts include Yang et al.,[33, 34] Humm et al.,[14] and Bergmann and Wilke.[2] Zarley[36] developed a case-based process planner for assembly operations, and Cser et al.[8] investigated case-based process planning in metal forming. Our work is essentially different from those mentioned previously because it is developed for general three-dimensional prismatic parts.

Although the work of Champati et al.[4] discusses casebased process planning for prismatic parts, the resulting process plans do not specify the machining process or tools to be used. A process plan in this work only corresponds to a sequence of fixture setups and a sequence of features to be machined within each setup. Our planning system accomplishes these tasks through a rule-based feature-sequencing mechanism (determining constraints on sequences of features) and a hierarchical plan merging mechanism, which determines a minimal number of fixture setups. Unlike this work, the casebased component of our system determines process selection (choosing one process or a set of processes) and sequencing for each feature of the part. The only other work in casebased process planning for prismatic parts is that of BenArieh and Chopra.[1] This work discusses an outline of an implementation in CLIPS, but lacks detail on how plan adaptation is to be carried out. Because plan adaptation is a crucial part of any case-based planning system, we view this work as preliminary and incomplete. Also, this work does not address the issues of tool and fixture selection. In contrast, we provide detailed descriptions of how tools and fixture setups are selected, including how the number of fixture setups and tool changes are minimized, as part of our complete planning and plan-merging methodology.

In the next section, we discuss the key aspects of the object-oriented knowledge representation used in our planner. Section 7.3 presents our case-based methodology for generating feature plans, whereas Section 7.4 discusses our hierarchical plan-merging methodology for combining feature plans into a global plan for the part. In Section 7.5, we discuss our developed prototype. We conclude the chapter and discuss future work in Section 7.6.

7.2 OBJECT-ORIENTED KNOWLEDGE REPRESENTATION

What types of knowledge are captured in the system, how the different types of knowledge are represented, and how the knowledge is retrieved and processed determine the domain and capabilities of the system. Three types of knowledge used by every knowledge-based process planner are the knowledge about the processes, the knowledge about the tools, and the geometric and engineering knowledge about the components. However, another important requirement for the development of a case-based approach to process planning is the capture, representation, and utilization of the knowledge contained in previous plans or solutions.

An advantage of object-oriented design includes structured and explicit representation of the knowledge of the system. A schematic rep-resentation of the knowledge of the system is shown in Figure 7.2. In an object-oriented process planning system, classes are created to represent the declarative knowledge of the system. The procedural knowledge is captured by the protocols in the created classes. A process plan for a part is generated by message passing among these classes.

7.2.1 Processes

The process knowledge needs to capture, represent, and utilize the geometric capabilities and the engineering constraints of each process. The geometric capabilities of a process include the shape, dimensions, and the features that a process can produce. The engineering constraints of a process include the geometric tolerances, surface finish, and the material hardness for which a process is applicable. In addition to these properties, we have also made use

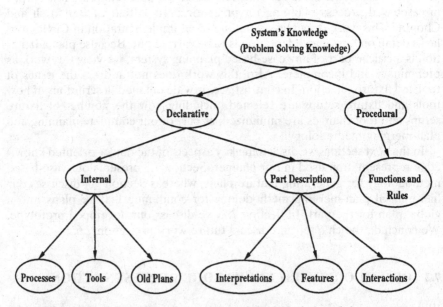

Figure 7.2 Schematic representation of the knowledge of the system.

of "the allowance for a process." The process allowance is intended to reflect, in a rough quantitative manner, the possibility that a process may not produce the anticipated end result. This reflection, which comes from both preferences and experience, may be related to the effectiveness of the process or the possibility of failure.

Tolerance specifications are abstracted to the symmetrical forms (dimensional, relative, and individual) that characterize the tightest constraints for use in process selection and sequencing decisions. This abstracted form matches the level of detail in which tolerances are typically specified in process capability lists (cf. Chang[5]). This is done to facilitate a direct match between feature requirements and process capabilities during the selection of processes in the process selection hierarchy. The actual design tolerance specifications can be maintained and used in plan verification.

A tolerance value is the maximum allowable symmetric (both increasing and decreasing) variation from the nominal value. The *dimensional tolerance* is the symmetric variation associated with the accuracy of dimensional quantities. The *relative tolerance* is the symmetric variation associated with the accuracy of the placement of two surfaces relative to each other. For example, two features may have a specific relative tolerance if a surface of one feature is used as a datum plane for the other feature. The *individual tolerance* represents the symmetric variation associated with a property of a feature by itself such as straightness or roundness. For example, if a slot has a straightness tolerance of $+0.005/-0.004$ and a flatness tolerance of $+0.005/-0.003$, then the abstracted individual tolerance value for this slot would be 0.003, because this is the tightest tolerance in this class of individual tolerances. The idea behind parameter abstraction is to include only enough detail to allow the planner to make the decisions it needs to make.

The process knowledge is represented by frames, which directly map to objects in the object-oriented sense. Hierarchical abstraction is effectively exploited in modeling and representing process knowledge. Figure 7.3 shows part of the process planning kernel process hierarchy. The processes are divided into two abstract classes of surface-generating processes and surface-finishing processes at the highest abstraction level. Each of these classes are further refined into milling processes, shaping and planing processes, broaching processes, and so on.

The developed system is capable of functioning with incomplete information. The incomplete information reflects the fact that all of the engineering information about a particular process may not be available for some particular processes. Therefore, the method has gone one step further by separating the essential information for a process, which includes its name, the features it can produce, the maximum hardness of the surface material on which it can be used, and the range of the attainable surface finish, from the rest of the information for a process. The planning may progress as long as the essential information for a process is available. Figure 7.4 shows an example instance of the class MILLING called face milling. The units of measure in our imple-

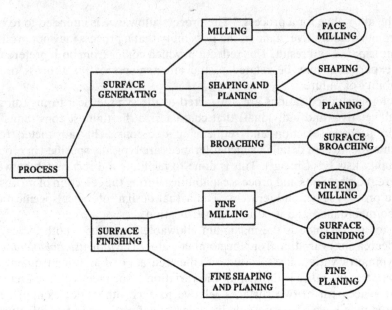

Figure 7.3   Part of the process planning kernel process hierarchy. Processes enclosed in rectangular boxes are abstract processes, whereas those in the ovals are real processes.

mentation are standard units, such as inches for dimensional and tolerancing parameters, microinches for surface finish parameters, and Brinell Hardness Number (BHN) for hardness. We will discuss how the process knowledge is utilized in plan generation in a later section on generating process plans.

7.2.2   Tools

Machine tools are used to apply a process. Tool knowledge can also be represented naturally using frames, as shown in Figure 7.5. The important considerations in decision making with tool knowledge are the types of processes in which the tool can be applied, the appropriateness of the tool dimensions for the dimensions of the feature being manufactured, and the kind of tool to be used. Similar to the process knowledge, in order to have the capability of planning even when complete knowledge about some tools is not available, the essential tool information is separated from the rest. Applying the tool knowledge to determine appropriate tools is discussed in a later section.

7.2.3   Old Plans

A significant part of the knowledge of a *case-based* process planner is the knowledge contained in old plans. Old plans capture the knowledge gained through previous experiences in a manner that makes this knowledge useful for the new problems. In order for the information in the previous experiences

```
instance         face-milling
class            GENERATING > MILLING
initialization
        type       milling        {type of the process}
        name       face milling   {name of the process}
        level      1              {level at which the
                                    process is applicable}
        hardness   369            {maximum hardness of a
                                    raw material}
        surFinMin  126            {minimum surface finish
                                    of a raw material}
        surFinMax  249            {maximum surface finish
                                    of a raw material}
        widthMin   3              {minimum width of the
                                    feature}
        widthMax   8              {maximum width of the
                                    feature}
        dimtol     0.01           {dimensional tolerance
                                    of a process}
        reltol     0.005          {relative tolerance
                                    achievable by a
                                    process}
        indtol     0.005          {individual tolerance
                                    of a process}
        features   (step island)  {features a process
                                    is applicable to}
        allowance  0.08           {allowance of a process}
```

Figure 7.4   Example instance of the class MILLING.

```
instance         carbide-tipped plain cutter
class            TOOLS
initialization
        type       milling        { type of tool }
        name       carbide-tipped plain cutter   { name
                                         of the tool }
        widthMin   129.032        { minimum width of the
                                    tool path }
        widthMax   258.064        { maximum width of the
                                    tool path }
        radiusMax  129.032        { maximum tool radius }
        processes  (End Milling,  { processes to which
                                    tool is applicable }
                   Fine End Milling, Peripheral Milling,
                   Fine Peripheral Milling, Profile
                   Milling)
```

Figure 7.5   Example instance of the class TOOLS.

to be useful for the new problems, the abstract knowledge in these experiences must be extracted from the detail information associated with the case. An abstracted plan contains only those aspects that are important for deciding its selection and determining its application to new situations. These aspects form the indices for storage, retrieval, and decision making with old plans. The old plans also need to contain the abstract solution to the situation, which is represented by their indices. In our planner, an old plan is an abstracted feature process plan. The important indices in the representation of old feature plans for a case-based process planner include the type of the feature, the dimensions of the feature, and the tolerances of feature for which a particular plan was used. The old plans contain a representation of the solution to the given situation in terms of the sequences of the processes to be used. Old plans are represented by frames whose slots (instance variables) correspond to the indices and are stored in the plan memory. An example instance of the class OLDPLANS is shown in Figure 7.6.

7.2.4 Part Representation

Part design is the input to the process planner. For process planning, the part representation should contain a detailed description of the features composing the part and their interdependence.

Defining a part as an integration of individual features does not sufficiently describe a part. Alternative courses of action may be taken in machining a

```
instance        anOldPlan
class           OLDPLANS
initialization
    featureName Slot           { type of the feature }
    width       4.5            { width of the feature }
    hardness    200            { hardness of the raw
                                 material }
    finish      159            { surface finish of the
                                 raw material }
    dimtol      0.002          { dimensional tolerance
                                 of the feature }
    reltol      0.0003         { relative tolerance of
                                 the feature }
    indtol      0.0003         { individual tolerance
                                 of the feature }
    interaction (parallel)     { types of interactions
                                 with other features }
    process     (end milling,
                 fine end milling,
                 jig grinding) { sequence of processes }
```

Figure 7.6 Example instance of the class OLDPLANS

part by considering a different set of features describing the same part. These new sets of features form alternative interpretations for a given part.[18] For instance, consider the part shown in Figure 7.7a. Five valid interpretations for the primitives involved (Fig. 7.7b) in this part are:

1. Interpretation 1 = Slot1 + Slot2 + Slot3
2. Interpretation 2 = Slot1 + Blindslot1 + Slot3
3. Interpretation 3 = Slot1 + Blindslot2 + Slot2
4. Interpretation 4 = Slot1 + Blindslot1 + Blindslot2
5. Interpretation 5 = Blindslot3 + Blindslot1 + Blindslot4 + Slot3

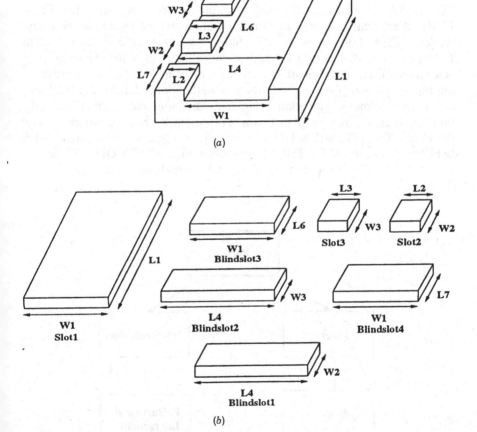

Figure 7.7 Alternate interpretations of a part: (a) example part, (b) different primitives forming the depressions in (a).

A hierarchical kernel similar to the one shown in Figure 7.8 is developed with classes PART, INTERPRETATION, FEATURE, INTERACTIONS, FACES, and EDGES under the category Part Representation to represent the part. As shown in Figure 7.8, a part is characterized by a list of interpretations and the faces that make up the part. Thus, PART has two instance variables: interpretations (a list of all the interpretations, where each interpretation is defined by a set of features and their interactions) and faceCollection (a list of all the faces that make up the part). Elements of these lists are objects representing an interpretation or a face. Another instance variable (processPlan) contains the generated process plan for the part. The processPlan slot is filled after the process plan for the part is generated. An instance of the class PART looks as shown in Figure 7.9.

Inheritance is advantageously exploited in representing the knowledge of the features. The abstract behavior common to all features is captured in the abstract class FEATURE, which has no instances but holds the common protocol of features. Each shape feature, such as slot, is a subclass of FEATURE. All the subclasses inherit the common attributes of the class FEATURE. New feature types can simply be implemented by adding new subclasses to FEATURE. New specializations of existing shape features (such as T-slot) can be added by creating appropriate subclasses of the existing classes (such as slot). The common attributes of a feature include the type of feature, the physical dimensions of the feature (length, width, depth), the tolerance information (dimensional, relative, and individual tolerances), and the interactions the feature has with other features of the interpretation. An instance of the class COUNTERSINK-HOLE, which is a subclass of the class HOLE (which is a subclass of FEATURE), is shown in Figure 7.10. COUNTERSINK-HOLE inherits both instance variables and methods from the class HOLE.

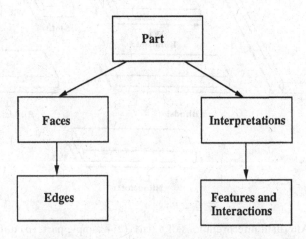

Figure 7.8   Part representation of the system.

```
instance              aPart
class                 PART
initialization
     partID           P3                    {ID of the part}
     interpretations  (Interpretation1,
                      Interpretation2,      {five
                      Interpretation3,       interpretations
                      Interpretation4,       of the part}
                      Interpretation5)
     faceCollection   (F1, F2, ..., Fn )    {collection of
                                             faces of the
                                             part}
     processPlan      ProcessPlan1          {process plan
                                             object for the
                                             part}
```

Figure 7.9   Example instance of the class PART.

Information about the relationships of a particular feature with other features in the same interpretation is declaratively represented using instances of the class INTERACTION, as shown in Figure 7.11. Two significant instance variables of the class INTERACTION are interactingFeature and typeOfInteraction, which represent the counterpart interacting feature and the interaction type, respectively. A table of currently implemented types of interactions[17] is shown in Table 7.1.

Given the previous discussion of the object-oriented representation of knowledge in our system, we can now describe the case-based approach to generating feature plans.

7.3   GENERATING FEATURE PROCESS PLANS

Most existing process planning systems use a description of the components as a group of features and plan for that one interpretation only (for one exception, see Nau et al.[27]). However, in our methodology, the component is treated as a collection of valid interpretations. Each interpretation consists of a set of features, their spatial relationships, and the engineering information (tolerances, dimensions, etc.) pertaining to the features. The methodology is to generate process plans based on the different valid interpretations, subsequently choose the most suitable process plan based on a measure of effectiveness and accuracy, and also present the possible alternatives to the user to aid in decision making. The approach for developing an appropriate process plan for machining a component can be summarized as follows:

```
class                   HOLE
superclass              FEATURE
instance variables      (radius, depth)
methods                 ()
    display-feature     { method to display a hole }
    add-feature         { add a hole to the workpiece }
    remove-feature      { method to remove a hole from
                          the workpiece }
    move-feature        { method to move a hole in the
                          workpiece }
    name-feature        { return "hole" }

class                   COUNTERSINK-HOLE
superclass              HOLE
instance variables      (outer-radius, total depth)
methods                 ()
    display-feature     { method to display a
                          countersink hole }
    add-feature         { add a countersink hole to the
                          workpiece }
    remove-feature      { inherited from class HOLE }
    move-feature        { inherited from class HOLE }
    name-feature        { return "countersink-hole" }
```

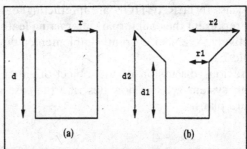

Figure 7.10   Example instance of the class COUNTERSINK-HOLE (*b*), which inherits much of its properties from its superclass HOLE (*a*).

```
class               INTERACTION
instance variables  (interactingFeature,
                     typeOfInteraction)
methods             ()
```

Figure 7.11   The class INTERACTION.

TABLE 7.1 Implemented Interaction Types

Type of Interaction	Description of the Interaction between Feature A and Feature B
Null interaction	No interaction between A and B
Parallel	A and B share area and length directions in parallel
Perpendicular	A and B share area and length directions perpendicularly
Intersecting	A and B have common volume
Contained-in	A is inside of B or B contains A

1. Generate a process plan for each interpretation. This involves:

· Sequencing the features in the order of machining based on the interactions among them and their accessibility.

· Determining machining processes for each feature using case-based reasoning.

· Computing a certainty value for the subplan corresponding to the feature.

· Selecting appropriate tools for each process.

· Generating a global plan for the part by merging feature subplans.

2. Select the most appropriate process plan for machining the part.

In a given interpretation, the relationships between the shape features of a part are used to sequence the order of their machining.

7.3.1 Sequencing the Features

The knowledge derived about the interaction between the features is important in sequencing them. As an example, let us consider the part shown in Figure 7.12a. A valid interpretation for this part is shown in Figure 12b, which consists of two pockets and a thru-slot. It is clear from the diagram that Pocket2 can only be machined with the given properties (i.e., dimensions) after the thruslot has been machined; otherwise, a pocket with different properties (dimensions) should be machined. In addition, the relationships between the features in the interpretation dictates which feature should be machined first, because a second feature may not be accessible until the first one is machined. In the preceding example, although the designed component has two pockets, the pockets do not have equivalent accessibility. Pocket2 is only accessible after the thrus-lot is produced. An alternate interpretation is one where Pocket2 is replaced by Alternative Pocket2 in the figure. In this case, Alternative Pocket2 and Pocket1 would have equivalent accessibility; hence, both Pocket1 and Alternative Pocket2 may be machined before Slot1.

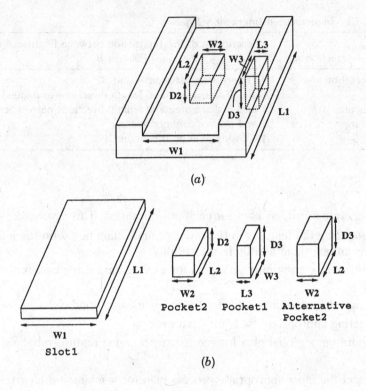

(a)

(b)

Machining Sequence = Slot1 ---> Pocket2 ---> Pocket1

(c)

Figure 7.12   Example part (a) along with its machining features (b) and the appropriate machining sequence (c).

It has been emphasized that a clear understanding and classification of feature relationships plays an important role in sequencing. In our reasoning methodology, we have exploited the following classification. The interaction between features may be nested or nonnested. Both of these classes are refined further. The nonnested relationships include different subclasses of parallel, perpendicular, and intersecting interactions. The difference between the perpendicular and the intersecting relationships is that in an intersecting relationship the two features share a common volume, whereas in the perpendicular case they share an area (two dimensional) which represents a common (virtual) face. A nested interaction, naturally, represents a "contained-in" relationship, where an open face of one feature is only accessible after the other feature has been machined. For example, features Pocket2 and Slot1 in Figure 7.12 have a containedin relationship because the open face of Pocket2 is only accessible after Slot1 has been machined. Finally, the null relationship is used to represent two features that do not have an interaction between them.

Our feature-sequencing methodology has its roots in accessibility, which is a hard constraint to be followed, and preferences, such as machining to reduce possible errors, which are soft constraints. In order to address accessibility, the features are divided into two categories: external features and internal features. External features are the features in which at least one of their opening faces is a boundary face of the component. This means that they can be directly accessed. On the other hand, in the internal features all the opening faces belong to the other features of the component. Hence, they can only be machined after the corresponding feature containing the opening face is produced. The overall methodology is to initially classify the features into the corresponding external and internal categories, address the external features, and reevaluate the remaining internal features. These remaining features are, once again, divided into the appropriate external/internal categories and addressed, and the process is repeated until all the features are handled. In order to sequence the features, a set of generic rules are used. These rules are based on the relationships between the external and the internal features, and they are derived based on the above previously developed classification for feature relationships. For simplicity, let us consider two features A and B. Then the heuristic followed by these rules can be summarized as follows:

1. If feature A is an external feature and feature B is an internal feature, then machine feature A first (this exemplifies accessibility).
2. If the relationship between features A and B is null, then the machining order is not affected by the interaction between these two features.
3. If feature A is parallel or perpendicular to feature B, then machine the feature covering a greater area on the component first (this is based on the ease of machining).
4. If feature A contains feature B (nested relationship) or vice versa, then machine feature A first or vice versa (this reflects accessibility).
5. If the relationship between features A and B is intersecting, then machine the feature with a larger volume first (this reflects better accessibility of larger volumes).

7.3.2 Feature Subplans

The problem of generating a process plan for a component is approached by dividing the problem into generating subplans for the individual features and integrating the subplans for the features to construct a global plan for the part. This approach maps directly to the described case-based methodology, which expects the solutions to be constructed from the solution pieces derived from previous experiences. The solution pieces or subplans for the features are constructed in a hierarchical methodology in which the plan for the feature is refined level by level. Before continuing with a description of how the old plans are retrieved and modified, it is important to first understand how the feature plans are generated. The hierarchical process of generating plans for

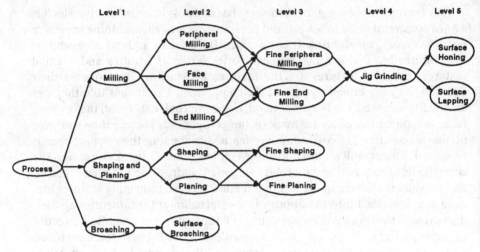

Figure 7.13   Part of the process selection tree.

features is like a tree, which becomes further refined at each level. At each level of the tree, the subplan is refined further by considering more properties of the feature. Thus, the properties used to choose a process for a feature depend on the level for which the process is being selected.

Figure 7.13 shows part of the process selection tree. Some of the important aspects of the knowledge about a process are the hardness, range of finish, range of width, tolerance values achievable, features that can be machined, and the degree of effectiveness (or allowance). At each level, a set of properties of the feature are of primary importance in selecting the appropriate processes. Table 7.2 summarizes the properties and the level of their primary importance in planning for the features (dimtol, reltol, and indtol represent the dimensional, relative, and individual tolerances, respectively).

In Table 7.2, the properties set in boldface type indicate the properties that are regarded as having primary importance at the indicated level. Therefore, to generate a subplan for a feature, if no old plan is applicable in the first level, an abstract (Milling, Shaping and Planing, or Broaching) operation is

TABLE 7.2   Important Properties of the Process Selection Hierarchy

Level	Properties of the Feature Used	Process Examples
1	**Hardness** and **surface finish**	Milling, shaping, and planing
2	**Type** and **width** of feature	End milling, shaping, and planing
3	Level 2 properties and **dimtol**	Fine end milling and fine shaping
4	Level 3 properties and **reltol**	Jig grinding and surface grinding
5	Level 4 properties and **indtol**	Surface lapping and surface honing

selected which best matches the hardness and surface finish characteristics of the feature, and then it is further refined in level 2 by considering the type and the width of the feature, and so on until all the desired constraints are satisfied. Clearly, when a process is selected to achieve the primary properties at a deeper level, it is ascertained that the process is also suitable for the properties that have been of primary importance in the previous levels, for example, a process selected to achieve a particular individual tolerance (level 5) is also suitable for the given hardness and surface finish, feature type and width, and so on. If there is more than one process at a level appropriate for machining the feature, the most appropriate process is selected based on the significant properties at that level. The criterion is to choose the process that has the parameter range whose center is closest to the required parameter value(s).

In the next section, we will show how process planning for features is done in a case-based approach. This implies retrieving an old plan and modifying it to fit the feature at hand.

7.3.3 Retrieving the Best Old Feature Plan

In order to generate a solution (process plan) for the new component, the subplans are generated by retrieving and modifying the solutions used in old feature plans. These solution pieces are the indexed and abstracted versions of the old plans generated by the system. When a new feature is encountered, the old feature plan that best suits the current feature is selected. The indices, which include the type of feature (slot, hole, pocket, etc.), the physical dimensions and the tolerances of the feature, and the type of interactions of the feature, are utilized to select the best old plan for a feature. Because the modification to generate new subplans for features is a hierarchical process in which the plan for the feature becomes further refined in a level-by-level basis, an old plan can only be useful, and changed to apply to the encountered situation, if, at the most abstract level, the important properties of the plan are applicable to the given situation. If, at the most abstract level, the characteristics of an old solution are not suitable (do not match) for the current problem, then it cannot be used at all. In such a case, a completely new solution must be generated from scratch.

At present, in order to select the feasibility of a particular feature plan to be modified and used for the case at hand, a 10 percent tolerance in the difference between the values of the properties in the old plan and the properties of the new feature is used. The actual number used is not of particular importance because the burden of repairing the selected old plan will be left with the repairer—a component that takes a given old solution and modifies it to generate a solution for the current problem. All the old feature plans within this tolerance are assumed feasible, but the best old plan is chosen based on how close the other attributes (lower weight indices of the old plan) match with the new feature's properties.

7.3.4   Modifying an Old Feature Plan

The best plan from the previous experiences is usually not completely suitable for the current problem. The repairer, which has the function of modifying a given old feature plan for application to the current situation, has to determine which portion of the old plan is usable in the current situation, and which part has to be changed. As described earlier, the subplans for the features are developed in a hierarchical level-by-level method, in which the plan is refined further in each level. At each level, the capabilities (of the processes) in the old plan are compared with the properties of the new feature. Because the solution becomes more refined at each level, once it is determined that the solution in the old plan at a particular level of detail is no longer appropriate for the characteristics of the current situation, the repairer discards the further details of the old solution. It uses the retrieved abstract portion and generates the further refinements by following the same level-by-level hierarchical plan development method described earlier together with the available knowledge about the process capabilities. The feature plan is refined further, and new processes are added to it until the sequence of the processes in the subplan satisfy all the engineering constraints (dimensions, tolerances, etc.) required for the feature. Thus, the system has the capability of selecting processes for a feature starting from any level of refinement. In the case that no old plan is retrieved, that is, no old plan meets the criteria for being suitable for the current feature even at the highest abstraction level, the system develops a new process plan fragment for the feature starting from scratch (i.e., level 1).

7.3.5   An Example

In this section, a simple example is used to briefly demonstrate certain aspects in plan selection, determination of applicable parts of a feature plan, and modification of the feature plan. Consider an instance of a slot shown in Figure 7.14a for which we would like to generate a process plan to be used as part of a complete process plan for a component containing the slot. Three old feature plans, shown in Figure 7.14b, have been found feasible for application to the situation pertaining to the new slot. This feasibility is determined using the most abstract characteristics of these plans, which is captured by the information indexed by the hardness, surface finish, the type of feature, and the type of interaction. Although all three are found to be feasible for application, it should be decided which one is to be used (first) to generate a new subplan. This decision is made by comparing the lower-weight indices, which relate to the other characteristics of the old plans and the new feature, such as width, dimensional tolerance (dimtol), and so on. For simplicity, let us assume that these more detailed characteristics have equal weight, and let us also assume that a binary decision is made about the proximity of the characteristics of the old plan and the detail feature properties using a 10 percent window on either side of the values for which the old plan was used.

name:	new_slot
length:	6.0 (1.0 0.0 0.0)
width:	6.0 (0.0 0.0 0.0)
depth:	2.0 (0.0 1.0 0.0)
.	
.	
.	
hardness:	250
finish:	165
dimtol:	0.01
reltol:	0.001
indtol:	0.0005
interaction:	parallel
process:	()
tool:	()
cerValue:	()

(a) Instantiated Slot

typeOfFeature:	slot
typeOfInteraction:	parallel
hardness:	260
finish:	160
width:	1.5
dimtol:	0.01
reltol:	0.001
indtol:	0.00054
process:	(milling, end milling, fine end milling, jig grinding)

(b) Oldplan1

typeOfFeature:	slot
typeOfInteraction:	parallel
hardness:	255
finish:	165
width:	6.2
dimtol:	0.011
reltol:	0.01
indtol:	0.01
process:	(milling, peripheral milling)

(b) Oldplan2

typeOfFeature:	slot
typeOfInteraction:	parallel
hardness:	265
finish:	170
width:	2.0
dimtol:	0.01
reltol:	0.01
indtol:	0.005
process:	(milling, end milling, fine end milling)

(b) Oldplan3

Figure 7.14 Instantiated slot (a) and the appropriate old feature plans (b).

Hence, for the example of Figure 7.14, Oldplan1 has three detail properties [dimensional tolerance (dimtol), relative tolerance (reltol), and individual tolerance (indtol)], which are in the proximity of the specifications of the new feature, while Oldplan2 has two, and Oldplan3 has one. It is important to note that although a particular percentage window is used to determine the proximity, this decision is not used to indicate that a given property constraint can be satisfied in the same manner as in the old plan. This decision is merely used to choose a feature plan from several with which to progress. Therefore, Oldplan1 is initially selected. The solution indicated by this plan contains the following sequence of processes: milling, end milling, fine end milling, jig grinding. It must now be determined which part of this old plan is useful, and how the plan should be changed for the new slot. Because the most abstract characteristics of the plan have already been found suitable for the new slot (otherwise the plan could not be used at all), the abstract process of milling is found applicable, and is used in the subplan for the new feature. At the next level, the width is of primary importance, and the new subplan must provide a suitable solution to this constraint as well as the previous characteristics. The old solution suggests the use of end milling, but checking the properties of this process with those of the current feature reveals that this process is not suitable, because the current knowledge of the system indicates that end milling can be used for features with a width in the range 0.3175 to 5.08, but the width of the new slot is 6.0. Using the knowledge of the current available processes, a new process, peripheral milling, which has an applicable width range of 0.3175 to 7.62 and satisfies all the previous characteristics, is found to be suitable for the given slot. Therefore, the current state for the subplan is: milling, peripheral milling. The developed subplan is checked to see whether it satisfies the specifications for the new slot. If this check is found to be satisfactory, no more refinement and/or detail processing is needed. However, in the preceding case, because the best relative tolerance achieved by peripheral milling is determined (from the representation of the knowledge associated with this process) to be 0.005, and the required relative tolerance (reltol) for the new slot is 0.001, further refinement of the previous subplan is needed. The process planner details the preceding subplan further to achieve the required relative tolerance; for example, it uses fine peripheral milling, which can produce a relative tolerance of 0.001. The final subplan obtained for the new slot is as follows: milling, peripheral milling, fine peripheral milling, jig grinding. It is worth mentioning that although, for example, the relative tolerance of the old plan was found to be in the proximity of the new feature in the initial assessment, the process used in the old plan, fine end milling, is not used as part of the new subplan. The reason is that the subplan needed to be modified at an earlier level, which made some of the later processes inappropriate for the situation at hand.

Without tools, a process plan is incomplete. Most of the existing computer-aided process planning systems only indicate the type of tool that should be used without providing much information. For example, to execute an end milling process, they simply recommend using an end mill. Because the tools are made with different dimensions, the extreme values for dimensions determine the appropriate range for application of the tool. For an axial tool, the maximum radius of the tool and the range of the width for the tool are two of the most important pieces of knowledge used in reasoning about the tool geometry. Examples of hard constraints in axial tool selection are as follows:

1. The maximum depth of the feature must be less than the maximum radius of the tool.
2. The process for which the tool is being applied must be among the list of applicable processes for the tool (the list of applicable processes is captured as part of the representation of knowledge about the tool).
3. The minimum width of the feature must be greater than the max imum width of the tool.

The second constraint is obvious, because not every tool can be used for every process. The first constraint models the need to avoid collision between the component and the noncutting portions of a tool. Most of the metal-cutting tools rotate on an axle, and this axle must stay outside the feature during the entire cutting operation. The third constraint models the physical constraint on the width of the cutter.

When all the feature subplans have been generated, it is necessary to combine these subplans into an efficient global plan for the part. The next section discusses our approach to merging feature plans such that the number of fixture setups and tool changes is minimal.

7.4 COMBINING FEATURE PLANS

The case-based process planner generates process plans for the individual features of a part. The output of this planner is a set of feature subplans and a set of orderings (sequencing) between the subplans. From this description, a process plan *could* be carried out by machining the part, feature by feature. The first feature would be machined, first by using one tool to do the rough cut, then by using other tools to do the finishing cuts. Then the next feature would be machined in a similar fashion and so on. This type of strict feature sequencing implicitly sequences the fixture positions and the tools used. Clearly, this would be an overcommitment of plan actions, which would lead to several disadvantages:

1. Planning decisions (plan coordination) in sequencing the part fixturing and application of tools would be done too early, eliminating the possibility of finding more suitable sequences.

2. Such early decisions in fixturing would make it hard to guarantee that tolerances will be met. In moving the part from one fixture to another, positional accuracy relative to features previously created would be reduced.

3. It is desirable to reduce the number of fixture positions because reorienting the part introduces delays in the manufacture of the part. With the previous strict sequencing, the number of fixture position changes would not be considered and hence would most likely not be optimal.

4. It is also desirable to minimize the number of tool changes within a given fixture positioning in order to save time. However, tool changes would most likely be frequent in the preceding overcommitment scheme because each feature would be machined from the rough cuts to the finishing cuts, one feature at a time.

To overcome these disadvantages, we utilize a hierarchical approach to merge each of the feature subplans into a more efficient global plan for the part, while minimizing fixture and tool setups.

The plan-merging problem is that of unifying separately generated plans into one global plan while obeying the constraints due to interactions within and between the individual plans. Very few works have explicitly addressed the plan merging problem. The work of Foulser et al.[12] presents a formal theory and treatment of the complexity of domain-independent plan merging. Foulser et al. developed an optimal algorithm utilizing dynamic programming methods for merging plans that consist of sets of independent linear sequences of actions. Because practical implementation of the optimal algorithm is infeasible for larger inputs, several greedy-based approximate (near optimal) algorithms were also developed along with their worst-case and average-case complexities for large inputs. It was also empirically shown that the approximate algorithms performed well for larger inputs. However, this work only considered actionmerging interactions[35]; that is, the only interaction considered was that which indicated that two or more actions could be profitably merged. Yang et al.[35] generalized a greedy-based algorithm in Foulser et al.[12] to handle a wider range of interaction types. However, restrictions were placed on the allowed interactions to deny the possibility of a cyclic constraint graph after plan merging. In this chapter, we relax these restrictions and address the cyclic conditions if and when they occur. We have developed a method for breaking such cycles which exploits the availability of alternatives in plan merging. Furthermore, in Yang et al.,[35] an admissible heuristic was developed for a best-first branch-and-bound algorithm, which searched through the space of mergings of alternative plans. In our methodology for decomposable domains, we explore alternatives without explicitly generating the alternative subplans.

In the remainder of this section, we characterize the domains for which our general plan-merging mechanism is applicable. Then we show how we utilize this method to merge feature process plans. More detail on the theoretical analysis and development of our domain-independent hierarchical plan-merging methodology can be found in Britanik and Marefat.[3]

7.4.1   Overview

We characterize *decomposable* planning domains as those in which operators can be hierarchically grouped by abstract types. An operator of a given abstract type must precede operators of successor abstract types. This type abstraction hierarchy yields a natural operator-type progression in the plans in the domain. For example, the process planning domain discussed in this chapter exhibits this decomposition property. Setup-type operations to hold a part naturally precede tooling-type operations, and tooling-type operations naturally precede material-removal-type operations. Plans in decomposable domains can be broken into plan fragments according to the abstract types. A *plan fragment* is a totally ordered group of actions that progresses through the type sequence, or some portion of the type sequence, in the decomposable domain. For example, a feature subplan may consist of the plan fragment: Setup$<+X,+Y> \rightarrow$ GetTool(MT1) $\rightarrow$ <NC Motions>. The shorter sequence, GetTool(MT1) $\rightarrow$

<NC Motions>, is also a plan fragment. Representing plans as collections of plan fragments is key to our methodology as shown later.

A *hierarchical plan graph* is used to develop a systematic methodology for uniform plan merging at different levels of abstraction. The hierarchical plan graph is broken into levels corresponding to the abstract operator-type hierarchy of the decomposable domain. Each level is a partially merged plan called a subplan sequencing graph. This approach decomposes the plan-merging problem in that, once external interactions between nodes on a given level have been established, the continued merging of the plan fragments in one node can take place independently of plan fragments in other nodes on that level. In the presence of the selection of alternative actions at some level of the hierarchical plan graph, this decomposition approach minimizes replanning effort. Only those plan fragments that are in the same branch as the alternative selection need be considered for replanning.[3]

7.4.2 Approach

Our approach to hierarchical feature process plan merging is a three-level approach, as shown in Figure 7.15. Level 0 is the output of the planner, a (partially) sequenced list of feature subplans. Level 1 is the grouping of feature subplans into common fixture setups. Each mergeability type (node) at this level is a fixturing (fixture setup). Each feature subplan that is a member of a given node in level 1 can be executed in the fixturing specified by that particular node. Directed edges between nodes imply that the fixturings represented by the source nodes of the edges must be sequenced before the fixturings represented by the destination nodes of the edges. Level 1 establishes a minimal set of fixturings and their appropriate sequence for executing the global plan. Each fixturing naturally decomposes into a set of toolings (tool setups) for that fixture setup. Level 2 is the grouping of feature subplan fragments into common tooling groups. Each mergeability type at level 2 is a tooling for the branch (fixturing parent from level 1) the node is in. Each subplan action that is a member of a node in level 2 can be executed using the tooling specified by that particular node. Directed edges between nodes in level 2 (toolings) imply that certain toolings must be executed before other toolings in the current fixture setup. Level 2 establishes a minimal set of toolings necessary to execute the plan fragments in the particular parent fixturing. The level after level 2 in the figure represents the actions or tool motions within a specific tooling.

7.4.3 Fixtures

For our planner, we will focus on generating fixture setups for a parallel vise clamp–type fixture such as that shown in Figure 7.16. However, the hierarchical plan-merging mechanism is not limited to the simplified fixture model presented. We use this model to facilitate a clear exposition of the methodology.

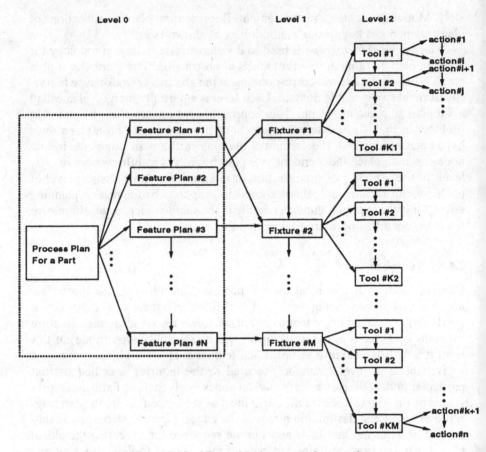

Figure 7.15   Hierarchical process plan graph.

Along with the clamp-type fixture, we also assume a vertical-type machine; that is, the tool chuck approaches the part from above the fixture as shown in Figure 7.16. Given this physical description, we can model a fixture setup as a two-tuple $\langle CA, O \rangle$, where CA is the clamping axis and O is the orientation of the part about the clamping axis. CA is specified as the direction parallel to the principle normal (in the part's local coordinates) of the faces to which the clamp is applied. O is the direction of the principle normal of the part surface that is facing up toward the tool chuck. The fixture setup in Figure 7.16 is described as $\langle +X; +Z \rangle$.

A feature is machinable in a given fixture if its tool access direction is opposite to O and it has a tool approach direction that is perpendicular to the clamping axis of the fixture, $\langle CA, O \rangle$. Table 7.3 shows sample tool access and approach directions for the various features. The base face normals

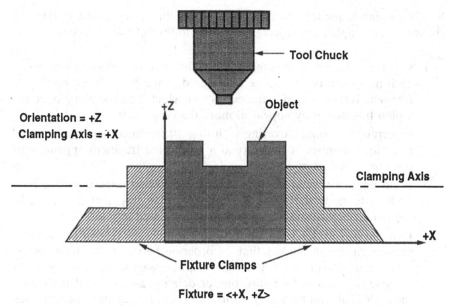

Figure 7.16 Clamp-type fixture used to hold part for machining.

can be found from feature extraction (cf. the cavity graphs in Marefat and Kashyap[24]).

7.4.4 Process Plan Merging

Process plan merging is accomplished through the construction of the hierarchical plan graph. The hierarchical plan graph is constructed by building each

TABLE 7.3 Tool Access and Approach Directions for Various Features

Feature	Tool Access	Tool Approach
Pocket	Base face normal	Opposite base face normal
Round pocket	Base face normal	Opposite base face normal
Prismatic hole	Hole axis	Along hole axis
Round hole	Hole axis	Along hole axis
Blind step	Side face normal	Opposite to normal of side face
Step	Side face normal	Opposite side face normal or Along direction perpendicular to both face normals
Blind slot	Base face normal	Opposite normal of back face
Slot	Base face normal	Along direction perpendicular to both side and base face normals

level's subplan sequencing graph and concatenating each level together. To construct the subplan sequencing graph, the following must be done:

1. Construct the set of maximal nodes. This involves creating a node for each mergeability type to be considered at the given level. Each plan fragment is then placed in the nodes with which it can be merged; hence, a plan fragment may appear in more than one node.
2. Generate a minimal covering such that all fragments are contained in a minimal number of nodes (and a given plan fragment appears only once in the set of nodes).
3. Build edges between nodes such that a directed edge is drawn from node A to node B if node A contains a plan fragment that must be sequenced before some plan fragment contained in node B.
4. Remove any cycles from the resulting graph. A cycle in the subplan sequencing graph implies that an ordering between the nodes in the graph cannot be obtained; therefore, it is necessary to remove any cycles. A cycle is removed by redirecting or deleting an edge that belongs to the cycle. This can be done by relocating plan fragments to other nodes.

First, we discuss interactions in the context of feature plan merging. Then we describe the subplan sequencing graph at the fixturing level and then at the tooling level. The part we will use as an example throughout this section is shown in Figure 7.17. This part has six features that need to be machined from the stock material. There is an independently generated subplan for each feature. Each of these subplans is shown in Figure 7.18 and corresponds to a plan fragment at level 0 of the hierarchical process plan graph. Before showing the merging process, we first need to describe how subplans (and plan fragments) can interact in process planning.

7.4.4.1 Subplan Interactions

There are two primary types of feature plan interactions in process planning: (1) contained-in relationships between two features and (2) tolerance constraints, which dictate that two features must be machined using the same fixture setup. These interactions correspond to *action-precedence* interactions and *identical-merge-action* interactions, respectively. These interactions are strong constraints that must be obeyed in the subplan merging process. For example, the fact that a containment interaction exists between two features, say, F_i and F_j, implies that the initial volume removal cut of F_i must be completed before the initial volume removal cut of F_j. There also may be weak constraints or *weak-precedence* interactions such as perpendicular or parallel feature interactions. These weak constraints can be used when obtaining a total ordering after the plan graph is generated.

We represent these interactions that provide strong and weak constraints in an interaction graph. An interaction graph is a mechanism by which we can represent complex interactions between feature subplans in process plan-

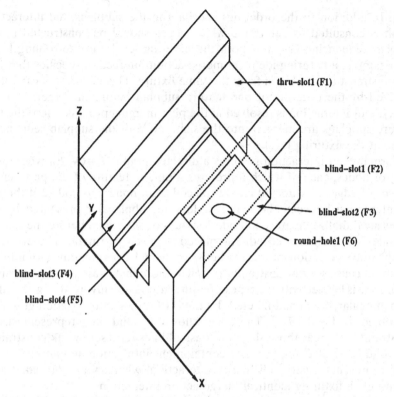

Figure 7.17 Example object with several interacting features.

F1: SetUp(<+X,+Z>) $\longrightarrow$ GetTool(MT1) $\longrightarrow$ <NC motions> $\longrightarrow$
$\qquad$ GetTool(MT2) $\longrightarrow$ <NC motions>
F2: SetUp(<+Y,+Z>) $\longrightarrow$ GetTool(MT1) $\longrightarrow$ <NC motions> $\longrightarrow$
$\qquad$ GetTool(MT2) $\longrightarrow$ <NC motions>
F3: SetUp(<+Y,+Z>) $\longrightarrow$ GetTool(MT1) $\longrightarrow$ <NC motions> $\longrightarrow$
$\qquad$ GetTool(MT2) $\longrightarrow$ <NC motions>
F4: SetUp(<-X,+Z>) $\longrightarrow$ GetTool(MT3) $\longrightarrow$ <NC motions> $\longrightarrow$
$\qquad$ GetTool(MT4) $\longrightarrow$ <NC motions>
F5: SetUp(<-X,+Z>) $\longrightarrow$ GetTool(MT3) $\longrightarrow$ <NC motions> $\longrightarrow$
$\qquad$ GetTool(MT4) $\longrightarrow$ <NC motions>
F6: SetUp(<+X,-Z>) $\longrightarrow$ GetTool(DT1) $\longrightarrow$ <NC motions> $\longrightarrow$
$\qquad$ GetTool(DT2) $\longrightarrow$ <NC motions>

Figure 7.18 Subplans to generate the part in Figure 7.17.

ning. In addition to the orderings specified in the subplans, the interaction graph is consulted to determine which edges should be constructed in the subplan sequencing graph at both the fixturing level and the tooling level. For example, a fixturing identical-merge-action interaction specifies that two features must be machined from the same fixture. This interaction cannot be resolved by the ordering of one feature subplan before the other. Instead, this type of interaction is resolved in the plan-merging process where the two feature subplans are merged into the same node of the subplan sequencing graph at the fixturing level.

Formally, an interaction graph is a directed graph $G = \langle N, E \rangle$, where N is a set of nodes, each of which represents a unique feature of the part, and E is a set of edges. A directed edge is placed from node 1 to node 2 if there is an interaction between the features (feature subplans) represented by the nodes such that the feature in node 1 should be machined before the feature in node 2. An undirected edge is placed between two nodes if there is an identical-merge-action interaction between them. Edges are drawn for interactions that cause strong constraints as well as those that cause weak constraints. Each edge is labeled with the type of feature interaction involved (e.g., parallel, perpendicular, contained-in, etc.). Figure 7.19 shows an interaction graph for the object in Figure 7.17. The edges shown as solid lines represent strong constraints, whereas those shown as dashed lines represent weak constraints. An edge label of C represents a containment interaction and an edge label of P represents a perpendicularity interaction, whereas an edge label of F represents a fixturing identical-merge-action interaction.

Consider the contained-in constraint directed from $F3$ to $F6$ in Figure 7.19. This strong constraint specifies that the rough cut for feature $F3$ must be completed before the rough cut for feature $F6$. If $F3$ and $F6$ are contained in different nodes of the fixturing subplan sequencing graph (i.e., they are not machined in the same fixturing), then an edge will be drawn between the corresponding fixturing nodes in the fixturing subplan sequencing graph. However, if both $F3$ and $F6$ are contained in the same node in the fixture subplan sequencing graph (i.e., they are machined during the same fixturing), then an

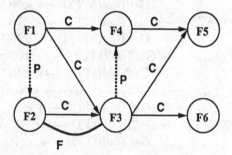

Figure 7.19 Interaction graph for the object of Figure 7.17.

edge will be drawn between the corresponding tooling nodes in the tooling subplan sequencing graph on the next level. Should the rough-cut actions for both $F3$ and $F6$ be contained in the same tooling node in the tooling subplan sequencing graph, then the rough-cut actions themselves will be ordered appropriately to satisfy the constraint. This example shows how the interaction graph is used to generate edges (orderings) between nodes in the subplan sequencing graphs in a manner dependent on how merging takes place between actions in the interacting subplans. Using the interaction graph, the subplan sequencing graphs are constructed such that they do not violate the constraints of the process planning problem.

The fact that there is a containedin interaction between two features, say, $F3$ and $F6$, implies that the initial volume removal process of $F3$ must be completed before the initial volume removal process of $F6$. Under this assumption, only the rough-cut processes of each feature subplan are sequenced by the sequencing of the feature subplans. In this manner, we can minimize the number of tool changes by doing the rough cuts for several features all in one tooling, then changing tools to do the finishing cuts. This is more efficient than the one-feature-at-a-time method, which would have one rough cut done, then a tool change followed by the finishing cut. Then the tool would be changed back to the rough-cut tool to do the next feature and so on.

7.4.4.2 Fixturing Level To generate the nodes of the fixturing level (i.e., level 1) subplan sequencing graph, we first build a list of feature subplan fragments applicable to each fixture. These lists can be regarded as *common fixture sets* (CFSs). The common fixture sets for the part in Figure 7.17 are shown in Figure 7.20. Notice that a feature subplan fragment may appear in more than one CFS, demonstrating alternative (i.e., different) fixtures from which the same feature may be machined. Common fixture sets correspond directly to the set of maximal nodes for this subplan sequencing graph.

The next step in building the fixture subplan sequencing graph is to generate a minimal covering of nodes, that is, a minimum set of fixturings. This corresponds directly to our desire to minimize the number of fixture changes while machining a part. Using a greedy approach (see Britanik and Marefat[3] for a detailed discussion of other approaches to generating a minimal covering of nodes), this is done by first selecting the largest CFS and removing the features it contains from all other CFSs. Then the next largest CFS is chosen and so on, until there are no nonempty CFSs remaining. For our example part, the

```
F<+X,+Z> : {F1,F4,F5,F6}    F<-X,+Z> : {F1,F4,F5,F6}
F<+Y,+Z> : {F2,F3,F6}       F<-Y,+Z> : {F2,F3,F6}
F<+X,-Z> : {F6}             F<-X,-Z> : {F6}
F<+Y,-Z> : {F6}             F<-Y,-Z> : {F6}
```

Figure 7.20 Common fixture sets for the part of Figure 7.17.

largest CFSs are $F\langle+X, +Z\rangle$ and $F\langle-X, +Z\rangle$. The set $F\langle+X, +Z\rangle$ is randomly chosen, which leaves only two nonempty CFSs remaining: $F\langle+Y, +Z\rangle = \{F2, F3\}$ and $F\langle-X, +Z\rangle = \{F2, F3\}$. Note that F6 has been removed from these two sets because, as mentioned previously, all features contained in the first chosen node have been removed from the remaining nodes. This is also why $F\langle+X, -Z\rangle$, $F\langle-X, -Z\rangle$, $F\langle+Y, -Z\rangle$, and $F\langle-Y, -Z\rangle$ are now empty. After randomly choosing the CFS $F\langle+Y, +Z\rangle$ in the second step, the generation of the minimal covering is complete since all six features are contained in chosen nodes. The minimal covering for the example part consists of the two nodes: $F\langle+X, +Z\rangle = \{F1, F4, F5, F6\}$ and $F\langle+Y, +Z\rangle = \{F2, F3\}$.

With this minimal set of fixture setups, we next construct the edges of the subplan sequencing graph. This corresponds to determining the sequence by which the fixtures are used in machining the part. This sequence is constrained by the strong interactions between features in different fixtures. Using the interaction graph in Figure 7.19, we have an explicit list of orderings (strong interactions), and we can generate the edges for the fixture subplan sequencing graph. We simply examine each strong interaction (represented by solid arrows in Fig 7.19) and include it in the fixture subplan sequencing graph as an edge if its source and destination features are in different fixture nodes. The fixture subplan sequencing graph after edge generation is shown in Figure 7.21a. Note that this graph has a cycle; hence, cycle removal is necessary.

The two general approaches for removing cycles from the subplan sequencing graph are: (1) move fragments between existing nodes and (2) move fragments to a newly created node. To remove cycles, break sets are constructed and ranked for suitability. A *break set* is a set of node components (feature subplan fragments) that, when moved to a different node, will break the cycle (remove an edge in the cycle). Break sets are ranked according to the following evaluation function:

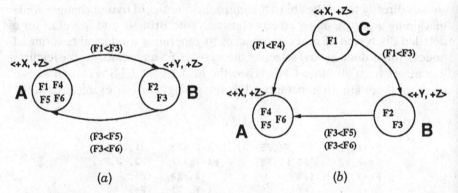

(a)   *(b)*

Figure 7.21 Fixture subplan sequencing graph for the object of Figure 7.17 before (*a*) and after (*b*) cycle removal.

F(Break set) = (Number of cycles edge is contained in)
- (Number of fragments to move)
- Σ(number of references per fragment moved)

The number of references per fragment is the total number of appearances of all the fragments in the break set in edge labels of the subplan sequencing graph. The edge in the cycle with the largest value of this function is the most favorable to break.

The list of break sets for the cycle in Figure 7.17a, ranked left to right, is as follows:

$$\{F1\} \quad *\{F3\}* \quad \{F5, F6\} \quad \{F2, F3\}$$

$\{F3\}$ is not a valid break set because it does not contain $F2$ (a fragment that has an identical-merge-action interaction with $F3$ as determined from the interaction graph shown in Fig. 7.19). Recall that an identical-merge-action interaction implies that the involved features must be merged into the same node; hence, they cannot be moved apart during cycle breaking. Because the break set $\{F2, F3\}$ already exists, $\{F3\}$ is simply removed from the list of break sets. The first approach of moving one or more plan fragments (feature subplans) to other nodes in the graph would be unsuccessful because only $F6$ can be moved to another node in the graph (this can be seen by observing the maximal nodes in Fig. 7.20), but there is no break set which contains only $F6$. Hence, the second approach of adding new nodes to the graph will be attempted. For the sake of exposition, let us pretend for the remainder of this paragraph that $\{F3\}$ is a valid break set. Note that the break set $\{F1\}$ has the highest ranking followed by $\{F3\}$. Even though they have the same cardinality, $\{F1\}$ has a higher ranking than $\{F3\}$ because $\{F3\}$ is referenced (appears in edge label orderings) two more times than $\{F1\}$. Also, it should be noted that the fragment $F3$ cannot be moved to any node, pre-existing or added, without generating another cycle. $\{F1\}$ is an eligible break set, and node A of Figure 7.21a is duplicated with the fragment $F1$ as its contents. Using the interaction graph of Figure 7.19, the edges of the fixture subplan sequencing graph are regenerated incorporating the new node, C. The resulting cyclefree fixture subplan sequencing graph is shown in Figure 7.21b.

Now that the fixturing subplan sequencing graph, that is, the fixturing level of the hierarchical process plan graph, is complete, it is necessary to build the tooling level by generating tooling subplan sequencing graphs for each fixture node in the fixturing level.

7.4.4.3 Tooling Level Figure 7.22 shows the partially merged subplans for the part in our example after merging takes place at the fixturing level. This figure is drawn to show the tooling level plan fragments, TF1–TF12, which are identified in Figure 7.23.

SetUp(<+X,+Z>) $\longrightarrow$ TF1 $\longrightarrow$ TF2

SetUp(<+Y,+Z>) $\longrightarrow$ TF3 $\longrightarrow$ TF4
$\qquad\qquad\quad\backslash \longrightarrow$ TF5 $\longrightarrow$ TF6

SetUp(<+X,+Z>) $\longrightarrow$ TF7 $\longrightarrow$ TF8
$\qquad\qquad\quad\backslash \longrightarrow$ TF9 $\longrightarrow$ TF10
$\qquad\qquad\quad\backslash \longrightarrow$ TF11 $\longrightarrow$ TF12

Figure 7.22 Partially merged plan after merging takes place at the fixturing level (level 1) of the process planning HPG.

It is these plan fragments that will be merged in the tooling subplan sequencing graph. With this information, we are now ready to generate the tooling subplan sequencing graph for node A of level 1 in Figure 7.24.

First, we need to generate the set of maximal nodes. One node is generated for each specific tool used (in the GetTool actions) in the plan fragments. A fragment is placed in a node if the required tool of its GetTool action matches that represented by the node or if the tool represented by the node is a suitable alternative to the required tool. (In this manner, alternative tools are considered to help minimize the number of tool nodes and hence the number of tool changes.) In this example, tools MT1 and MT3 are suitable alternatives of each other, as are tools MT2 and MT4. The following is the set of maximal nodes for the tooling subplan sequencing graph for node A of level 1 in Figure 7.25:

$T1$: TF7, TF9	$T2$: TF8, TF10	$T3$: TF11
$T4$: TF12	$T5$: TF7, TF9	$T6$: TF8, TF10

TF1: GetTool(MT1) $\longrightarrow$ <NC motions>
TF2: GetTool(MT2) $\longrightarrow$ <NC motions>
TF3: GetTool(MT1) $\longrightarrow$ <NC motions>
TF4: GetTool(MT2) $\longrightarrow$ <NC motions>
TF5: GetTool(MT1) $\longrightarrow$ <NC motions>
TF6: GetTool(MT2) $\longrightarrow$ <NC motions>
TF7: GetTool(MT3) $\longrightarrow$ <NC motions>
TF8: GetTool(MT4) $\longrightarrow$ <NC motions>
TF9: GetTool(MT3) $\longrightarrow$ <NC motions>
TF10: GetTool(MT4) $\longrightarrow$ <NC motions>
TF11: GetTool(DT1) $\longrightarrow$ <NC motions>
TF12: GetTool(DT2) $\longrightarrow$ <NC motions>

Figure 7.23 Set of plan fragments for the tooling level (level 2) of the process planning HPG. MT1–MT4 and DT1–DT2 are constants representing tools.

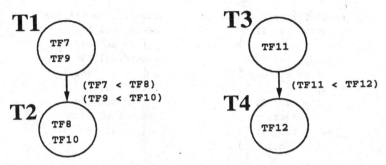

Figure 7.24 Tooling subplan sequencing graph for node *A* of level 1 in Figure 7.25.

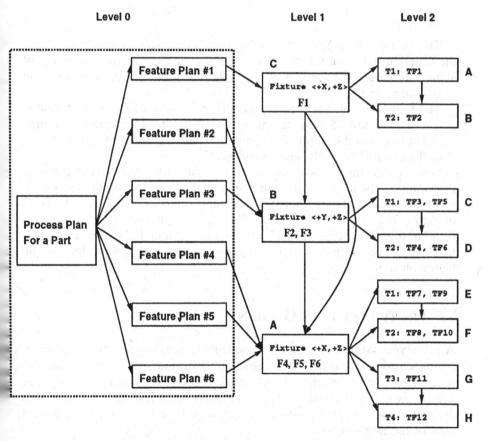

Figure 7.25 Hierarchical process plan graph incorporating the tooling subplan sequencing graphs into the tooling level (level 2).

```
      setup fixture <+X,+Z>          setup fixture <+X,+Z>
        setup tool T1                  setup tool T1
          { do actions }                 { do actions }
        setup tool T2                  setup tool T3
          { do actions }                 { do actions }
      setup fixture <+Y,+Z>            setup tool T2
        setup tool T1                    { do actions }
          { do actions }               setup tool T4
        setup tool T2                    { do actions }
          { do actions }
```

Figure 7.26 Global plan extracted from the hierarchical plan graph in Figure 7.25.

The corresponding minimal covering is:

$T1$: TF7, TF9 $T2$: TF8, TF10
$T3$: TF11 $T4$: TF12

To generate the edges in the tooling subplan sequencing graph, we utilize the orderings between the fragments as shown in Figure 7.22. The result of adding the appropriate edges is the cycle-free tooling subplan sequencing graph shown in Figure 7.24.

The tooling subplan sequencing graphs are generated for nodes B and C of level 1 in Figure 7.25 in a similar manner as before. These graphs are then added at level 2 to the hierarchical plan graph (HPG) representing the process plan. The final HPG is shown in Figure 7.25.

Now that the hierarchical process plan graph is complete, we can generate the outline of the final plan (it is an outline in the sense that we do not include specific motions and dimensions). This plan, which is extracted from the HPG of Figure 7.25, is shown in Figure 7.26. The execution sequence is from top to bottom and then left to right. This is one of multiple possible sequences that can be generated through a topological sort on the orderings in the hierarchical plan graph.

7.5 PROTOTYPE DEVELOPMENT

A prototype system has been developed and tested on a SUN workstation in an object-oriented environment (Smalltalk–Objectworks, Version 4.1). The process planner is interfaced with the Pro/ENGINEER solid modeler. A feature extractor that generates all the valid interpretations of the part is integrated with the planner via the database. Figure 7.27 shows the top-level view of the user interface.

The process planning interface (see Fig. 7.28) depicts the process plan of a part and the machining processes and tools that have been used to generate

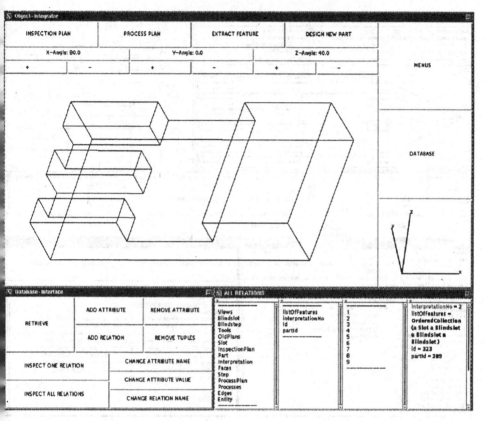

Figure 7.27 Main view of the graphical interface of the integrated system (top) along with the database interface views (bottom).

the process plan. The old plans that have been changed to obtain the current process plan are also shown. A hierarchical (tree) and a textual representation of the process plan are included in the interface. The sequence of the features is reflected in the tree (vertically across the second level in the tree) as well as the listing of the processes and tools associated with the individual features. A graphical display of the part is also included whose features and faces can be selected and highlighted while interrogating a particular aspect of the part. Comprehensive tools to interact with the model and system knowledge allow the user to examine the existing process and tool capabilities, to display the process plan for an alternative part, to invoke graphic commands, and to return to the main view at any stage. Options for generating an alternate course of actions from any detail level (using an alternative set of features, an alternative process, or alternative tools) in the plan tree are also provided. Such a request modifies and updates everything in the subtree below the

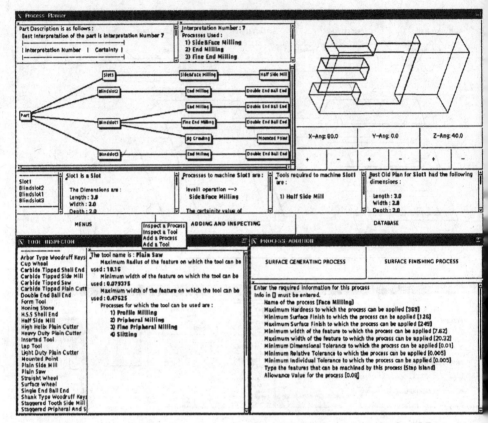

Figure 7.28 Process plan interface (top) along with the process addition (lower-right) and the tool inspection (lower-left) windows.

desired change. A switch for the database graphical interface enables queries to the database, such as finding out all the process plans that use a particular machining process.

7.6 CONCLUSIONS AND FUTURE WORK

The research discussed in this chapter presents a casebased approach to process planning. The major contributions of our research may be summarized as follows:

· A process planning system that utilizes case-based techniques for multi-level process selection (feature process planning) for three-dimensional prismatic parts has been developed. The case-based methodology involves retrieving old feature plans generated from past experiences, modifying them to fit the part at hand, and abstracting and storing the new plan for future use.

· A hierarchical method for merging feature subplans into a global plan for the part has been presented. The resulting global plan is efficient because the number of fixtures and tool changes is minimized.

· A prototype case-based process planning system has been developed and it has been used for experimentation to validate the discussed techniques.

Future work includes adding design error detection to the planner to minimize the failures in plan generation and integrating information from the hierarchical plan graph into the user interface. This would allow the user to modify fixture selections interactively, with the planner giving advice on the allowable (or most efficient) alternatives.

ACKNOWLEDGMENTS

The support of this work by the National Science Foundation under Grant DDM9210018 to Dr. Marefat is gratefully appreciated. Additional support was provided by the Department of Electrical and Computer Engineering at the University of Arizona.

REFERENCES

1. Ben-Arieh, D., and M. Chopra, 1995, A Case Based Process Planning System for Prismatic Parts, *Proceedings of the Fourth Industrial Engineering Research Conference (IERC Proceedings)*, IIE, Nashville, pp. 827–835.

2. Bergmann, R., and W. Wilke, 1995, Building and Refining Abstract Planning Cases by Change of Representation Language, *J. Artificial Intelligence Res.*, Vol. 3, pp. 53–118.

3. Britanik, J., and M. Marefat, 1996, Hierarchical Plan Merging, Technical Report TR-ISL-20, Department of Electrical and Computer Engineering, University of Arizona.

4. Champati, S., W. Lu, and A. Lin, 1993, A Case-Based Process Planning for Prismatic Parts Machining, *PED: Manufacturing Science and Engineering*, ASME, New York, Vol. 64, pp. 299–305.

5. Chang, T. C., 1990, *Expert Process Planning for Manufacturing*, Addison-Wesley, Reading, MA.

6. Chang, T., and R. Wysk, 1985, *An Introduction to Automated Process Planning*, Prentice-Hall, Englewood Cliffs, NJ.

7. Cho, H., A. Derebail, T. Hale, and R. A. Wysk, 1994, A Formal Approach to Integrating Computer-Aided Process Planning and Shopfloor Control, *J. Engineering for Industry, Trans. ASME*, Vol. 116, pp. 108–116.

8. Cser, L., M. Geiger, W. Greska, and M. Hoffman, 1991, Three Kinds of Case-Based Learning in Sheet Metal Manufacturing, *Computers in Industry,* Vol. 17, pp. 195–206.

9. Davies, B., and I. Darbyshire, 1984, The Use of Expert Systems in Process Planning, *Ann. CIRP,* Vol. 33, pp. 303–306.

10. Descotte, Y., and J. Latombe, 1985, Making Compromises among Antagonist Constraints in a Planner, *Artificial Intelligence,* Vol. 27, pp. 183–217.

11. Dong, J., H. R. Parsaei, and A. Kumar, 1995, Intelligent Feature Extraction for Concurrent Design and Manufacturing, in *Design and Implementation of Intelligent Manufacturing Systems,* H. Parsaei and M. Jamshidi (eds.), Prentice-Hall, Upper Saddle River, NJ, Vol. 13, pp. 301–323.

12. Foulser, D. E., M. Li, and Q. Yang, 1992, Theory and Algorithms for Plan Merging, *Artificial Intelligence,* Vol. 57, pp. 143–181.

13. Hammond, K. J., 1988, Case-Based Planning, *Proceedings: Case-Based Reasoning,* J. Kolodner (ed.), DARPA/ISTO, Washington, DC.

14. Humm, B., C. Schulz, M. Radtke, and G. Warnecke, 1991, A System for Case-Based Process Planning, *Computers in Industry,* Vol. 17, pp. 169–180.

15. Hummel, K., 1989, Coupling Rule-Based and Object-Oriented Programming for the Classification of Machined Features, *Proceedings of the ASME Computers in Engineering Conference,* San Francisco, pp. 409–418.

16. Irani, S. A., H. Y. Koo, and S. Raman, 1995, Feature-Based Operation Sequence Generation in CAPP, *Internat. J. Production Res.,* Vol. 33, pp. 17–39.

17. Joshi, S., N. N. Vissa, and T. C. Chang, 1988, Expert Process Planning System with Solid Model Interface, *Internat. J. Production Res.,* Vol. 26, pp. 863–885.

18. Karinthi, R. R., and D. S. Nau, 1989, Using a Feature Algebra for Reasoning about Geometric Feature Interactions, *Proceedings of the International Joint Conference on Artificial Intelligence,* Morgan Kaufmann, Detroit, pp. 1219–1224.

19. Kolodner, J., 1993, *Case-Based Reasoning,* Morgan Kaufmann, San Mateo, CA.

20. Korde, U. P., B. C. Bora, K. A. Stelson, and D. R. Riley, 1992, Computer-Aided Process Planning for Turned Parts using Fundamental and Heuristic Principles, *J. Engineering for Industry, Trans. ASME,* Vol. 114, pp. 31–40.

21. Leake, E. D., 1996, *Case-Based Reasoning. Experience, Lessons, and Future Directions,* AAAI Press, Menlo Park, CA.

22. Mani, J., and S. Raman, 1996, A Methodology for Manufacturing Precedence Representation and Alternate Task Sequencing, *Trans. NAMRI/SME,* Vol. 24, pp. 251–256.

23. Marefat, M., and J. Britanik, 1994, Case-Based Process Planning, *Proceedings of the NSF Design and Manufacturing Systems Conference,* Cambridge, MA.

24. Marefat, M., and R. Kashyap, 1992, Automatic Construction of Process Plans from Solid Model Representations, *IEEE Trans. Systems Man Cybernet.,* Vol. 22, pp. 1097–1115.

25. Marefat, M., R. Kashyap, and S. Malhotra, 1993, Object Oriented Intelligent Computer Integrated Design, Process Planning, and Inspection, *IEEE Computer,* Vol. 26, 54–65.

26. Nau, D., and M. Gray, 1986, SIPS: An Application of Hierarchical Knowledge Clustering to Process Planning, in *Integrated and Intelligent Manufacturing,* C. Liu and T. Chang (eds.), ASME, New York, pp. 219–225.

27. Nau, D., S. Gupta, and W. Regli, 1995, AI Planning versus Manufacturing Operation Planning: A Case Study, *Proceedings of the International Joint Conference on Artificial Intelligence*, Morgan Kaufmann, Montreal, pp. 1670–1676.

28. Prabhu, P., S. Elhence, and H. Wang, 1990, An Operations Network Generator for Computer Aided Process Planning, *J. Manufacturing Systems*, Vol. 9, pp. 283–291.

29. Tsatsoulis, C. and R. Kashyap, 1988, A Case-Based System for Process Planning, *Robotics and Computer-Integrated Manufacturing*, Vol. 4, pp. 557–570.

30. Tsatsoulis, C., and R. Kashyap, 1988, A System for Knowledge-Based Process Planning, *Artificial Intelligence in Engineering*, Vol. 3, pp. 61–75.

31. Tsatsoulis, C., and R. Kashyap, 1993, Case-Based Reasoning and Learning in Manufacturing with the TOLTEC Planner, *IEEE Trans. Systems Man Cybernet.*, Vol. 23, pp. 1010–1022.

32. Usher, J. M., 1995, Object Oriented Approach to Feature-Based Process Planning, in *Design and Implementation of Intelligent Manufacturing Systems*, H. Parsaei and M. Jamshidi (eds.), Prentice-Hall, Upper Saddle River, NJ, Vol. 12, pp. 275–300.

33. Yang, H., and W. F. Lu, 1993, PROCASE: A Prototype of Intelligent Case-Based Process Planning System with Simulation Environment, *Proceedings of the ASME Computers in Engineering Conference*, San Diego, pp. 571–577.

34. Yang, H., W. F. Lu, and A. C. Lin, 1992, A Framework for Using Case-Based Reasoning in Automated Process Planning, *Proceedings of the ASME Winter Annual Meeting—Concurrent Engineering*, New Orleans, pp. 101–114.

35. Yang, Q., D. S. Nau, and J. Hendler, 1992, Merging Seperately Generated Plans with Restricted Interactions, *Computational Intelligence*, Vol. 8, pp. 648–676.

36. Zarley, D. K., 1991, A Case-Based Process Planner for Small Assemblies, *Proceedings: Case-Based Reasoning*, DARPA/ISTO, Washington, DC, pp. 363–373.

8

AN INTEGRATED METHODOLOGY FOR PRODUCT AND PROCESS DEVELOPMENT IN AUTOMATED MANUFACTURING CONTROL SYSTEM DESIGN

FRANK S. CHENG
Central Michigan University

ERNEST L. HALL
University of Cincinnati

8.1 INTRODUCTION

All knowledge about the physical world may be classified into four types: observable facts, procedures, concepts, and functional relationships. A fact is a report of a direct sensory input experienced by someone at some time. A procedure is a sequence of actions that produce a desired outcome. Both facts and procedures are often learned through observation and memorization. Concepts are ideas that refer to classes or categories in which the idea members share common features or characteristics. Concepts can be represented and

Integrated Product and Process Development, Edited by John Usher, Utpal Roy, and Hamid Parsaei
ISBN 0-471-15597-7   © 1998 John Wiley & Sons, Inc.

organized into a hierarchy that illustrates the logical relationships among the classes. The final type of knowledge, functional relationships, represents the highest level of abstraction. Functional relationships bridge concepts and recognize commonality between the concepts. Principles, mathematical models, rules, and generalizations are examples of functional relationships. These relationships encourage development beyond replication into the arena of the design, creation, and synthesis of new outcomes.

As a representation of the state-of-the-art technologies, an automated manufacturing system (AMS) is a man-made system that is composed of intelligent machines, sensors, automated devices, computers, and networks. It is designed to achieve improved productivity and quality of manufacturing processes. The key parts of an AMS are automated manufacturing cells and associated automated material-handling systems.

An automated manufacturing cell usually contains a cell controller and several computer-controlled machines such as industrial robots and computer numerical controlled (CNC) machines. The automated material-handling systems associated with the work cell include conveyors, automated storage and retrieval systems (AS/RS), and automated guided vehicles (AGVs).[14] Figure 8.1 shows such an automated work cell. Here, an AS/RS is responsible for sending requested raw materials to the cell according to production orders. The synchronous moving conveyor delivers the materials to each workstation through pallets. The solenoids at the workstation hold and fix the pallet when the sensor detects its presence. The robot performs specified tasks on the materials on the pallet and releases them to the next workstation as the operation is finished. Necessary inspection procedures may also be applied to the materials and the pallet before the AS/RS stores them.

Through engineering practice, "specification or design" has acquired precise meanings in the classical disciplines of engineering. A design can be viewed as both a product and a process of problem solving, which takes place in a task environment with particular properties. The process of design is an interactive process of capturing information, analyzing the captured information against the criteria for a good design, and either returning to the first step for refinement or releasing the design for implementation. A more detailed model of a design process could be based on the following activities that provide an abstract solution to satisfy the function, performance, resource, and other constraints. They are: (1) analysis of the problem, (2) conceptual design, (3) embodiment of schemes, and (4) detailing.

A key component of an AMS design is the ability to specify and implement the required control functions of the controllers. Usually, the control design at the machine control level is to specify motion types and motion sequences of the controlled machine using programmed steps. At the work cell level, control objectives are to supervise the interactions between a group of related machines or processes. Both design tasks require the ability to precisely address the temporal and spatial constraints of the operations, as well as the behavior of the controlled devices in real-time processes. In addition, data

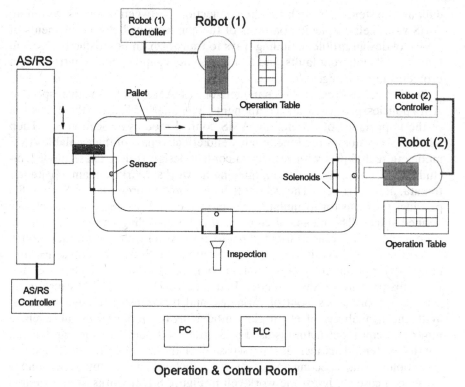

Figure 8.1 Example of an automated manufacturing work cell.

must be available for integrating the control functions of each component into a practical AMS. Control design practice shows that the specification stage is an important process in which control requirements are imposed and expressed in convenient mediums such as texts, schematics, diagrams, tables, or graphs. Good control specifications play the role of communication mediums (e.g., contracts and documents) between users and control system designers to aid the verification of candidate control solutions. However, due to a system's complexity in components, processes, functions, constraints, and changes, designers may not easily develop precise criteria to evaluate the candidate control solutions with respect to the overall specifications. Also, separated design specifications provide further difficulties for designers to undergo a stepwise refinement process in which each step decomposes a (sub)specification into several component specifications arranged in a hierarchy. Consequently, the design of AMS controls may result in high cost, long design periods, and inflexible or unsafe systems.

This chapter discusses the reasons for these limitations and presents an integrated approach for reducing these limitations. This is achieved by conducting a systematic search for the connections between elements within an AMS work cell, displaying patterns of the connections between elements of a control design problem, finding ways to transform an unsatisfactory system to remove its inherent faults, and dividing complex and unclear control design problems into manageable parts.

Section 8.2 addresses the characteristics of AMS control with respect to AMS policies, processes, and components. This analysis leads to the discussion of the importance of developing AMS control models in Section 8.3. Then Section 8.4 evaluates the functionality and effectiveness of the available AMS modeling techniques with respect to control design issues. These models include discrete-event simulation, queuing networks, Markov chains, finite automata, and Petri nets. The system theory approach presented in Section 8.5 provides the basis for formulating the modeling and design process as clearly defined stages. The means of constructing the modeling and design stage is proposed by using object-oriented paradigms. As a result, a reference model, called the *event class,* is developed in Section 8.6. Section 8.7 presents integrated descriptions for control system components. The products generated from this process are the extended Petri net–based models that capture the subsystems, processes, control variables, and functions of the event classes. With the availability of event class models at the required detail levels, a multilevel control structure is used in Section 8.8 to design an integrated cell control system. This includes a discussion of the methods for decomposing subproblems and selecting and designing cell coordination inputs. Section 8.9 presents a case study for the work cell in Figure 8.1. It works as an example of the integrated method in systematically developing work cell controllers. Finally, some conclusions are presented in Section 8.10.

8.2   CHARACTERISTICS OF AMS CONTROL

The AMS policies, processes, and components provide fundamental relationships to AMS control, as shown in Figure 8.2.[3] A policy is a plan of a system to meet its goals. The AMS policies provide positive guidance about the goals of the designed system and the means of how to achieve the goals, or they set constraints limiting the way of achieving the goals.

The processes of an AMS are time-based and discrete-event-based hybrid procedures. First, the progression of the processes is driven by the discrete events that occur at discrete times; each event has states with discrete values that are assumed to be logical or symbolic rather than numeric. Second, a process generally has internal continuous-time dynamic behavior that represents interactions and responses to the system's environment. Third, the processes operate and share resources both concurrently and sequentially. "Hard" real-time deadlines must often be met for safe operations. As a result, system correctness depends not only on the logical results of the system behavior, but also on the time at which the results are produced.

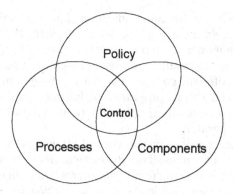

Figure 8.2 Integrative representation of AMS controls.[3]

The hardware and software components of an AMS are the resources used to realize both the policy and the processes. The hardware components represent the controlled resources, including computers, intelligent machines, sensors, and automated devices. Usually, the associated machine controller controls the operations of an intelligent machine. Figure 8.3 shows the structure and components of a machine controller. In addition to an operating system, memory, and a user interface, today's machine controller is often equipped with some communication interface such as RS-232 and input/output (I/O). A programmable logic controller (PLC) is a common type of controller primarily designed to interface I/O devices and perform logic control on them. A PLC has a central processing unit (CPU), memory, a user interface, and communication interface. To allow information and instructions to be communicated between PLCs and more powerful computers, today's PLCs also provide RS-232 and local-area network communication capabilities as a part of their

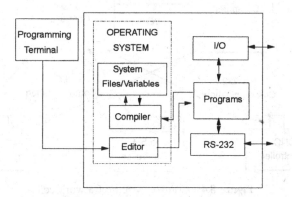

Figure 8.3 Structure of a machine controller.

specifications. Thus, operator communication, data acquisition, and programming of PLCs can be done using a personal computer (PC) interface.

The software components often represent the control programs or subroutines that are largely the results of enforced languages and designed systems. For example, the software component of a machine controller has a well-defined public interface and a programming language as shown in Figure 8.3. Both the visible and inaccessible implementations of the software components can be compiled separately.

Automated manufacturing system control integrates the operational policies, processes, and resources. Two distinct control activities are observed. The first control activity is to create and interpret a policy. Policies need to be specified in ways that enable them to be applied to the resources. The second control activity is to perform control operations for monitoring or controlling the behavior of the resources related to a control function. A controlled resource must provide a control interface as well as support the operation related to its normal functionality. Control decisions must often be made quickly from both control activities to meet "hard time" deadlines for safe operations. For example, the time scale of a controller's decision can range from minutes to seconds in a work cell. Also, it is possible in a distributed manufacturing process that the control decisions affecting an operation may be made at one physical location and sent to control the devices in other locations. This distribution of decision making, as shown in Figure 8.4, may create interface constraints due to the limitations imposed by either the hardware or the software of the controllers. These limitations may be further complicated by temporal constraints and the inability of the controllers to execute control functions or communicate with each other in real time.

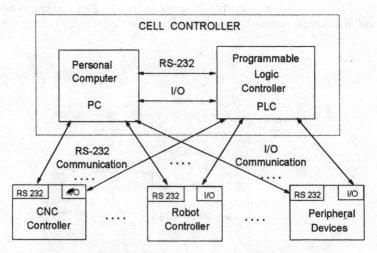

Figure 8.4 Control system of a work cell.

8.3 DEVELOPMENT OF AMS MODELS

The need for developing useful models arises as one is dealing with an inherently complicated system. This is because models are manageable specifications, devised to aid the understanding of the system. Models have many applications, among them is the ability:

- To clearly and concisely describe existing systems.
- To prescribe the main characteristics of future systems.
- To design a system to given requirements.
- To predict system properties reliably.
- To closely define a system and analyze it rigorously.

Modeling methods may be differentiated by noting the methods used to capture the vocabulary of the system (the static dimension) or the methods used to capture the behavior of the system (the dynamic dimension). A static domain describes a static structure with a collection of variables. These variables represent the useful attributes of the system. The dynamic dimension of modeling is less specific in definition. As a consequence, the decomposition of the system highly influences the selection of models.

A "good" model usually conveys the characteristics chosen by the designer to fit the objectives of the design and analysis. This implies the following:

1. The model must suit the designer—the sender of the message.
2. The model must suit the modeled system in its context, that is, be capable of capturing the chosen characteristics so as to carry the message.
3. The model must suit the intended user—the recipient of the message.

8.4 MODELING TECHNIQUES FOR AN AMS

It has been conjectured that a formal mathematical approach would be useful for the specification and implementation of complex AMS control systems. Differential or difference equation-based models provide successful control engineering systems with closed-form specifications and solutions for the state at all times. To use these mathematically convenient models, the system must be a continuous-state system and must be "time driven" in its state transition mechanism. In contrast, models for the discrete-event problems of an AMS are not agreed upon. So far, no discrete-event models have been provided to permit closed-form solutions similar to those available for continuous- and discrete-time systems.

8.4.1 Discrete-Event Simulation

Discrete-event simulation has been recognized as a useful tool for analyzing a variety of manufacturing system design problems.[10] A simulation approach depends on the availability of appropriate simulation software packages and computing power. Usually, the modeling language used in simulation software contains structures and constructs that best serve high-level system modeling. Using such a simulation tool may introduce significant model uncertainties with respect to lower-level control specifications (e.g., concurrence and synchronization). This problem limits the control system designer to address control issues such as control function distribution and feedback requirements. Without correct and detailed control information (e.g., control structure and variables), designing controllers becomes very difficult.

8.4.2 Queuing Networks and Markov Chains

For the modeling and performance analysis of an AMS, queuing networks (QNs) and Markov chains (MCs) are among the most widely used techniques. As performance modeling tools, MCs are similar to QNs.[1] Both modeling techniques are mathematical approaches that deal with only a limited range of problems. Queuing network models have inherent shortcomings in their lack of descriptive power in the presence of phenomena such as synchronization and blocking. Also, the distributed features of a manufacturing process destroy the product form characteristics of the overall system specifications. Thus, even a simple QN model must be translated into its corresponding MC model to generate a solution.[9] However, deriving MC models requires the ability to account for all the possible states of the system. For a practical-size problem, MCs often suffer from an exponential state space explosion. This increase in size and complexity makes it difficult to construct and analyze the MC models. In addition, correcting or modifying these models is not an easy task.

8.4.3 Finite Automaton

The theory of automata[2] provides one of the formal ways to model the logical behavior of a discrete AMS. An automaton is a model that is capable of generating a sequence of events according to well-defined rules. Figure 8.5 shows the graphical representation of a finite automaton. This is simply a directed graph consisting of circles and arrows. Circles denote states and arrows denote events. In this automaton, a discrete system is assumed to be affected by four different events described by an event set $E = \{e_1, e_2, e_3, e_4\}$. The model is also endowed with a state set $S = \{s_0, s_1, s_2, s_3\}$, where $s_0 \in S$ is designated as the initial state. The automaton obeys a set of rules of the form: If the state is s and an event e or a sequence of events is observed, then the state must change to the next state s'.

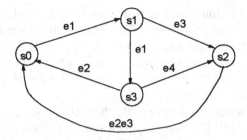

Figure 8.5 State transition diagram of a finite-state automaton.

An attractive feature of a finite-state automaton is its *decidability* to event sequences and states. However, as the system becomes more complex, the modeling with a finite-state automaton becomes more and more difficult because the state of the total system must be known. Meanwhile, each finite-state automaton represents a fixed number of states and must be modified every time the state information changes. In addition, the combination of multiple systems or asynchronous processes rapidly increases the complexity of the automaton and hides some of the intuitive structures involved in this combination.

8.4.4 Petri Net Basics

An alternative to automata is provided by Petri nets (PNs).[2, 6, 11] A Petri net, described by a Petri net graph as shown in Figure 8.6, consists of *transitions, places, arcs,* and *tokens.* Generally, transitions represent the events driving a discrete-event system and are normally denoted by the set $T = \{t_j\}$ for $j = 1, 2, \ldots, m$. Places describe the conditions related to the occurrence of events and are normally denoted by the set $P = \{p_i\}$ for $i = 1, 2, \ldots, n$. The *input* places to a transition t_j, denoted by the set $I(t_j)$, represent the associated conditions required for this transition to occur. The *output* places of a transition t_j, denoted by the set $O(t_j)$, represent the associated conditions that are affected by the occurrence of this transition. A typical arc is of the form (p_i, t_j) or (t_j, p_i), which graphically represents the connection of the input place p_i to

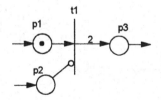

Figure 8.6 Petri net graph.

transition t_j or the output place p_i to transition t_j. Multiple arcs are allowed to connect place p_i and transition t_j. Or, equivalently, a weight, $w(p_i, t_j) = k$ or $w(t_j, p_i) = k$, can be assigned to a single arc representing k numbers of arcs. For example, a weight, $w(t_1, p_3) = 2$, is assigned to the arc that connects t_1 and p_3 in Figure 8.6. If no weight is shown on an arc of a Petri net graph, it is assumed to have weight 1.

A *token* is essentially used to indicate the fact that the condition described by a place is satisfied. The way in which tokens are assigned to a Petri net defines a marking m. The *state* of a Petri net is its marking $m = [m(p_1), m(p_2), . . ., m(p_n)]$, where n is the number of places in the Petri net and the ith entry of the marking, $m(p_i) \in \{0, 1, 2, . . .\}$, is the (nonnegative integer) number of tokens in place p_i. A Petri net changes its states by moving tokens through the net as transitions become enabled and fired. To enable and fire a transition, a token must be present in each input place of the transition. Considering the case of arc weight, enabling and firing transition t_j requires that the number of tokens in its input place p_i must be at least as large as the weight of the arc connecting p_i to t_j. After the firing of the enabled transition t_j, tokens are removed from its input places and placed in its output places. The number of tokens placed in the output place p_i is determined by the weight of the arc connecting t_j to p_i. For example, in Figure 8.6, with the weight, $w(t_1, p_3) = 2$, there are two tokens placed in p_3 after t_1 is fired. An inhibitor arc is a zero-weighted arc, with which the transition is enabled when the input place has no token. In Figure 8.6, an inhibitor arc connects p_2 and t_1. An "enabled transition" in a Petri net is equivalent to the idea of a "feasible event" used in state automata. Unlike state automata, a Petri net includes explicit conditions under which an event can be enabled. Using this feature of a Petri net, one may represent a very general discrete-event system whose operation depends on complex control schemes.[1, 14] For example, a sequential Petri net, as shown in Figure 8.7a, represents a series of successive operations. Transition t_2 can be fired only after the firing of t_1. This imposes the precedence constraint "t_2 after t_1." Similarly, a parallel Petri net, as shown in Figure 8.7b, represents the concurrency and synchronization of activities. The firing of t_1 leads to the conditions that enable transitions t_2 and t_4 concurrently. Transition t_5 captures the synchronization of activities represented by transitions t_3 and

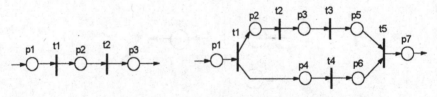

Figure 8.7   (a) Sequential PN and (b) parallel PN.

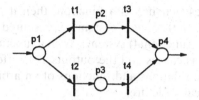

Figure 8.8 Choice Petri net.

t_4. A choice Petri, as shown in Figure 8.8, represents the conflict and merging formed by successive processes. A choice must be made at place p_1 for firing either transition t_1 or t_2. Merging occurs at place p_4 with the tokens arriving into p_4 from both transition t_3 and t_4.

Compared to finite automata, a Petri net model is more easily expended. For example, the addition of a component to the system only affects a Petri net model locally, which is achieved by simply adding a few places and/or transitions that represent the coupling effects between the component and the original system. Moreover, by looking at a Petri net graph, one can conveniently see the individual components, discern the levels of their interactions, and, ultimately, decompose a system into logical distinct modules. Therefore, a Petri net has the ability to *decompose* or *modularize* a complex system. In addition, it has also been proven that a finite-state automaton can always be represented by a Petri net. This supports the claim that Petri nets are indeed very general modeling tools. However, it is not always easy when one deals with the event sequences and states of a Petri net model. This reflects a natural trade-off between the decidability and the richness of a Petri net model.

8.5   SYSTEM THEORY APPROACH TO AMS MODELS AND MODELING PROCESS

According to system theory, a general system S may be defined as an abstract relation of the Cartesian products of set X and Y:

$$S \in X \times Y \tag{8.1}$$

As a relation defined in (8.1), the component set of a system S is referred to as its objects: X contains the input objects; Y contains the output objects. The representation of a system as a relation is, therefore, an input/output representation.

To study the behavior and properties of a given system, specific kinds of auxiliary functions are defined in the form:

$$S: X \to Y \tag{8.2}$$

Equation (8.2) implies that if S is a function, then it may be defined as a mapping from X to Y. In this sense, systems represented by (8.2) are referred to as function-type (or functional) systems. With a function, the system inputs X can be considered as causes and the outputs Y as effects.

More precisely, the relations and functions of an arbitrary system are defined as a mathematical structure:

$$S = (E, R_E) \qquad (8.3)$$

where $E = \{e_1, e_2, . . ., e_n\}$ is a set of elements that may be finite or infinite, and $R_E = \{r_{E1}, r_{E2}, . . ., r_{Em}\}$ is a set of relations on the element of E.

The system definition expression in (8.3) is general and makes clear the essential distinction between a system and a set. Without the system forming relation set R_E, the elements of set E would remain unrelated. In a real-life system, elements of E can be related to each other in a variety of ways. R_E carries the responsibility of providing the set of "system constructor" relations, which connect the elements of E into the coherent entering of the system as a unity.

The system theory approach provides system designers with organizing principles for coping with system complexity. First, any system may be viewed as an indivisible and enclosed whole. It is characterized accordingly as a "black-box" with attributes and properties. The way in which the referent system appears to an outside observer at its boundary is a behavioral model. Second, any system may be regarded as a divisible composition. A structural model describes the internal organization of the referent system by showing the parts of its composition. It also allows the deduction of the referent system's black-box model from the system structure. Although the system structural models can be irrelevant from the viewpoint of end users, they are essential for designers, manufacturers, or problem solvers who are concerned with the development, implementation, and serviceability of a system.

With the system theory approach, modeling becomes a process that forms a relationship between the referent system S and the model M. Both the referent system and the model are definable as some forms of system definitions or functional expressions. For example, the referent system S may be a tangible entity such as a single object or a finite or infinite collection. It may also be a natural phenomenon or an abstraction such as a concept, a theory, or a computer program. The model M is always a specification, preserving specifically chosen features of the referent system S. In practical situations, the specification is achieved by selecting relevant features of a system and omitting irrelevant attributes of the system. Often the model M is not a direct product of a referent system as an object, but a product of some earlier modeling process. In such cases, the referent system itself is a model and the modeling process becomes a "chain," each new link of which creates a new model of the previous model.

The formulation of a four-stage modeling process is then illustrated as follows:

Stage 1. Formalize the relationship between the referent system and the model. This is based on the assumption that the referent system S belongs to the real world and the system model is a part of an artificial domain. The relationships of these two can be represented in two ways: either by the modeling process that creates M from S by abstraction or by the refinement process of moving from the general to the more specific.

Stage 2. Model the internal organization of each system. The referent system S will have a network of elaborate interactions with the numerous entities in its world. As a process of modeling, such interactions are simplified to define only the referent system and its immediate environment.

Stage 3. Focus on the referent system and regard it as a unity (an atomic entity). The referent system is then modeled as the system of related properties as defined by a designer.

Stage 4. Examine the referent system as a composition of parts. These parts represent the way in which it is constructed from other systems.

Abstractions represent an important technique used to manage system complexity. Procedural abstractions represent a form of abstraction used extensively by requirement analysts, designers, and programmers. It is often characterized as the "function or subfunction" analysis of a system. The functional analysis method offers a means of considering essential functions and the level at which the problem under consideration is to be addressed. For example, the essential functions are often those that the system must satisfy, regardless of the physical components required. The problem level is frequently described by establishing a "boundary" around a coherent subset of functions. The functional analysis method may be formulated as a five-stage procedure:

Stage 1. Express the overall function of a design in terms of the conversion of inputs and outputs. A simple black-box model may be used to represent all the functions that are necessary for converting the inputs to the outputs. In this stage, the system boundary is used to define the function of the product. It is important to ensure that all the relevant inputs and outputs are identified. They can usually be classified as flows of either materials, energy, or information.

Stage 2. Break down the overall function into a set of essential subfunctions. Usually, this includes the conversion of the set of inputs into a set of outputs. This conversion is a complex task inside the black-box model, requiring that the task be broken down into subtasks or subfunctions. In specifying subfunctions, it is helpful to ensure that they are all expressed in the same form. Each subfunction should be a statement including a verb and a noun, for example, "amplify signal," "move part," and so forth. Each subfunction has its own input(s) and output(s), and the compatibility between these should be verified.

Stage 3. Draw a block diagram showing the interactions between the subfunctions. A block diagram consists of all the separate subfunctions identified by enclosing them in graphic boxes. These boxes are linked together by their inputs and outputs so as to accurately describe the overall function of the product or system that is being designed.

Stage 4. Draw a system boundary. The system boundary defines the functional limits of the system. The functional limits will be a function of client requirements.

Stage 5. Search for appropriate components for performing the subfunctions and the interactions between the subfunctions. Only when the subfunctions have been defined and divided at an appropriate level will it be possible to identify a suitable component for each subfunction. The identification of the specific components will depend on the nature of the product or system that is being designed.

8.6 OBJECT-ORIENTED REFERENCE MODEL

Several reasons exist for using object-oriented paradigms[8, 13] to achieve an integrated representation or modeling for AMS controls. First, in engineering terms, an object is a component of a system. More specifically,

- An object is a real-world identifiable item to which attention can be directed and to which characteristics can be ascribed.
- An object is a component of a system that performs some individual function as its contribution toward the function of the system as a whole.
- An object has an unambiguous boundary. It is clearly understood what is internal and what is external to the object.
- An object has an internal design structure that is independent of everything external to the object.

Second, the operations of an object are the characteristics that represent how the object reacts to actions directed toward it. Many different terms are used to refer to the operations of the objects. They can be called functions, operations, methods, responsibilities, or service. However, it is generally agreed that they all represent responses, either directly or indirectly, to the actions directed at one object by other objects. The process requirements of an object are represented by the list of operations that are assigned to the object. Each operation represents one process requirement, and the sum of the process requirements represents the requirements of a system. The attributes of an object are the set of properties to which values can be assigned to describe the object and thus to establish its identity. In practical terms, attributes are data elements that can be accessed in an object.

Third, objects are packages of operations and attributes. Objects are identified by the names of items in the domain and by the groups of operations and attributes of which they are composed. Operations and attributes are identified by the description of the actions to be taken (process operations) and the information needed to carry out the action (data attributes).

Finally, identifying objects, operations, and attributes is often recursive. Objects suggest operations and attributes. Operations suggest other attributes. Attributes suggest other operations. Together, these new attributes and operations suggest other objects. Each object represents one requirement of the system.

An integrated perspective of AMS control to its processes and resources' is formulated from the notion of an event class.[4] An event class in an AMS is a block-box model, as shown in Figure 8.9. It consists of a material object set $M = \{m_i\}$ for $i = 1, 2,. . .,n$, an equipment object set $E = \{e_j\}$ for $j = 1, 2,. . .,k$, and a controller that controls the equipment objects to perform operation q on the material objects. The input/output relationships of an event class in real time can be expressed as

$$Q_{out}(t + \delta t) = q \ (Q_{in} \ (t)) \tag{8.4}$$

where $Q_{in}(t) = \{E_{in}(t), M_{in}(t)\}$ is the input set of the event class at time t, and $Q_{out}(t + \delta t) = \{E_{out}(t + \delta t), M_{out}(t + \delta t)\}$ is the output set of the event class at time $t + \delta t$.

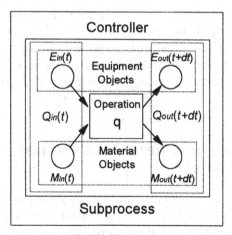

EVENT CLASS

Figure 8.9 Object-oriented reference model.

TABLE 8.1   Work Cell Event Classes in Figure 8.1

	AS/RS Event Class	Robot (1) Event Class	Robot 2 Event Class	Inspection Event Class
Equipment E	AS/RS, solenoids	Robot (1), sensor, solenoids	Robot (1), sensor, solenoids	Vision, sensor, solenoids
Materials M	Pallets, parts	Pallets, tools, parts	Pallets, tools, parts	Pallets, parts
Operations q	Retrieve and store pallet	Pick and place parts	Pick and place parts	Inspect parts and pallet

Using the event-class-based decomposition, the operations or tasks of overall production can then be logically grouped as autonomous subprocesses controlled by event class controllers. For example, the operations of the work cell in Figure 8.1 can be decomposed into four event classes. They are the AS/RS event class, the robot (1) event class, the robot (2) event class, and the inspection event class. Table 8.1 lists the equipment and material objects and the key operations of each event class.

As far as control units are concerned, the function or operation q needs to be decomposed only if the outputs of q are not solutions of the associated decision problems. In general, as a complement of functional analysis, the function specification of the operation q in an event class might be given in the following ways:

- Define q procedurally, as a controlled sequence, such that any valid instance of $Q_{in}(t)$ should yield a valid instance $Q_{out}(t + \delta t)$.
- Give an informed or semiinformed textual description of q, complemented by drawings, graphs, and so on.
- Define q operationally, by constructing a simulator or prototype artifact to serve as a dynamic model of q, and generate any model instance of $Q_{out}(t + \delta t)$ in response to any valid instance of $Q_{in}(t)$.

8.7   INTEGRATED DESCRIPTIONS OF CONTROL SYSTEM COMPONENTS

In reality, truly complex control systems can almost evade completed and detailed design in a single step. Often a control solution specification is first created from a problem. The design brief is described through high-level languages such as flowcharts, reflecting the requirements of all users and undertaking to achieve a convergence of viewpoints. The specifications written in such a high-level language clearly identify the structure and the operation modes of the system for the successful development of a graphical system model. For example, the rectangular boxes of the flowchart in Figure 8.10 describe the basic operation modes of the work cell in Figure 8.1. The diamond boxes describe the conditions that cause the shift of the operation modes.

There often exists the dilemma between the abstraction of the descriptions in higher levels and the need to take into account a system's numerous behavioral (i.e., input/output) aspects in lower levels. A resolution of this dilemma is sought in an integrated family of models, each of which is concerned with the behavior of the system as viewed from a different level of abstraction.

As discussed in Section 8.6, formulating the control specifications of an AMS may start with the black-box descriptions of event classes. The modeling language used in this design stage should provide a set of concepts and notions that allow the designer to define the interface and the associated input/output properties of a system or a component. Refining the black-box model into a transparent-box model becomes the next phase of the control design, which decomposes an event class into lower-level components and defines their interactions in terms of both data and control rules. Conducting this refinement process requires two additional requirements to the modeling language:

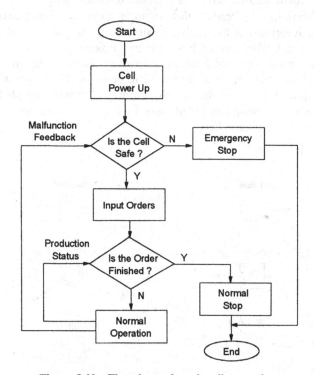

Figure 8.10 Flowchart of work cell operation.

· The notation encompasses both the representations of the specifications and the representations of the hierarchy among subspecifications.

· The notation used to represent the hierarchy can be mapped directly or indirectly into structural constructs of the target programming language.

One modeling tool that may meet these requirements is extended Petri nets (EPNs).[6, 12] Extended Petri nets were developed from the original Petri nets to include extended classes of places, arcs, and tokens. These extended constructs greatly enhance the ability of Petri nets to model the flow of control, resources, and information in complex systems such as an AMS. For example, Figure 8.11 shows some extended place constructs commonly used in EPNs for aiding the specification of conditions or procedures that may arise in a system. A *status place* is equivalent to a place in an original Petri net. The procedure of a status place is the enable check for the associated transitions. A *simple place* has a simple procedure associated with it, in addition to the transition-enable check. An *action place* is used to represent procedures that take a long time to be executed. Usually, these procedures are spawned off as subprocesses that are executed externally in parallel with the Petri net–based model, for example, on other control computers or by other equipment. A *subnet place* denotes a subnet that represents a system abstraction, a subsystem area, and so forth. In this way, it is possible to model a complex system in a hierarchical fashion. A *decision place* denotes a decision being taken. It may use external information to resolve a conflict. A special case of a decision place is the *switch place,* which has a binary choice.

Using EPN constructs and Petri net stepwise refinement technologies,[6] the behavior and structure of an event class can be refined into a graphical EPN model at required detail levels. The design may start with the abstraction of important processes within an event class. Figure 8.12 shows one of the initial EPN models for an event class.

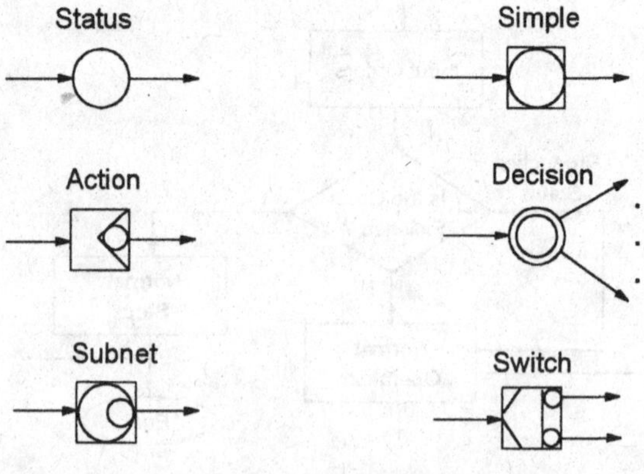

Figure 8.11   Constructs commonly used in EPN design.

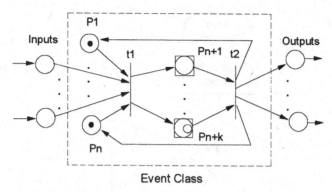

Event Class

Figure 8.12   Initial EPN model for an event class.

In this model, status places $p_1, p_2, \ldots, p_n$ are the *resource places,* which represent the conditions of the resources such as the operating equipment and the operated materials of an event class. Some specified resource places are the interface input and output places of the event class indicating the flow of external resources between event classes. Other resource places are the internal resources of an event class. The initial tokens in these places represent the initial availability of the resources. Action places or subnet places p_{n+1}, $p_{n+2}, \ldots, p_{n+k}$ are k *operation places* representing k concurrently working modules, and they are initially not marked. Two transitions in the event class model represent the start and the end of the working of the event class. The firing of the start transition t_1 lets the operation places occupy both the internal and the external resources for a process. The end transition t_2 is to be enabled and fired as the event class finishes all the working modules. Consequently, the internal resources return to their original places and the external resources flow to other event classes. Each operation place in an event class model can be further refined by some special Petri net module as introduced in Section 8.4.4, describing a specific type of operation in an event class.

As an example of using the EPN approach, Figure 8.14 shows an EPN model that describes the operations of the AS/RS event class in Figure 8.13. A token initially at place p_1 represents the single AS/RS resource shared by both the retrieving and the storage processes. This situation is also called parallel mutual exclusion. As the AS/RS starts an application program, the firing of transition t_1 moves the token from place p_1 to p_2, indicating that the AS/RS positions the solenoids for holding a coming pallet. Then a decision has to be made at p_3 regarding the execution of either "retrieving" or "storage" subroutines according to a predefined priority. For example, the work cell may give "storage" a higher priority than "retrieving." This means that the AS/RS will start the "storage" subroutine as long as a pallet is conveyed to the AS/RS workstation from the previous station (i.e., a token flows into p_5).

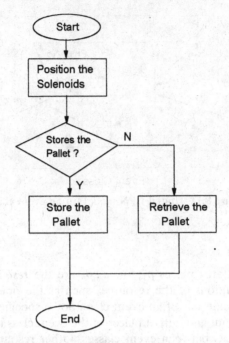

Figure 8.13 Flowchart of the AS/RS event class in normal operation mode.

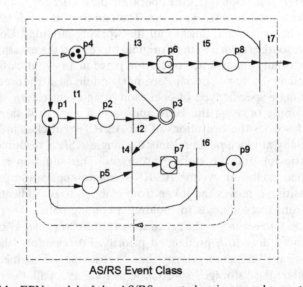

AS/RS Event Class

Figure 8.14 EPN model of the AS/RS event class in normal operation mode.

TABLE 8.2 Description of the Places and Transitions of the AS/RS Event Class in Figure 8.14

Places		Transitions	
p_1	AS/RS is ready	t_1	Start program
p_2	Position the solenoids	t_2	Solenoids in position
p_3	Decide which subroutine is to be executed	t_3	Start "retrieving" subroutine
p_4	Pallet is ready to be retrieved	t_4	Start "storage" subroutine
p_5	Pallet is ready to be stored	t_5	End of "retrieving" subroutine
p_6	Execute "retrieving" subroutine	t_6	End of "storage" subroutine
p_7	Execute "storage" subroutine	t_7	Start to release the retrieved pallet
p_8	Ready to release the retrieved pallet		
p_9	Stored pallet is ready		

Otherwise, it is ready to execute the "retrieving" subroutine. Table 8.2 gives more detailed descriptions of the places and transitions in Figure 8.14.

Similarly, the flowchart in Figure 8.15 describes the required normal operations of the robot event class. A more detailed EPN description is shown in Figure 8.16. A token at place p_1 represents that the robot is ready to run an application program. The firing of transition t_1 indicates the start of the program. The decision place p_2 represents the selection of operation subroutines according to the production order. If the robot chooses to execute the "pick/place" subroutine denoted by subnet place p_6, it has to position the solenoids (p_3) for holding the coming pallet and fix the pallet detected by the sensor (i.e., a token is present at p_4). In the case of no request for conducting an operation to the coming pallet, the robot executes the "bypass" subroutine (p_5), which deenergizes the solenoids to bypass the pallet through the station. Table 8.3 gives the description of the places and transitions in Figure 8.16.

8.8 MULTILEVEL APPROACH FOR AMS CONTROL STRUCTURE

An AMS may consist of multiple event classes, each of which has a recognized explicitness. Conflicting goals often exist among event classes. For example, the operation interactions between subprocesses and the fact that each piece of equipment in one event class is unaware of the actions taken by other event classes may generate intraorganizational conflict. This conflict appears not only as the result of the composition of the system, but also as a necessity for the efficient operation of the overall system. In addition, there is always a need for the event class controller to act without delay, while there is an equal need for the event class controller to understand the situation better before

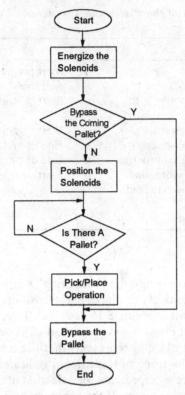

Figure 8.15 Flowchart of the robot event class in normal operation mode.

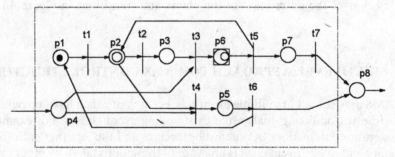

Robot Event Class in Normal Operation Mode

Figure 8.16 EPN model of the robot event class in normal operation mode.

TABLE 8.3 Description of the Places and Transitions of the Robot Event Class in Figure 8.16

Places		Transitions	
p_1	Robot ready to run program	t_1	Start the program
p_2	Execute either "pick/place" subroutine or "bypass" subroutine	t_2	Start to position the solenoids
p_3	Position solenoids	t_3	Start the "pick/place" routine
p_4	Pallet is ready at the station	t_4	Start the "bypass" routine
p_5	Execute "bypass" subroutine	t_5	End of the "pick/place" routine
p_6	Execute "pick/place" subroutine	t_6	End of "bypass" routine
p_7	Release the pallet	t_7	Pallet has been released
p_8	Pallet is ready at the next station		

taking any action. One of the resolutions for solving the conflicting goals is to arrange the decision units in a multilevel, multigoal structure. In the design of a multilevel, multigoal system, the higher-level decision units condition, but do not completely control, the goal-seeking activities of the lower-level decision units. Providing such a freedom of actions at lower levels makes the economical use of the resources available for decision making.

The hierarchy approach for AMS controls[5, 7, 15] is one example of multilevel system design. In this design, the complex tasks of planning and controlling manufacturing processes are distributed among various levels in an overall functional hierarchy. This includes the organization level, the coordination level, and the execution level, as shown in Figure 8.17. Each level is characterized by the length of the planning horizon and the type of data required for the decision-making process. Higher levels of the hierarchy typically have

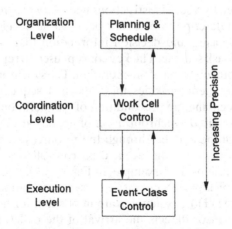

Figure 8.17 Multilevel approach for AMS controls.

long horizons and use highly aggregated data, whereas lower levels have shorter horizons and use more detailed information. The function of each level is closely related to its corresponding time scale. For example, the time scales could range from years and months at the organization level to minutes and seconds at the coordination level and execution level. Harmony between these goals can be formalized as a logical proposition. The overall goal set at the organization level has to be realized through the action of the decision units at the execution level because the decision units at the execution level are the only subsystems in direct contact with the overall process. This also requires that the decision units at the execution level must be coordinated relative to the given overall decision problem. Indeed, there are many ways of decomposing a given overall decision problem into subproblems. The real design question, however, is how to coordinate decomposed subproblems; this requires the modification of the subproblems, the selection of coordination principles, and the design of methods to define coordination inputs.

8.8.1 Decomposition at the Execution Level

The execution level of an AMS can be further decomposed into two lower levels: the Euclidean level and the logical level. The subproblems at the Euclidean level are characterized by the locations, orientations, and movements of the objects in continuous-time processes. The control issues are usually local ones with continuous-time data that are often expressed in coordinates, transformations, kinematics constraint equations, differential equations, and so on. The logic level is an abstraction of the Euclidean level and is mainly concerned with logical conditions and events that appeared in the discrete-event processes.

The main objective of an event class controller at the execution level is to execute the actuation defined in the Euclidean level according to the conditions abstracted at the logical level. Thus, an equipment control program becomes a task plan described by a set of instructions necessary to achieve some defined actions. A detailed description of a task plan can be developed by explicitly including the sequencing and decision information that is "hidden" within the behavior of the task scheme. The key concept used to represent sequencing information is the precondition of an operation. The synchronization of actions under a cell environment provides the important sequencing and decision information for specifying the precondition of an operation. Generally, there are two kinds of required synchronization of actions at the execution level.

The first kind can be defined through the resources within the domain of an event class. In this case, the event class controller is able to make the synchronization decisions. For example, in Figure 8.15, before the robot controller starts the "pick/place" operation, a synchronization of actions is required to make sure: (1) the solenoids are in position for holding the coming pallet and (2) the sensor detects the arrival of the pallet. Because both the solenoids and the sensor at the robot workstation are the equipment resources within the robot event class, the controller of the robot event class has full authority to make such a decision. In Figure 8.16, this synchronization of actions is easily modeled by the transition t_3 and its two input places p_3 and p_4.

The second kind of required synchronization of actions may require the use of resources outside the control domain of a single event class. For example, one required synchronization of actions for the AS/RS event class in Figure 8.14 concerns when to release a retrieved pallet to the next robot event class. Obviously, this synchronization of actions relates to the processes of the robot event class in the work cell. In this case, the AS/RS event class controller cannot make its own decision without the assistance of the controller at the coordination level.

8.8.2 Cell Coordination Principles

At the coordination level, coordination is the task of the cell controller, whereby it attempts to cause a harmonious functioning of the work cell. To make correct synchronization decisions for all event classes, the cell controller must be able to access both the local operational state and the global operational requirements. To this end, the functional relationship of message passing between the cell controller and the event class controllers must be established. Actually, the design of the cell control system could be defined as a structure problem. A hierarchical cell control structure is shown in Figure 8.18.

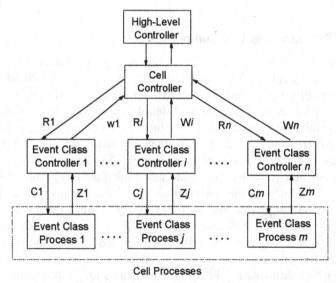

Figure 8.18 Message-passing requirements in the cell.

8.8.3   Message-Passing Requirements in the Work Cell

The data signal interactions shown in Figure 8.18 can be classified into two types. One type is the downward transmission of the command signals. Signals from the event class controller to its operation process are termed operational control inputs C_i, whereas signals from the cell controller to each event class controller are called cell coordination inputs R_i. The other type of vertical signal interaction is an upward transmission of the feedback signals. This includes the feedback Z_i from a subprocess to the event class controller and the feedback W_i from the event class controller to the cell controller. As part of the communication interface specification, message passing in the cell may relate to the communication capabilities of the hardware and software components used to implement the controllers.

8.8.4   Performance Balance Control Criterion

Considering both the operation processes and the control requirements for an AMS cell, a closed-loop control criterion, called the performance balance principle, is feasible and effective for the AMS cell operation realization. This criterion states that the operational control inputs C_i from an event class controller to the overall cell process q solve the overall cell control problem when (1) the event class controller has a solution to its own control problem and (2) the operation performance of process q in real time balances with the desired operation requirements of the cell in real time. This balance occurs when the control inputs C_i are applied to the process q.

8.8.5   EPN Conceptual Cell Control Model

In the case of EPN modeling, a general representation of the control requirements in a work cell is shown in Figure 8.19. The event class model models the relationships of the required resources, processes, and operations within an event class at a required level of detail. The input/output places of the materials establish the operational interfaces between event classes. The feedback signals from the event class controller provide information about the operational status of the controlled equipment. Sending a feedback signal to the cell controller requires the event class controller to initialize the action; thus, this is modeled as the output place P_f of some transition in the event class. All the output places from the ith event class specify the feedback signals W_i in Figure 18.8. The cell controller consists of control modules, each of which specifies a sequence of control actions required at the coordination level. The progression of the sequence generates the cell coordination input place P_c, which controls the firing of some transition in the event class. All the control input places to the ith event class specify the command input signals R_i in Figure 8.18.

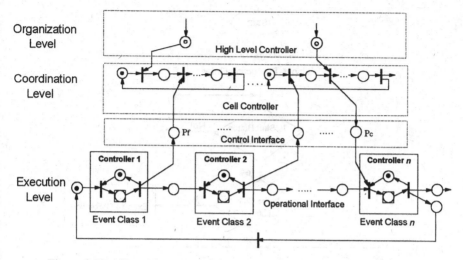

Figure 8.19   General representation of cell level control requirement.

8.9 CASE STUDY: DESIGN OF AUTOMATED WORK CELL CONTROLLERS

A case study is presented to illustrate the design of the control system for the work cell in Figure 8.1 using the developed method. To simplify the illustration, only the AS/RS event class and the robot (1) event class of the cell are used to conduct the operation. The design starts with the overall cell control specification described by the flowchart in Figure 8.20. This is a simplified cell control specification, which does not take into account the malfunction check and emergency stop functions in Figure 8.10.

8.9.1 Developing EPN Models for Event Classes

Based on the EPN models of both the AS/RS event class in Figure 8.14 and the robot event class in Figure 8.16, Figure 8.21 shows the newly developed event class models. This is done by adding calibration procedures and the required feedback signals and cell coordination inputs. A description of the places and transitions is given in Tables 8.4 and Table 8.5.

8.9.2 Developing Cell Control Model

Based on the flowchart in Figure 8.20 and the control requirements specified for each event class in Figure 8.21, the functions of the cell controller are specified as an EPN model in Figure 8.22. A description of the places and

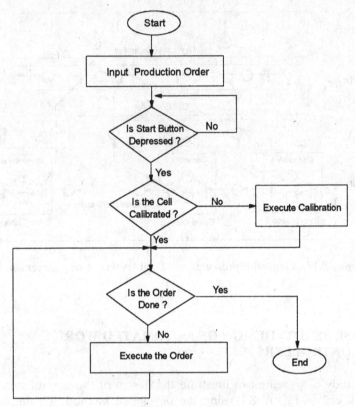

Figure 8.20   Flowchart of the overall cell control function.

transitions of the cell control model is given in Table 8.6. There are four modules specified to achieve the overall cell control specification.

The "Order & Start" module is designed to make sure that the cell cannot start the operation without the availability of an input production order. This is simply achieved in the model through place p_2, which connects transition t_1 and transition t_2 sequentially.

The "Calibration & Start" module is designed to regulate the execution of the calibration/start procedures required by the event classes. This is modeled as the choice Petri nets in Figure 8.22. For example, in the case of the AS/RS event class, place p_6 has two output arcs. This indicates that the token at p_6 may be used by either of two paths according to the availability of feedback Pf_1 from the AS/RS event class controller. If the AS/RS has not calibrated, no Pf_1 signal has been sent from the AS/RS controller as shown in Figure 8.21; thus, no token appears at place Pf_1. In this case, transition t_6 is enabled and fired, moving the token from its input place p_6 into its output place Pc_1. Then the token at place Pc_1 will be used as the cell coordination

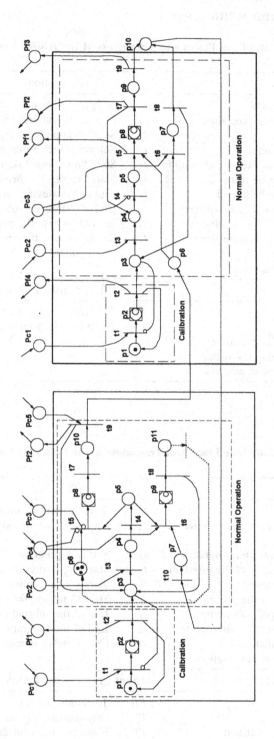

AS/RS EVENT CLASS

ROBOT (1) EVENT CLASS

Figure 8.21 Newly developed event class models.

TABLE 8.4 Description of the Places and Transitions of the AS/RS Event Class in Figure 8.21

Places		Transitions	
p_1	AS/RS is powered up	t_1	Start the calibration of AS/RS
p_2	Execute calibration procedure	t_2	End of the calibration
p_3	AS/RS is ready	t_3	Start an application program
p_4	Position the solenoids	t_4	Solenoids are in the position
p_5	AS/RS is ready to execute subroutine	t_5	Start "retrieving" subroutine
p_6	Pallet is ready to be retrieved	t_6	Start "storage" subroutine
p_7	Pallet is ready to be stored	t_7	End of "retrieving" subroutine
p_8	Execute "retrieving" subroutine	t_8	End of "storage" subroutine
p_9	Execute "storage" subroutine	t_9	Start to release the pallet
p_{10}	AS/RS is ready to release the retrieved pallet	t_{10}	Sensor detects the presence of the pallet
p_{11}	Pallet has been stored		

Cell Control Input Places		Event Class Feedback Output Places	
Pc_1	Calibration	Pf_1	Calibration is finished
Pc_2	Cycle start	Pf_2	Pallet has been released
Pc_3	Input order has been finished		
Pc_4	Request for storing the pallet		
Pc_5	Release the pallet to next station		

TABLE 8.5 Description of the Places and Transitions of the Robot (1) Event Class in Figure 8.21

Places		Transitions	
p_1	Robot is powered up	t_1	Start the calibration of robot
p_2	Execute calibrating and homing procedure	t_2	End of the calibration
p_3	Robot is ready to run program	t_3	Start the program
p_4	Start the execution of the program	t_4	Start to position the solenoids
p_5	Position the solenoids	t_5	Start the subroutine
p_6	Pallet is ready at the station	t_6	Start to release the pallet
p_7	Execute "bypass" subroutine	t_7	End of the subroutine
p_8	Execute "pick/place" subroutine	t_8	End of the program
p_9	Release the pallet	t_9	End of pallet release
p_{10}	Pallet is ready at next station		

Cell Control Input Places		Event Class Feedback Output Places	
Pc_1	Calibration	Pf_1	Solenoids are in the position
Pc_2	Cycle start	Pf_2	One request is done
Pc_3	Order has been finished	Pf_3	Pallet has been released
		Pf_4	Calibration is done

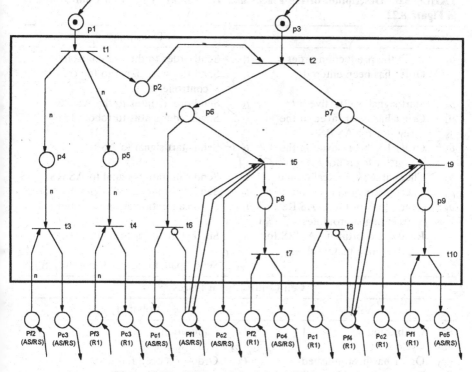

Figure 8.22 EPN model of the cell controller.

input to the AS/RS controller to trigger its calibration. If the AS/RS has finished the calibration, a token has been sent from the AS/RS controller to the cell controller through place Pf_1. In this case, t_5 is enabled and fired. The token moved from place p_6 to place Pc_2 is then used as the start signal that starts the operation of the AS/RS event class. The token that is moved back to place Pf_1 after the firing of t_5 maintains the "start" status of the AS/RS event class. A similar choice Petri net is used for regulating the "Calibration & Start" case of the robot (1) event class.

Once all the event classes have been started, the cell controller coordinates them to conduct a normal operation cycle. The cell control module called "Pallet Release" is designed for two kinds of cell coordination. Given that the "storage" operation in the AS/RS event class has a higher priority, then one of the cell coordinations is to coordinate the AS/RS event class controller to resolve the decision conflict between "retrieving" and "storage." To make this coordination correctly, the cell controller needs to know when the robot (1) event class is ready to release a pallet to the AS/RS event class. In the cell control model, this is modeled by place p_8, transition t_7, feedback place

TABLE 8.6 Description of the Places and Transitions of the Cell Control Model in Figure 8.22

Places		Transitions	
p_1	Input the production order	t_1	Send order to the cell controller
p_2	Order has been entered	t_2	Send the start signal to the cell controller
p_3	Start signal is effective	t_3	Send order status to the AS/RS
p_4	Order has been saved in the register for AS/RS	t_4	Send order status to robot (1)
p_5	Order has been saved in the register for robot (1)	t_5	Send start signal to AS/RS
p_6	Check if AS/RS is calibrated	t_6	Send calibration signal to AS/RS
p_7	Check if robot (1) is calibrated	t_7	Send "storing pallet" signal to AS/RS
p_8	Ready to coordinate AS/RS for executing subroutines	t_8	Send calibration signal to robot (1)
p_9	Ready to coordinate AS/RS for performing pallet release	t_9	Send start signal to robot (1)
		t_{10}	Send "pallet release" signal to AS/RS

Cell Control Input Places			
AS/RS		Robot (1)	
Pc_1	Calibration	Pc_1	Calibration
Pc_2	Cycle start	Pc_2	Cycle start
Pc_3	Order has been finished	Pc_3	Order has been finished
Pc_4	Request for storing pallet		
Pc_5	Release pallet to next station		

Event Class Feedback Output Places			
AS/RS		Robot (1)	
Pf_1	Calibration is finished	Pf_1	Solenoids are in the position
Pf_2	Pallet has been released	Pf_2	One cycle is done
		Pf_3	Pallet has been released
		Pf_4	Calibration is done

Pf_2 from the robot (1) event class, and cell input place Pc_4 to the AS/RS event class. Another cell coordination to the AS/RS event class controller concerns when to release a retrieved pallet from the AS/RS event class to the robot (1) event class. This depends on when the robot (1) event class controller is ready to accept an incoming pallet. In the cell control model, this is modeled by place p_9, transition t_{10}, feedback place Pf_1 from the robot (1) event class, and cell control input place Pc_5 to the AS/RS event class.

The fourth cell control module is to coordinate the event classes to perform the "Normal Stop" procedure after they have finished an input order. This is modeled by using weighted arcs. In Figure 8.22, after t_1 fires, n tokens flow into places p_4 and p_5, respectively, representing the n number of operation cycles required by each event class. Once the event classes have finished one cycle, cycle feedback signals are sent to the cell controller. This is represented by adding a token into the feedback places [i.e., Pf_2 (AS/RS) or Pf_3 (R1)]. When n tokens appear in the feedback places, associated transitions (i.e., t_3 or t_4) are fired and cell control inputs [i.e., Pc_3 (AS/RS) or Pc_3 (R1)] are sent to the associated event classes.

8.9.3   PLC Implementation of the Cell Control Model

A PLC ladder diagram is developed from the EPN cell control model in Figure 8.22. The data definition of the ladder diagram is given in Table 8.7. In the ladder diagram shown in Figure 8.23a and b, p_1 and p_3 are external contacts representing the "Enter Order" and "Start" control buttons of the operator panel, respectively. The internal contacts used by the PLC include p_2, p_6, p_7, p_8, and p_9. They work as indicators that sequence the functions of the cell control modules. The symbols p_4 and p_5 represent the integer files that store the input number of cycles n for event classes. They are used as the preset values of the PLC counter 1 and counter 2, respectively. The symbols Pc_1 (AS/RS) to Pc_5 (AS/RS) represent the external contacts of the PLC digital output module that interfaces with the digital input module of the AS/RS controller. The symbol Pf_1 (AS/RS) and Pf_2 (AS/RS) represent the external contacts of the PLC digital input module that interfaces with the digital input module of the AS/RS controller. Similarly, Pc_1 (Robot 1) to Pc_3 (Robot 1) and Pf_1 (Robot 1) to Pf_4 (Robot 1) are PLC external contacts that interface with the robot (1) event class controller for coordination and feedback purposes. In addition, data transfer instruction MOV and counter instruction CTU are also used in the PLC ladder diagram to implement the "Normal Stop" cell control function. Counter resets (RES) are controlled by the "Start" button p_3.

8.10   CONCLUSIONS

The specification, implementation, and maintenance of an AMS control system are important problems in AMS designs. Confused requirements, conflicting demands, time pressure, and insufficient resources may make the problem exceedingly complex. This is especially true when new control functions are required for an AMS or existing control functions of an AMS need to be analyzed and changed. Consequently, the incompletely described characteristics and behavior of an AMS control may build communication barriers among designers and users and may cause AMS control designers to overlook important control functions. The integrated approach and associated reference models presented in this chapter target these common problems by providing

TABLE 8.7 Data Definition of the PLC Ladder Diagram in Figure 8.3

Symbol	Function	Type	Source Name	Location
P_1	Enter order	Boolean	Push button	Operator panel
P_2	Order indicator	Boolean	Boolean file	PLC memory
P_3	Start operation	Boolean	Push button	Operator panel
P_4	Store order for AS/RS	Integer	Integer file	PLC memory
P_5	Store order for robot (1)	Integer	Integer file	PLC memory
P_6	Cell start indicator	Boolean	Boolean file	PLC memory
P_7	Cell start indicator	Boolean	Boolean file	PLC memory
P_8	AS/RS start indicator	Boolean	Boolean file	PLC memory
P_9	Robot (1) start indicator	Boolean	Boolean file	PLC memory
n	Number of cycles	Integer	Integer file	PLC memory
Pc_1 (AS/RS)	AS/RS calibration control	Boolean	Output file	PLC memory
Pc_2 (AS/RS)	AS/RS cycle start control	Boolean	Output file	PLC memory
Pc_3 (AS/RS)	AS/RS normal stop control	Boolean	Output file	PLC memory
Pc_4 (AS/RS)	AS/RS release pallet control	Boolean	Output file	PLC memory
Pc_5 (AS/RS)	AS/RS pallet storage control	Boolean	Output file	PLC memory
Pc_1 (Robot 1)	Robot 1 calibration control	Boolean	Output file	PLC memory
Pc_2 (Robot 1)	Robot 1 cycle start control	Boolean	Output file	PLC memory
Pc_3 (Robot 1)	Robot 1 normal stop control	Boolean	Output file	PLC memory
Pf_1 (AS/RS)	AS/RS calibration done feedback	Boolean	Input file	PLC memory
Pf_2 (AS/RS)	AS/RS cycle done feedback	Boolean	Input file	PLC memory
Pf_i (Robot 1)	Position solenoids done feedback	Boolean	Input file	PLC memory
Pf_2 (Robot 1)	Robot 1 pallet done feedback	Boolean	Input file	PLC memory
Pf_3 (Robot 1)	Robot 1 cycle done feedback	Boolean	Input file	PLC memory
Pf_4 (Robot 1)	Robot 1 calibration done feedback	Boolean	Input file	PLC memory

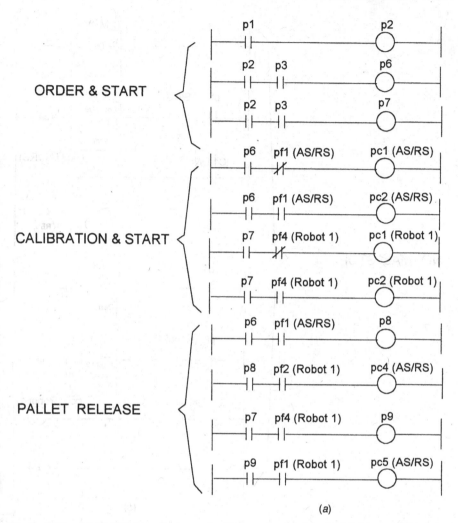

Figure 8.23 *(a)* PLC ladder diagram of cell control modules. *(b)* PLC ladder diagram of cell control modules.

control designers with concepts, procedures, models, and methods. System theories, object-oriented paradigms, and Petri nets are used as tools for developing efficient top-down and bottom-up design approaches and solutions. As a result, process/product models are developed to establish a consistent and refined cell control specification. This methodology shows that even the complex AMS control problem could be approached in a logical manner using a conquer strategy. It also shows that the integrated system approach can enhance the quality of the AMS control system by permitting the capture of system requirements, providing a method for representing the requirements, and formulating evaluation criteria for an improved system design.

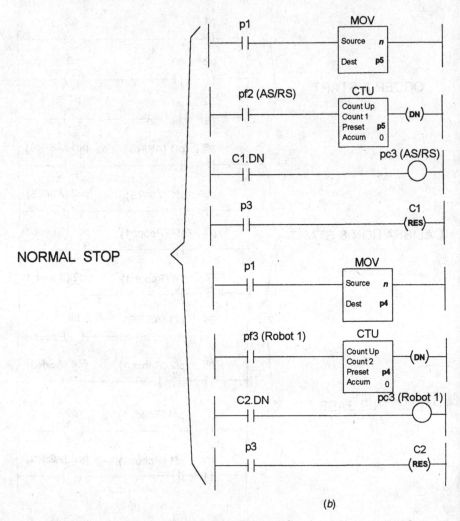

NORMAL STOP

(b)

Figure 8.23   *(Continued).*

REFERENCES

1. Al-Jarr, R. Y., and A. A. Desrochers, 1990, Performance Evaluation of Automated Manufacturing System Using Generalized Stochastic Petri Nets, *IEEE Trans. Robotics and Automation,* Vol. 6, pp. 621–639.

2. Cassandras, C. G., 1993, *Discrete Event Systems: Modeling and Performance Analysis* Richard D. Irwin and Aksen Associates, Homewood, IL, pp. 1–140.

3. Cheng, F. S., 1995, A Hybrid Method for the Modeling and Design of Automated Manufacturing Control Systems, Ph.D. Dissertation, University of Cincinnati.

4. Cheng, F. S., and E. L. Hall, 1994, Modeling and Design of Automated Manufacturing Control Systems Using Object-Oriented Timed Petri Nets, in *SME Transactions on Robotics Research,* B. O. Nnaji Robotics International Association of SME, Dearborn, MI, Vol. 3, pp. 109–124.

5. Combacau, M., and M. Courvoisier, 1990, A Hierarchical and Modular Structure for FMS. Control and Monitoring, *Proceedings of the AI, Simulation and Planning in High Autonomy Systems,* pp. 80–88.

6. Desrochers, A. A., and R. Y. Al-Jaar, 1995, *Applications of Petri Net in Manufacturing Systems,* IEEE, New York, pp. 1–306.

7. Gershwin, S. B., 1989, Hierarchical Flow Control: A Framework for Scheduling and Planning Discrete Events in Manufacturing Systems, *Proc. IEEE* Vol. 77, pp. 195–209.

8. Graham, I., 1994, *Object Oriented Methods,* Addison-Wesley, New York, pp. 193–326.

9. Jothishankar, M. C., and H. P. Wang, 1992, Performance of Optimal Number of Kanban Using Stochastic Petri Nets, *J. Mfg. Sys.,* Vol. 11, pp. 449–461.

10. Law, A. M., and W. D. Kelton, 1982, *Simulation Modeling and Analysis,* McGraw-Hill, New York, pp. 1–266.

11. Peterson, J. L., 1981, *Petri Net Theory and the Modeling of the System,* Prentice-Hall, Englewood Cliffs, NJ, pp. 1–100.

12. Ramaswamy, S., and K. P. Valavanis, 1994, Modeling, Analysis and Simulation of Failures in a Materials Handling System with Extended Petri Nets, *IEEE Trans. System Man Cybernet.,* Vol. 24, pp. 1358–1373.

13. Sullo, G. C., 1994, *Object Engineering: Designing Large-Scale Object-Oriented Systems* Wiley, New York, pp. 194–215.

14. Viswanadham, N., and Y. Narahari, 1992, *Performance Modeling of Automated Manufacturing Systems,* Prentice-Hall, Englewood Cliffs, NJ, pp. 1–579.

15. Wu, B., 1992, *Manufacturing System Design and Analysis,* Chapman and Hall, New York, pp. 303–400.

9

A GRAPH-THEORETIC APPROACH WITH FUZZY CRITIQUE TO MANUFACTURING DIAGNOSIS

KHOO LI-PHENG and ANG CHENG-LEONG
Nanyang Technological University

ZHANG JIAN
Hewlett-Packard Singapore

9.1 INTRODUCTION

The success of a product in the marketplace is dictated by its cost, performance, and reliability, which are predetermined, to a very large extent, by the design and manufacture of the product. One of the ways to reduce the product manufacturing cost and improve the product quality is to implement a fast and accurate fault diagnosis system for the manufacture of the product. This would also lead to a decrease in unforeseen processes or machine downtime, reducing unnecessary scraps and improving the overall production throughput. Traditionally, fault diagnosis is performed by operators and, in many instances, has been confined to the individual machine level. With the advent of technology, shopfloor automation and real-time monitoring of manufacturing processes becomes possible. This

Integrated Product and Process Development, Edited by John Usher, Utpal Roy, and Hamid Parsaei
ISBN 0-471-15597-7 © 1998 John Wiley & Sons, Inc.

permits better efficiency and flexibility in meeting production schedules and, as a result, leads to lower cost and higher-quality products. However, the success of a highly automated manufacturing system is very much dependent on trouble-free operation of the manufacturing processes. It is therefore important to troubleshoot the root cause and carry out appropriate corrective actions as soon as a failure or a fault symptom has been detected in any manufacturing process. Many techniques have been developed to assist operators in troubleshooting. These techniques can be broadly classified into two categories: the artificial intelligence (AI) approach and the non–artificial intelligence (non-AI) approach.

The AI approach utilizes techniques such as knowledge-based systems (KBSs) and artificial neural networks (ANNs) for manufacturing diagnosis. A KBS diagnostic system uses heuristic rules to reason about the possible causes of process malfunctioning. Graham et al.[4] developed a KBS for the diagnosis of a computer-integrated manufacturing (CIM) system. More recently, Png[16] implemented a KBS for the maintenance and troubleshooting of the auxiliary power unit of a weapons system. However, it is difficult to ensure that the knowledge captured in a KBS is complete and correct in practice.[23] Artificial neural networks are distributed information-processing systems comprising a number of simple computational neurons, like processing elements, locally interacting through a set of weighted connections. Knowledge is internally represented by the values of the weights and the topology of the connections. The inputs and outputs of the network are simply additional connections originating or terminating in the external environment. Artificial neural networks can learn and adapt themselves to the inputs from actual processes, thus allowing the representation of complex engineering systems that are difficult to model either with traditional physical engineering relations or KBSs. Venkatasubramanian and Chan[22], and Naidu et al.[14, 15] applied ANNs to the diagnosis of a chemical process and a control system, respectively. However, ANNs require a comprehensive set of training data for network training. This makes ANNs unsuitable to be used for the diagnosis of large or complex manufacturing processes.

In the domain of model-based reasoning, Rojas-Guzman and Kramer[18, 19] applied a model-based approach to the diagnosis of a large chemical plant. Bayesian belief function and genetic algorithms[7] were used to predict probable process failures. Such an approach would require the exact prior probability of failure of each component to be determined as subjective estimation of uncertainty is not suitable for the Bayesian approach.[21] Advocated by Zadeh,[25] fuzzy-set theory has become an effective tool to deal with uncertainties. Its applications are widespread.[10, 11] Qian[17] employed fuzzy sets to deal with the fault propagation of a chemical process. In a similar application, Chang and Yu[2] used fuzzy sets augmented with signed directed graphs to improve the resolution of diagnosis. Depth-first search and fuzzy-set manipulation were employed to identify the possible cause of a process failure.

The non-AI approach applies conventional techniques, such as failure modes and effects analysis (FMEA) and fault tree analysis (FTA), to diagnosis. Safoutin and Thurston[20] employed FMEA to minimize the impact of failure in the configuration of design teams. Hu et al.[8] used FTA to assess the failure probabilities of rocket motors. In general, these techniques are process specific and have limitations in dealing with dynamic processes.

This chapter presents a graph-theoretic approach with a fuzzy critique for the diagnosis of manufacturing processes. Graph theory is used to describe the functionality and model the processes used for the manufacture of a product. A hybrid approach based on fuzzy-set theory and graph theory has been developed to deal with fault propagation. Section 9.2 gives an overview of the fuzzy sets and graph theory. Section 9.3 presents the details of the hybrid approach. Section 9.4 describes a prototype fuzzy-directed graph (FDG)-based system for manufacturing diagnosis. Using a simple manufacturing system for the production of mechanical parts and assemblies as an example, Section 9.5 briefly discusses the performance of the prototype system. Section 9.6 summarizes the conclusions reached in this work.

9.2 FUZZY SETS AND GRAPH THEORY

9.2.1 Fuzzy Sets and Possibility Measure

As already mentioned, fuzzy-set theory is widely used to deal with the uncertainty and ambiguity in human expression. A fuzzy set is a class of objects characterized by a membership function that assigns to each object a "0 to 1" grade of membership.[9, 26] The basic notions of a crisp set, such as union, intersection, complement and associativity, can be extended to fuzzy sets.

Definition 1. A fuzzy subset **A** of a universe of discourse U having n objects $(u_1, u_2, . . .,u_n)$ is characterized by a membership function $\mu_A(u_i)$. This function associates with each object in U a real number in the interval $[0, 1]$, which indicates the membership of this object. Mathematically, **A** can be expressed in Zadeh form as follows:

$$\mathbf{A} = \sum_{i}^{n} \mu_A(u_i)/u_i \tag{9.1}$$

where $u_i \in [0, 1]$.

Using such an expression, the objects are written on the right-hand side of the slash, and the membership on the left. The symbol + means *or*. For example, the fuzzy set of what could be called "big" numbers in the range of integers from 1 to 6 is expressed as

"big" = 0.1/1 + 0.2/2 + 0.3/3 + 0.4/4 + 0.9/5 + 0.4/6

and the fuzzy set of "quite big" numbers can be represented as

$$\text{"quite big"} = 0.3/1 + 0.5/2 + 0.7/3 + 0.6/4 + 0.3/5 + 0.1/6$$

The mathematical notions of fuzzy-set theory provide a natural basis for possibility theory[21,22] which is different and yet analogous to probability theory.[3] The importance of possibility theory in manufacturing diagnosis is that much of the information on which human decisions are based is possibilistic rather than probabilistic in nature. In particular, the intrinsic fuzziness of natural language such as *this is likely to be the cause of delay in the manufacturing process* is possibilistic in origin.

Definition 2. Let **A** be a fuzzy subset of a universe of discourse U, which is characterized by its membership function μ_A, and let X be a variable taking values in U. The possibility distribution function associated with X is denoted by π_X and is expressed as follows:

$$\pi_X = \mu_A$$

Thus, the possibility that $X = u_i$, that is, $\pi_x(u_i)$, is postulated to be $\mu_A(u_i)$.

Definition 3. Let **A** be a fuzzy subset of U and let the possibility distribution function π_X take values in U. The possibility measure $\pi(\mathbf{A})$ of **A** is defined as follows:

$$\text{Poss}\{X \text{ is } \mathbf{A}\} = \pi(\mathbf{A}) \tag{9.2}$$
$$= \sup_{u \in U} \{\mu_A(u) \wedge \pi_X(u)\}$$

where sup and $\wedge$ are the supremum and minimum operations, respectively.

For example, let $U = \{1, 2, 3, 4, 5\}$ be the severity of the deviation from the nominal value, that is, *very small deviation, small deviation, medium deviation, large deviation,* and *very large deviation* in linguistic terms, for a process component. If **A** is the fuzzy set of deviations that are large, such that

$$\mathbf{A} = 0.1/1 + 0.2/2 + 0.5/3 + 0.7/4 + 0.5/5$$

then let the possibility distribution function π_X of the proposition "X is a large deviation" be

$$\pi_X = 0.0/1 + 0.3/2 + 0.5/3 + 0.9/4 + 0.3/5$$

Then

$$\text{Poss}\{X \text{ is a large deviation}\}$$
$$= \sup_{u \in U} \{\mu_A(u \wedge \pi_X(u)\}$$

$$\begin{aligned}
&= \sup_{u \in U} \{(0.1 \wedge 0.0)/1 + (0.2 \wedge 0.3)/2 + (0.5 \wedge 0.5)/3 \\
&\qquad + (0.7 \wedge 0.9)/4 + (0.5 \wedge 0.3)/5\} \\
&= \sup_{u \in U} \{0.0/1 + 0.2/2 + 0.5/3 + 0.7/4 + 0.3/5\} \\
&= 0.7
\end{aligned}$$

This shows that "X is a large deviation" is a *strong* proposition. If *large deviation* is associated with a fault, the condition of the process component can thus be inferred.

9.2.2 Graph theory

Graph theory[5] has been widely used to represent the deep knowledge of systems or manufacturing processes.[6, 12] It provides a rigorous framework for analyzing the causal relations of process components and locating the possible root cause of a process failure. Nodes (vertices) and edges (arcs) are two common elements of a graph for modeling the structures of a manufacturing process.

In general, the nodes represent the objects, concepts, process components, or propositional variables of the manufacturing process. The edges relate the possible cause–effect relations between two nodes. When the directions of the edges are indicated, the resulting graph or, model is known as a directed graph or, in short, digraph (Fig. 9.1).

Definition 4. Let V be a nonempty set and $E \subseteq V \times V$. The pair (V, E) is defined as a digraph on V, where V and E are the sets of nodes and edges, respectively. Mathematically, a digraph can be expressed as follows:

$$G = (V, E) \tag{9.3}$$

To enhance their capability, digraphs can be modified to incorporate qualitative and quantitative information. Signed directed graphs (SDGs) and Bayesian networks are examples of modified digraphs. A node of an SDG represents a process component. It takes the value of $+$, 0, or $-$, which means positive deviation, no deviation, or negative deviation from the nominal condition. An edge shows the immediate influence between the nodes. It also takes the

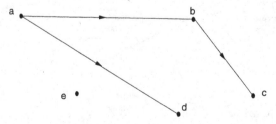

Figure 9.1   Example of a directed graph with $V = \{a, b, c, d, e\}$ and $E = \{(a, b), (a, d), (b, c)\}$.

value of $+$, 0, or $-$, which denotes the direction of influence. Typically, Chang and Yu[2] expressed an SDG as follows:

$$SDG = (V, E, \Lambda, \Delta) \tag{9.4}$$

where V is the set of nodes, E is the set of edges, Λ represents the forward edges, and Δ is the fault manner of a node.

Uncertainties can also be attached to the nodes or the edges of an SDG using probability or possibility values. The Bayesian network is an example of a probabilistic graphical representation with its nodes representing process components and the edges denoting the probabilistic causal relationships between nodes. Prior and conditional probabilities are included in the network. This inclusion enables the network to model the uncertainties and the interactions of a complex manufacturing process.[18, 19]

9.3 THE HYBRID APPROACH

A hybrid approach[1] based on fuzzy-set theory and digraphs (FDGs) has been developed to perform real-time process diagnosis. As with SDGs (Eq. 9.4), an FDG can be defined as follows:

$$FDG = (V, E, \varphi, \delta) \tag{9.5}$$

where V is the set of nodes, E is the set of edges of causal influence, φ represents the possibility of having fault, and δ represents the membership functions of a node.

Figure 9.2 shows a simple manufacturing process comprising four process components, O, A, B, and C. Assume that a symptom has been detected at O by an operator. To determine the fault propagation path, that is, OA, OB, or OC, the operator would intuitively carry out the following procedure:

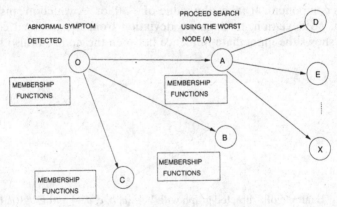

Figure 9.2 Worst-first research using the worst node on an FDG.

1. Assess the conditions of nodes *A, B,* and *C*.
2. Identify the node that has the worst condition, that is, the worst-first search.
3. Deal with the node, that is, the "faulty process component."

This manual fault-tracing process is repeated until the root cause has been identified. This procedure can be implemented using an FDG (Fig. 9.2). Each of the nodes is tagged with a set of membership functions (Fig. 9.3), which are expressed in terms of triangular fuzzy numbers (TFNs).

The condition of a node can be linguistically described using phrases such as *highly impossible to be faulty, impossible to be faulty, may be faulty, possible to be faulty,* and *highly possible to be faulty.* For simplicity, five different overlapping states, states 1–5, are employed to represent these linguistic variables (Fig. 9.3). Thus, the universe of the condition of a node contains these five overlapping states as objects. Intuitively, the state of a node is related to its deviation from the nominal condition, that is, the difference between the observed and the nominal values. For ease of implementation, an index known as the normalized degree of deviation X is used to indicate the condition of the node indirectly. Mathematically, X can be expressed as follows:

$$X = \frac{\text{Observed value} - \text{Nominal value}}{\text{Expected maximum deviation}} \tag{9.6}$$

The input (normalized degree of deviation) to a node is fuzzified using the membership functions attached to the node. As a result, the condition C of the node can be readily expressed in terms of the five states using Zadeh's formulation as follows:

$$C = m_1/1 + m_2/2 + m_3/3 + m_1/4 + m_5/5$$

where m_1, m_2, m_3, m_4, and m_5 are the grades of membership of each of the states.

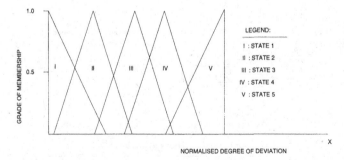

Figure 9.3 Typical membership functions expressed in terms of TFNs.

The possibility φ of having a faulty node can be determined using (9.2), that is, the possibility measure. This enables the condition of the node to be inferred. The value obtained is used to completely define the FDG (Eq. 9.5). Subsequently, ranking of the possibility measures is performed. Tracing of the possible fault propagation paths and the identification of the root cause for process failure will then be carried out using the worst-first search strategy.

9.4 PROTOTYPE FDG-BASED SYSTEM

The structure of a prototype FDG-based system for manufacturing diagnosis is depicted in Figure 9.4. It comprises three modules: the modeling module, the input module, and the reasoning module. The modeling module enables an engineer to model the manufacturing processes used to manufacture a product. Membership functions are attached to all the nodes at this stage. The input module gathers the on-line real-time process performance data from all the nodes and forwards them to the reasoning module. The reasoning module is equipped with a control mechanism, a deviation assessor, a predictor, and a fault searcher. The control mechanism is essentially a traffic controller. It decides the timing to fire the other submodules. The deviation assessor is employed to estimate the normalized degree of deviation of each of the nodes using the on-line real-time process performance data. Based on the normalized degree of deviation, the predictor carries out two tasks:

1. Assess the conditions of all the nodes.
2. Compute the possibility measures to predict the states of all the nodes.

The fault searcher then applies the worst-first search strategy to identify the possible fault propagation paths and the root cause for the process failure.

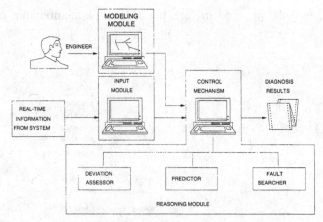

Figure 9.4 Prototype FDG-based diagnosis system.

9.5 DIAGNOSIS OF A SIMPLE MANUFACTURING SYSTEM

A simple manufacturing system (MS) for the production of mechanical parts is employed to illustrate the performance of the prototype system (Fig. 9.5). The MS comprises a number of manufacturing cells (MCs), each, in turn, comprising some machines and material-handling equipment to carry out the specific manufacturing processes or tasks.

For example, MC 1 consists of a machining center, a lathe, a pick-and-place robot, a process robot, and an automated guided vehicle (AGV). The operation of MC 1 is dependent on the smooth running of all the component machines or processes. The three production machines, namely, the machining center, the lathe, and the process robot, are used to produce the part components. The pick-and-place robot is used to serve these machines.

Preprocessed parts or subassemblies are picked from the AGV and placed on the conveyors serving the three production machines. The pick-and-place

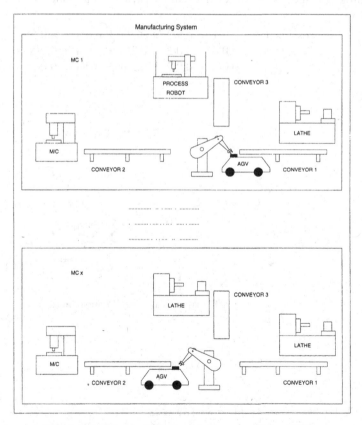

Figure 9.5 Simple MS comprising a number of MCs.

robot is triggered by scanning the bar code registered on the pallet loaded onto the AGV during docking. The throughputs of the various machines, the status of pallets, the availability of the AGVs, and the performance of the docking facilities are tracked using sensors. The information obtained is stored in the PLCs, which control and coordinate the activities. MC 1 is controlled by a cell controller, which constantly polls relevant information from the PLCs. It forwards the information gathered to the prototype system for analysis. The manufacturing activities of the MS is modeled by the engineer using the prototype system. Subsequently, it is translated into an FDG representation (Fig. 9.6). Assume that a lower-than-expected throughput is detected at node 11, that is, Manufacturing 1 (Fig. 9.6). The prototype system begins fault diagnosis by identifying the descendant nodes of node 11.

Based on the real-time process performance data captured, the prototype system applies (9.6) to compute the normalized inputs. Using the membership functions (Table 9.1) attached to the descendant nodes, the conditions and the possibility measures of the descendant nodes are estimated. As an example, the conditions of nodes 21, 22, and 23, that is, Processing 1, Machining 1, and Machining 2, respectively, at level 2 can be expressed as follows:

$$\text{Processing 1} \quad 0.0/1 + 0.9/2 + 0.2/3 + 0.0/4 + 0.0/5$$
$$\text{Machining 1} \quad 0.0/1 + 0.0/2 + 0.2/3 + 0.8/4 + 0.0/5$$
$$\text{Machining 2} \quad 0.0/1 + 0.9/2 + 0.2/3 + 0.0/4 + 0.0/5$$

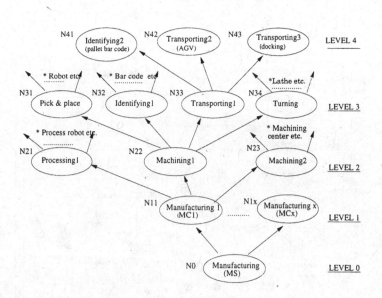

*Further decomposition of nodes (activities) is possible.

Figure 9.6   FDG representation of the MS.

TABLE 9.1 Membership Functions for the MS

Node Number		Membership Functions	
N0			
		Range	Function
	State 1	[0.00, 0.25]	$-4x + 1$
	State 2	[0.00, 0.25]	$4x$
		[0.25, 0.50]	$-4x + 2$
N11, N12, N41	State 3	[0.25, 0.50]	$4x - 1$
		[0.50, 0.75]	$-4x + 3$
	State 4	[0.50, 0.75]	$4x - 2$
		[0.75, 1.00]	$-4x + 4$
	State 5	[0.75, 1.00]	$4x - 3$
		Range	Function
	State 1	[0.00, 0.25]	$-2x + 1$
	State 2	[0.00, 0.25]	$4x$
		[0.25, 0.50]	$-2x + 1.5$
N21, N22, N23, N31, N32, N33, N34,	State 3	[0.25, 0.50]	$4x - 1$
N42, N43		[0.50, 0.75]	$-4x + 3$
	State 4	[0.50, 0.75]	$4x - 2$
		[0.75, 1.00]	$-4x + 4$
	State 5	[0.75, 1.00]	$4x - 3$

Notes: (1) For simplicity, two groups of membership functions are used. (2) The normalized input value outside the stipulated range is assumed to have a "0" membership.

In this work, once a node has attained state 4 or 5, it is considered to be a *possible faulty* node. The possibility measures (Eq. 9.2) are given as follows:

Processing 1   Poss{Processing 1 is at faulty} = 0.0
Machining 1   Poss{Machining 1 is at faulty} = 0.8
Machining 2   Poss{Machining 2 is at faulty} = 0.0

A series of worst-first searches is then carried out until the root cause is identified. The results of the diagnosis are summarized in Table 9.2. The worst node is therefore node 22 at level 2, that is, Machining 1. The preceding computation process is then repeated for the descendants of node 22. Two fault propagation paths, namely, N11–N22–N33 and N11–N22–N34, can be obtained. It can be inferred that the drop in the throughput of Manufacturing 1 is due to nodes 33 and 34, that is, Transporting 1 and Turning, respectively. For simplicity, it is assumed that level 4 is the lowest level and further decomposition is not possible. Physical checking on the descendants of nodes 33 (nodes 41, 42, and 43) and 34 (not shown and assumed to be operational) is then performed.

Table 9.2 shows that the AGV (node 42) is not able to dock properly at Conveyor 1. As a result, the bar code device cannot read the bar code posi-

TABLE 9.2   Results of Diagnosis

Level	Node Number	Node Name	Normalized Input X_i	Possibility Measure (State 4 or 5)	Status of Node
0	N0	Manufacturing	—	—	—
1	N11	Manufacturing 1	0.65	0.60	√
1	N12	Manufacturing 2	0.55	0.20	×
2	N21	Processing 1	0.30	0.00	×
2	N22	Machining 1	0.65	0.60	√
2	N23	Machining 2	0.30	0.00	×
3	N31	Pick & Place	0.55	0.22	×
3	N32	Identifying 1	0.55	0.22	×
3	N33	Transporting 1	0.65	0.60	√
3	N34	Turning	0.78	0.88	√
4	N41	Identifying 2	*	*	×
4	N42	Transporting 2	*	*	√
4	N43	Transporting 3	*	*	√

Notes: √—possible faulty nodes. *—physical checking is carried out at this level.

tioned on the pallet. Hence, the pick-and-place robot cannot be activated to serve the lathe. Eventually, the throughputs at nodes 22, 33, and 34 are affected.

The example has demonstrated the possibility of using the hybrid approach to perform real-time diagnosis of MS. With this, the on-line monitoring of process performance, which provides an early warning about process failure, becomes possible. As compared to the knowledge-based approach, human involvement has been largely reduced. In this case, the involvement is limited to the physical checking of suspicious process components such as nodes 42 and 43 as shown in the example. Furthermore, the hybrid approach is also more flexible and adaptable to changes. It is able to provide more than one fault propagation path to assist diagnosis using the worst-first search strategy. Knowledge about the probability of failure, a prerequisite of the Bayesian network approach, is not required here. The membership functions used in the proposed approach are derived by intuition. Based on the membership functions and the real-time information obtained from the physical system, the condition of a node, which changes dynamically with time, can be easily assessed using the possibility measure. As the FDG model is stored as an ASCII file, new knowledge about the manufacturing system, once acquired, can be easily included. The FDG model is essentially activity based. This is similar to the way in which an IDEF0[13] (ICAM DEFinition Zero) model of a system is built. Work is currently underway to integrate the prototype system with DESIGN IDEF, a commercially available IDEF0 modeling tool. With this integration, once an IDEF0 model of a manufacturing system has been successfully built, it is possible to package a real-time diagnostic program if the behaviors of the process components, such as the membership functions, are known.

9.6 CONCLUSIONS

This chapter outlines a hybrid approach based on graph theory and fuzzy-set theory for on-line real-time manufacturing diagnosis. To deal with the uncertainties in diagnosis, membership functions are incorporated into the digraph. Using the hybrid approach, the condition of a node can be easily estimated using its possibility measure. By ranking the possibility measures of the various nodes located at the same level, a worst-first search strategy for tracing the possible fault propagation paths has also been formulated. As the hybrid approach does not require prior knowledge on the probability of failure, the condition of a node, which changes dynamically with time, can be easily assessed. A prototype FDG-based diagnosis system, which utilizes the hybrid approach for reasoning, has been developed. A simple manufacturing system for the production of mechanical parts was employed to illustrate the performance of the prototype system. It has been shown that the prototype system is able to trace the fault propagation path and locate the possible root cause in a manner similar to that of human experts. The FDG model is essentially activity based. This is similar to the way in which an IDEF0 model is built. The integration of the prototype system with DESIGN IDEF would allow a real-time diagnostic program to be packaged upon the completion of system modeling.

REFERENCES

1. Ang, C. L., L. P. Khoo, and J. Zhang, 1996, Model-Based Approach to Fault Diagnosis of FMS, *IEEE Symposium on Emerging Technologies & Factory Automation, ETFA, Vol. 1,* pp. 254–260.
2. Chang, C. C., and C. C. Yu, 1990, On-Line Fault Diagnosis Using the Signed Directed Graph, *Industrial Eng. Chem. Res.,* Vol. 29, pp. 1290–1299.
3. Dubois, D., and H. Prade, 1989, *Possibility Theory—An approach to Computerized Processing of Uncertainty,* Plenum, New York.
4. Graham, J. H., J. Guan, and S. M. Alexander, 1993, A Hybrid Diagnostic System with Learning Capabilities, *Engineering Applications of Artificial Intelligence,* Vol. 6, No. 1, pp. 21–28.
5. Grimaldi, R. P., 1985, *Discrete and Combinatorial Mathematics,* Addison-Wesley, Canada.
6. Han, C. C., R. F. Shih, and L. S. Lee, 1994, Quantifying Signed Direct Graphs with the Fuzzy Set for Fault Diagnosis Resolution Improvement, *Industrial Eng. Chem. Res.,* Vol 33, No. 8, 1943–1954.
7. Holland, J., 1992, Escaping Brittleness: The Possibilities of General-Purpose Learning Algorithms Applied to Parallel Rule-Based Systems, in *Genetic Algorithms,* B. P. Buckles and F. E. Petry (eds.), IEEE Computer Society Press, pp. 48–78.

8. Hu, B., Y. Yang, and W. Xie, 1993, Method of Fault-Tree Quantitative Analysis for Solid Rocket Motor, *J. Propulsion Technol.* in Chinese, Vol. 4, pp. 55–59.

9. Kaufmann, A., 1988, *Fuzzy Mathematical Models in Engineering and Management Science,* North-Holland, New York.

10. Khoo, L. P., N. C. Ho, and T. W. Chao, 1994, A Fuzzy Management Decision Support System for Scenario Analysis, *J. Electron. Mfg.,* Vol. 4, No. 10, pp. 1–9.

11. Khoo L. P., and Y. H. Yeong, 1995, A Prototype Fuzzy Resistance Spot Welding System, *Internat. J. Production Res.,* Vol. 33, No. 7, pp. 2023–2036.

12. Kokawa, M., S. Miyazaki, and S. Shingai, 1983, Fault Location Using Digraph and Inverse Direction Search with Application, *Automatica,* Vol. 19, No. 6, pp. 729–735.

13. Kusiak, A., T. N. Larson, and J. Wang, 1994, Reengineering of Design and Manufacturing Processes, *Computers and Industrial Engineering,* Vol. 26, No. 3, pp. 521–536.

14. Naidu, S. R., E. Zafiriou, and T. J. McAvoy, 1990, Use of Neural Networks for Senior Failure Detection in a Control System, *IEEE Control Systems Mag.,* Vol. 10, No. 3, pp. 49–55.

15. Naidu, S. R., E. Zafiriou, and T. J. McAvoy, 1989, "Application of Neural Networks on the Detection of Sensor Failure during the Operation of a Control System, *Proceedings of the 1989 American Control Conference, Pittsburgh.*

16. Png, L. H., 1997, PC-Based Expert System for the Troubleshooting of Auxiliary Power Unit System, M.S. Dissertation, Nanyang Technological University.

17. Qian, D. Q., 1990, An Improved Method for Fault Location of Chemical Plants, *Computers Chem. Eng.,* Vol. 14, No. 1, pp. 41–48.

18. Rojas-Guzman, C., and M. A. Kramer, 1993, Comparison of Belief Networks and Rule-Based Expert Systems for Fault Diagnosis of Chemical Process, *Engineering Application of Artificial Intelligence,* Vol. 6, No. 3, pp. 191–202.

19. Rojas-Guzman, C., and M. A. Kramer, 1994, Remote Diagnosis and Monitoring of Complex Industrial Systems Using a Genetic Algorithm Approach, *Proceedings of the 1994 IEEE International Symposium on Industrial Electronics,* pp. 363–367.

20. Safoutin, M. J., and D. L. Thurston, 1993, Communications-Based Technique for Interdisciplinary Design Team Management, *IEEE Trans. Eng. Management,* Vol. 40, No. 4, pp. 360–372.

21. Terano, T., K. Asai, and M. Sugeno, 1992, *Fuzzy Systems Theory and Its Applications,* Academic, Boston.

22. Venkatasubramanian, V. R., and K. Chan, 1989, Neural Network Methodology for Process Fault Diagnosis, *Amer. Inst. Chem. Eng. J.,* Vol. 35, No. 12, pp. 1993–2002.

23. Vinson, J. M., and L. H. Ungar, 1995, Dynamic Process Monitoring and Fault Diagnosis with Qualitative Models, *IEEE Trans. System Man and Cybernet.,* Vol. 25, No. 1.

24. Yager, R. R., S. Ovchinnikov, R. M. Tong, and H. T. Nguyen, 1987, *Fuzzy Sets and Applications: Selected Papers by L. A. Zadeh,* Wiley, New York.

25. Zadeh, L. A., 1988, Fuzzy Logic, *Computer,* Vol. 21, No. 4, pp. 83–93.

26. Zimmerman, H. J., 1985, *Fuzzy Set Theory and Its Applications,* Kluwer Nijhoff, Boston.

SECTION III
CASE STUDIES

10

INTEGRATED PROCESSES IN DEFENSE MANUFACTURING*

GERALD C. SHUMAKER
Air Force Research Laboratory

RICHARD E. THOMAS
Texas A&M University

10.1 INTRODUCTION

The challenges associated with designing and building airplanes began, naturally, with the early pioneers who were successful in their efforts to fly. Both Otto Lilienthal and the Wright Brothers were proficient with gliders and the Wrights, of course, achieved the first sustained flight. Beyond the task of creating sufficient lift, thrust, and stability to enable flight, they were confronted with the development of structures that were light enough to fly but strong enough to withstand the loads encountered. In contrast to the design of other products, weight minimization was then and continues to be a major concern of aircraft, missile, and spacecraft designers—many are not aware that the engine used by the Wrights had the highest ratio of horsepower to weight in existence at the time. Moreover, the structural integrity of the lifting surfaces had to be such that the proper shape was maintained during flight in order to produce the desired aerodynamic properties. At the time, these issues were, for the most part, new, never having been encountered with other early engineering products.

* The views expressed here are solely those of the authors and not those of the U.S. Air Force or the Department of Defense.

Integrated Product and Process Development, Edited by John Usher, Utpal Roy, and Hamid Parsaei
ISBN 0-471-15597-7 © 1998 John Wiley & Sons, Inc.

As aircraft became more a part of our culture, as our knowledge and understanding of them increased, and as they began to be produced in substantial numbers, there evolved an array of new design techniques for both the aircraft and the manufacturing methods used for their fabrication. In the early days, airplanes were made from wood and cloth. Over the intervening years, light but strong metals have come into use; today, they are made from exotic metal alloys and composite materials, which impose special requirements on the product/process design team.

The primary objective of this chapter is to describe how integrated processes in defense industries differ from those in the civil sector. By raising the awareness of these differences, it will become clear that some of them bespeak a unique foundation of information and data and, as a result, pose different requirements on research and development efforts intended to support the design and manufacture of weapons systems.

Although one would inherently expect that the detailed aspects of integrated product and process development (IPPD) operations would be different for an automobile or a television set marketed to civilian buyers than for a complex fighter plane purchased by the government, when examined at a sufficiently high level, some of the tenets are common to the two environments. In 1995, then Secretary of Defense William Perry described the Pentagon view of IPPD[10]; those principles are presented in Figure 10.1. Although framed in the perspective of the Department of Defense (DoD), most of the ideas offered there could be applied to IPPD activities in the commercial world. Clearly, some of the parameters, such as life cycle planning, have a somewhat different meaning to the military.

Despite any commonality among the basic tenets of IPPD, the different environments will produce divergent objectives, organizations, and methodologies. It is these differences that we seek to explore, mainly in the context of their sources, those factors in the weapons business and in government acquisition that cause the military IPPD rubric to be unique. The details of that uniqueness will be described in subsequent sections.

Now, there are many weapons systems and each one will exert its own influences on the related IPPD system, so it is worthwhile narrowing the range of this discussion somewhat. Therefore, we shall focus on a modern military fighter plane, arguably one of the most complex products made by man, a result of the manifold requirements that it must meet. The integration of its product and process design operations is similarly complex, made more so by the organizational and acquisitional frameworks in which they take place. The goal here is to describe those special concerns that confront the experts in engineering analysis, materials, manufacturing, information systems, and other disciplines who operate in that arena.

The integration of product and process activities, of course, involves the commingling of two separate design functions, one related to the product and one related to the manufacturing system that will be used to make the product. Both design tasks involve trade-offs that are separately driven, that is, that

Integrated Product and Process Development (IPPD) Tenets

IPPD is an expansion of concurrent engineering utilizing a systematic approach to the integrated, concurrent development of a product and its associated manufacturing and sustainment processes to satisfy customer needs.

IPPD Defined: A management process that integrates all activities from product concept through production/field support, using a multi-functional team, to simultaneously optimize the product and its manufacturing and sustainment processes to meet cost and performance objectives. Its key tenets are as follows:

1. *Customer Focus.* The primary objective of IPPD is to satisfy the customer's needs better, faster and at less cost. The customer's needs should determine the nature of the product and its associated processes.

2. *Concurrent Development of Products and Processes.* Processes should be developed concurrently with the products which they support. It is critical that the processes used to manage, develop, manufacture, verify, test, deploy, operate, support, train people, and eventually dispose of the product be considered during product development. Product and process design and performance should be kept in balance.

3. *Early and Continuous Life Cycle Planning.* Planning for a product and its processes should begin early in the science and technology phase (especially advanced development) and extend throughout a product's life cycle. Early life cycle planning, which includes customers, functions and suppliers, lays a solid foundation for the various phases of a product and its processes. Key program events should be defined so that resources can be applied and the impact of resource constraints can be better understood and managed.

4. *Maximize Flexibility for Optimization and Use of Contractor Unique Approaches.* Requests for Proposals (RFP's) and contracts should provide for maximum flexibility for optimization and use of contractor unique processes and commercial specifications, standards and practices.

5. *Encourage Robust Design and Improved Process Capability.* Encourage use of advanced design and manufacturing that promote achieving quality through design, products with little sensitivity to variations in the manufacturing process (robust design) and focus on process capability and continuous process improvement. Utilize such tools as Six Sigma process control and lean/agile manufacturing concepts to advantage.

6. *Event-Driven Scheduling.* A scheduling framework should be established which relates program events to their associated accomplishments and accomplishment criteria. An event is considered complete only when the accomplishments have been completed as measured by the accomplishment criteria. This event-driven scheduling reduces risk by assuring that product and process maturity are incrementally demonstrated prior to beginning follow -on activities.

Figure 10.1   Pentagon view of IPPD.

7. *Multidisciplinary Teamwork.* Multidisciplinary teamwork is essential to the integrated and concurrent development of a product and its processes. The right people at the right place at the right time are required to make timely decisions. Team decisions should be based on the combined input of the entire team (e.g. engineering, manufacturing, test, logistics, financial management, contracting personnel) to include customers and suppliers. Each team member needs to understand their role and support the roles of the other members, as well as to understand the constraints under which the other team members operate. Communication within teams and between teams should be open with team success emphasized and rewarded.

8. *Empowerment.* Decisions should be driven to the lowest possible level commensurate with risk. Resources should be allocated at levels consistent with authority, responsibility and the ability of the people. The team should be given the authority, responsibility, and resources to manage their product and its risk commensurate with the team's capabilities. The team should accept responsibility and be held accountable for the results of their effort.

9. *Seamless Management Tools.* A framework should be established which relates products and processes at all levels to demonstrate dependency and interrelationships. A single management system should be established that relates requirements, planning, resource allocation, execution and program tracking over the product's life cycle. This integrated approach helps ensure teams have all available information thereby enhancing team decision-making at all levels. Capabilities should be provided to share technical and business information throughout the product life cycle through use of acquisition and support databases and software tools for accessing, exchanging, and viewing information.

10. *Proactive Identification and Management of Risk.* Critical cost, schedule and technical parameters related to system characteristics should be identified from risk analyses and user requirements. Technical and business performance measurement plans, with appropriate metrics, should be developed and compared to best-in-class industry benchmarks to provide continuing verification of the degree of anticipated and actual achievement of technical and business parameters.

William J. Perry
Secretary of Defense, 1995

Figure 10.1   *(Continued).*

are controlled by widely divergent parameters. It therefore seems reasonable to conclude that what we are talking about here is, in fact, the integration of two activities, each of which involves an integration of many factors and which are hierarchical in nature.

Ideally, of course, it would be desirable if this integration could occur in the mind of one person and, indeed, when a product and its manufacturing system are relatively simple, such "single-mindedness" might be achieved. There would seem to be no way that an individual who is steeped in the business of designing something as complex as a fighter could possibly come to grips with everything that is necessary to the design of the manufacturing system for the airplane. He or she would be doing well to effectively integrate all of the disparate elements of aircraft system design. As a matter of fact, one notes in the research literature a substantial number of investigations directed more or less exclusively to the integration of product design. Those efforts focus, naturally, on the first P in IPPD.

Time was when there were relatively few ways of making components and assembling them—now there are many different techniques as conveyed by the notion of manufacturing systems. They are just as complex and are just as difficult to design optimally as the products they yield. Saying it another way, the integration of the design of the process system, the second P, is, by itself, somewhat daunting. Pulling both P's together in an interlocking, cooperative, mutually reinforcing, truly integrated design operation is a challenge of the first order.

10.1.1 General Differences

Let us now describe in very general terms some of the factors that cause IPPD activities in the aerospace defense manufacturing sector to differ from those in the commercial world. Some of the more important and unique of these will be examined in more detail in later sections. Incidentally, we define "commercial world" to mean the usual mass-produced products such as automobiles and television sets—specifically excluded is the design and production of commercial aircraft. In a later section, the recent advances by Boeing in applying IPPD to its new 777 will be described in terms of its influence on the defense establishment.

The most fundamental difference lies in the goals—the objectives that the aerospace defense team pursues are not defined by the usual commercial marketplace parameters but rather by government agencies, which will specify the performance and other attributes the aircraft must have. It is important to understand how those different goals will shape the design and operation of the IPPD system and the metrics by which its efficacy is evaluated.

In addition to the complexity of our fighter plane and the fact that it is designed and manufactured in a setting wholly different from the usual civil-sector operations, there are other important elements that will affect the associated IPPD system. A rough rule of thumb holds that 70 percent of a contemporary fighter will be made and supplied by subcontractors, so supply chain dynamics are quite complex. Some of the fighter's subsystems such as the engine and avionics units are not simple elements that are easily integrated into the assembly of the plane. Product design and manufacturing experts from different companies, perhaps even from different nations, must be brought into the system at some level. The task of information sharing frequently poses difficulties in database interfaces and access, complicated sometimes by proprietary concerns.

The character of the defense industrial base has changed substantially since the end of the Cold War. Defense budgets have been reduced, and, with the prospects of even more cuts, aerospace defense firms are continuing to downsize and consolidate. The impact is great. There are, today, vastly fewer companies dedicated to weapons design and manufacturing so, in the future, the military must consider more extensive utilization of elements of the civil industrial base for the manufacture of military components. Additionally, cost pressures are leading to the use of more commercially made parts.

To maintain a state of military readiness, the Department of Defense is following a policy of acquiring weapons in smaller numbers and stretching their production out over longer periods of time in order to keep some production lines "warm." Therefore, again in contrast to the civil sector, wherein high-volume, high-rate production is more the norm, in the defense arena the environment is one of low-volume, low-rate production. This, too, has an effect on the configuration and functioning of the IPPD system.

During the days of the Cold War, wherein it was essential to maintain a technological edge over the Soviet Union, superior combat capability and performance was the dominant requirement. As a result, it was not unusual for military leaders to specify relatively unproven technologies in the descriptions of the attributes of new weapons systems. This understandably had a significant effect on both the design and the manufacture of those systems and frequently accounted for a substantial amount of redesign, rework, and cost in the process of bringing new systems into operational status. The United States still desires to maintain a technological edge over potential enemies but cost is now as important as performance. Therefore, the pressures to use unproven, high-risk, and potentially expensive advances in weapons are reduced. Whereas the unknowns associated with designing cutting-edge technologies into new weapons are lowered, other factors increase the difficulties of decision making by the IPPD team.

The long duration of the Cold War gave rise to an entire culture; one that saw generations of military leaders and weapons designers matured in an atmosphere where the issue of cost was secondary to performance. No longer! The U.S. military must assiduously consider the trade-off between performance or benefit and cost—that factor is now referred to as affordability. A great effort is underway to develop a new mindset, a new framework for decision making about new technologies and new weapons. A major part of the problem here is the absence of appropriate tools for making judgments of affordability, which, in the current mode, is leading defense manufacturers to adopt a *minimum* cost approach. This is not necessarily the condition for maximum affordability.

These new requirements call for the study of financial data, much of which resides in the records of industrial firms, and for analytical research based thereon, which will evolve new tools and methodologies. Because of the proprietary nature of that information, investigators must be prepared to take unusual steps to safeguard the interests of the firm yet evolve credible techniques for estimating the cost and value of performance and other attributes in order to arrive at good measures of affordability. IPPD teams must have these tools so that they can properly assess the impact of their product and process design decisions.

Turning to another matter, it has been pointed out repeatedly in the literature that one of the major problems in the implementation of IPPD activities is the diversity of the expertise that must be brought to bear on the design problems. This translates into the issues associated with information needed by each party involved, information associated not only with its own concerns but also that needed to deal with other elements of the team. Some researchers have offered the view that the development of readily accessible and properly designed databases is, in fact, the key to making a true IPPD team work.[8] There are others who describe the need for decision support systems of one type or another in order to achieve full integration in these operations.[13]

There have even been some researchers who have investigated ways in which artificial intelligence software can be used to select leaders, decision makers, for these teams. The issue here is this. The more diverse the item being designed, the greater the number of subsystems involved, the greater the number of disciplines and expertise that is required, the more severe these problems are. In the area of defense products, particularly something as complicated as a fighter aircraft, these issues are enormous and tend to transcend those similar problems that occur in the civil world. In considering the integration of aircraft and manufacturing systems design in the aerospace defense sector, one must also keep in mind that there are certain manufacturing issues that are unique to that sector. For example, stealth requirements impose special fabrication techniques. Some components are made from advanced materials such as titanium, aluminum–lithium alloys, and carbon composites, all of which pose unusual challenges to the designer of the manufacturing system. A fighter is a small vehicle that is crammed with sophisticated components; therefore, the issue of dense assembly is a major concern.

In the remainder of the chapter, it will be worthwhile to revisit all of these topics in more detail and to describe their impact on IPPD systems in an aerospace defense manufacturing world that is in a considerable state of flux and that will likely continue to be so over the next decade.

Indeed, it is difficult, at this point, to conjecture what the ultimate configuration of that industry will be by, say, the 2010 time frame. It is obvious, however, that on the civil side the IPPD movement is gathering speed rather quickly and that there will be contained therein some methods and tools that can be used by the defense establishment. In 1992,the Air Force set in motion the Lean Aircraft Initiative, which was designed to probe these matters. The Lean Aircraft Initiative will be described later.

10.2 THE IPPD ENVIRONMENT IN AEROSPACE DEFENSE MANUFACTURING

In this section, the framework in which IPPD operations are conducted in the aerospace defense manufacturing sector will be described in more detail. Under normal conditions, where one examines the environments in two different but relatively stable systems, the task would be comparatively easy. However, as described previously the defense establishment is currently undergoing massive changes, which some have likened to a revolution! There are several crosscurrents making the situation more than a little cloudy.

10.2.1 The Changing Defense Industrial Base

First, there is the substantial downsizing and realignment of the corporate entities that for years have comprised significant components of the defense industrial base. Some organizations, such as General Dynamics and Rockwell, are no longer in the business of making military airplanes. Now we have Lockheed–Martin, Grumman–Vought, and Boeing–McDonnell–Douglas operating as merged entities. As a result, engineering design and manufacturing staffs have been combined with a consequent blurring of previously well-defined organizational frameworks and procedures that have been years in the making.

As a result of the reduced manpower needs associated with the downsizing of the defense establishment, many firms have offered a variety of early retirement plans to their employees. The exodus has been substantial and most of those choosing to leave have been the more experienced people, including especially both weapons and manufacturing systems designers. An enormous amount of "corporate memory" has gone out the door. One industrial representative said that in the past the practice for gearing up to manufacture a new airplane involved selecting individuals who had been through at least two "new starts," but that now it was difficult to find those among the remaining employees who had the experience associated with participation in a single new effort!

Nor will there be as much opportunity as there was in the past for a young engineer to acquire the requisite design experience. During the years of the Cold War, the Department of Defense would have, at any given time, several different aircraft in various stages of design, development, and deployment. With the substantial decrease in defense spending, there will be fewer new projects through which young engineers can hone their analysis and design skills. This may ultimately force universities to drastically overhaul course structure and content in order to improve their ability to supply industry with engineers who are better equipped to "hit the ground running" and who will need less seasoning to become mature vehicle and manufacturing systems designers. Then, too, new AI and expert systems software, properly integrated into computer-aided design (CAD) and computer-aided engineering (CAE) systems, can offer some assistance to those with less experience, but this will surely change the dynamics of design and will affect the manner in which trade-offs are assessed and decisions made.

Another change that is occurring relates to the manner in which new technologies are matured and inserted into new products. This is, of course, a major parameter confronting IPPD teams in both the civil and the defense arenas. In the commercial world, advanced concepts are usually developed inside a given company or its "family," so described to include closely related suppliers, and are designed and manufactured into products wholly within that extended organization. This is, of course, a salient aspect of competitiveness.

The defense establishment has a substantially different system for delivering new technologies for insertion into advanced weapons and components, one in which government laboratories and government-sponsored R&D programs play a major role. That system, too, has been severely perturbed by the end of the Cold War and the associated changes in what might be called the defense science and technology framework.

It is significant that government laboratories are themselves initiating the use of IPPD teams—their activities will clearly impact the subsequent functioning of contractor teams in the industry. Moreover, through the use of models which, in part, will come from industry, judgments about the cost and affordability of new technologies, weapons components, and systems will have been initiated before the industry IPPD teams begin to operate. In fact, those industry teams will likely include government representatives who will insert customer concerns directly into the design/manufacturing deliberations. This is clearly at variance with the practice of most firms in the civil sector.

As will be examined in more detail in the next section, the issues of cost and affordability are having a major impact on the defense industry, even as it copes with reductions and realignment. The Air Force–sponsored Lean Aircraft Initiative is an effort to bring some of the more efficient manufacturing practices from the commercial sector into defense. This is causing industrial firms to reexamine all of their organizational structures along with their engineering and manufacturing practices in order to seek out more cost-effective ways of operating. IPPD is perceived as a major force in these efforts.

Other aspects of "the new paradigm" will be examined in the following discussion, but it seems clear that the aerospace defense industry is at a defining moment in its history. Things will never again be as they were. The changes that are occurring are irreversible and the future is by no means clear. With that in mind, the reader is cautioned that the snapshot of the aerospace defense industry that is presented here may not be an accurate description, given the nature of the "moving target." It will likely be even more inaccurate in another five or ten years.

10.2.2 The Acquisition Framework

To state again our objective, we seek to describe the current environment in which IPPD efforts are being mounted in the U.S. aerospace defense industry and to contrast that environment with that found in the civil sector. Although there are many small differences that are important to understand, perhaps the most significant factor is the overall objective. In the commercial world, IPPD is intended to help firms be more competitive by enhancing their capabilities to get advanced, high-quality products to market quickly in response to customer demands. The goal is to minimize the time to market without compromising other vital consumer-oriented characteristics such as quality, cost, and performance.

In the military world, competition does, of course, exist but it is constrained in various ways. There, the product and process design teams seek to develop and produce a weapon component or system that meets or exceeds a set of performance characteristics or other attributes given to all competitors by an agency of the Department of Defense. Moreover, the time lines are also defined—the competitors are required to submit their proposals by a specified deadline and so there is no advantage to being first. Usually, teams will continue to refine their designs right up to the deadline for submission.

However, one should not assume that the proposal teams dawdle along, taking their time, expending funds while they tinker with their planned submission. Defense firms are greatly concerned about proposal "cycle time" and seek to minimize it along with its cost. Again, though, the situation is different from a commercial firm trying to capture a share of a market by being the first to offer a new product.

What, then, are the primary objectives for IPPD in the aerospace defense manufacturing framework? Cost and affordability. Because manufacturing affects both of them, its impact is substantial.

Therefore, new requirements, new parameters have been thrust upon the IPPD teams. No longer can they focus on designing and manufacturing an aircraft, missile, or component in such a way as to "simply" maximize its performance. They must now weigh the trade-off between performance and cost. Nor can they look simply at acquisition cost, rather they must concern themselves with the cost of ownership, the total life-cycle cost. This will necessitate the development of a new set of tools and the assessment of a new set of trade-offs but actually the challenge is even broader.

Fundamentally, we are talking about the creation of a new culture, a new way of thinking that will permeate the entire defense structure from administrative elements in the Pentagon through the government's science and technology system and the industrial firms as well. Teams, both cross-disciplinary and cross-functional, will seek total concurrency in an effort to ensure that potential problems, the sources of unexpected costs, are anticipated and solved as early in the process as possible. The span of that effort, stretching across both government and industry, is broader than is typically found in the commercial world. Effecting these changes will require a great deal of education and training—as we all know, altering thought patterns is a daunting challenge, indeed.

10.2.3 Life-Cycle Issues

There are still more differences in the goals of IPPD operations in the civil and the defense sectors. That all-encompassing term, *quality,* is another major objective of IPPD activities in both manufacturing arenas, but here, again, when one looks at the details, weapons systems issues differ from those relating to the quality of mass-produced civil-sector products.

Consider operational lifetime, for example. Whereas a customer purchasing an automobile or a computer expects it to operate with "reasonable" reliability for a "reasonable" period of time, the consumer would not likely expect or want the item to be in service 20 years from now. The venerable B-52 bomber has undergone several rejuvenations, has been in service for more than 30 years, and is expected to remain so for at least another decade. Although that may be an extreme example, today's declining defense budgets and the absence of a substantial threat to national security will combine to require both weapons and manufacturing systems designers to focus more on an extended service life.

This has other implications. During its operational life, a fighter plane will need to have its combat capability upgraded with new engines, electronic components, and other parts. Because it is viewed as a "platform" housing components, and supporting several different subsystems, the basic design of the vehicle must be such as to render as easy as possible its later upgrading. Thus, designers must attempt to envision the characteristics of those items and plan for them—the manufacturing specialist must do the same in order to ensure that the processing barriers to future modernization are minimized.

In the defense business, quality has other different facets than those applied to commercial products. In the design of weapons systems, one must consider not only performance but parameters such as reliability and maintainability, factors that take on different dimensions when dealing with weapons compared with normal civil products. Maintainability is an especially important factor because it directly relates to force readiness; that is, if a combat aircraft is out of service because of maintenance, the available combat force has been reduced and its ability to fulfill its military mission has been compromised to a degree.

Some specialists have suggested that the definition of maintainability should be broadened and be called "availability," thereby encompassing the notions of whether or not a repair mechanic is available to do the necessary maintenance and whether or not a given part is available from the logistics system.

In the defense arena the vendor's service center is not just down the street. Moreover, whereas commercial products are designed to be repaired, they do not have to contend with the matter of combat damage. This, of course, goes somewhat beyond the issue of maintainability, but it directly affects the availability of a weapons system, which are quality concepts that must be considered by the IPPD team.

For the most part, IPPD and concurrent engineering have been viewed as a philosophy and methodology for improving productivity through a systematic approach to design. However, there are those[3] who believe that even in the civil sector future teams of multidisciplinary engineers and designers will collaborate in the development, production, and *support* of products *throughout* the product life cycle. The objective here is a continued manufacturer commitment to meeting the ongoing requirements of the customer well beyond the traditional warranty period. This factor is clearly applicable in the defense arena.

In the civil world, the task of responding to consumer requirements, especially if we are talking about a member of the public, would seem to be a comparatively simple process. Although his decision may be affected by an array of unspoken "delights," John Q. Citizen would be expected to enumerate perhaps three or four major requirements.

Government requirements, reflecting complex military mission objectives, will likely define and describe the characteristics of each of the subsystems in our fighter plane. Their number and detail are such that some researchers working for aerospace defense manufacturers[1] have investigated the use of computer-based artificial intelligence techniques simply to "manage" these requirements. Obviously, because each of them would interact, to some degree, with the others, trade-offs can be expected to be abundant and complex, posing substantial challenges to the team of aircraft and manufacturing systems designers.

10.2.4 Other Issues

Let us turn to some more detailed aspects of the IPPD environment in the aerospace defense industry. A fighter plane is generally not a large structure yet it contains an incredible number of components, some of them quite complex. Therefore, the issue of dense assembly is one that must be addressed early on in any IPPD operation. It is essential that the CAD system being used enable design engineers to perform collision detection analyses of complicated assemblies on a routine basis rather than waiting until the end of the design cycle is near. This is essential for identifying problems at a time when they are relatively easy and inexpensive to fix. Some of the latest generation of integrated CAD programs provide such capabilities.[6]

As one further considers the detailed aspects of aerospace defense design and manufacturing activities, the challenges associated with the use of advanced materials have already been mentioned, as has been the matter of low-volume, low-rate manufacturing. However, there must be concern for the capabilities of those systems to respond to national emergencies, to be able to "surge" production to meet requirements related to armed conflict. In those situations, the defense industrial base must be "mobilized" in order to ensure the proper defense of the nation.

Therefore, we have seen that IPPD in the aerospace defense manufacturing arena varies rather substantially from that in the commercial world. Researchers working in the areas of concurrent engineering or IPPD generally content themselves to focus on the problems of the civil sector and, as a result, there is not a great deal of fundamental data and information out of which new IPPD methods and technologies suited to defense activities can spring. This void needs to be filled.

The important issues of cost and affordability will now be examined in more detail.

10.3 COST AND AFFORDABILITY

Cost is surely not a new issue in the world of design and manufacturing of commercial products. After all, accurate estimates of all costs related to a product are significant steps along the road to profitability. Increased competitiveness leads to a continuous search for better methods for making these judgments. Little wonder that the current research literature contains several articles on the shortcomings of industrial accounting systems and related problems in cost modeling.[5, 12]

The problem here is that accounting systems are designed by and for accountants and they well serve the development of reports for managers and tax officials. Cost estimation models must be based on the proper financial data, but they need above all to be useful to engineers; thus, there is a cultural mismatch. The research issue that is implied here will be explored later.

Neither is cost a new parameter in the military arena although, as was explained earlier, during the years of the Cold War it took a back seat to the matter of weapons system performance. The United States desire to maintain an edge in combat capability vis-à-vis the Soviet Union and to use advanced technologies in its weapons to achieve that advantage made it necessary to relegate cost to a secondary position. Of course, that is no longer the case.

Affordability? Now that is a different matter! Return on investment and benefit/cost analyses have been around for some time and are well understood as a basis, for example, for decisions to invest in advanced manufacturing systems. Typically, however, one does not refer to such analyses as affordability assessments, but that is indeed what they are.

In the lexicon of the military services, few words have come to have greater impact than has affordability. It is one of those rather strange words that has languished in obscurity, connoted a rather restricted meaning, and has had only marginal significance. Then events combine to cause it to be used in a new framework and suddenly it becomes invigorated, assumes a new, much stronger identity, and brings with it an array of concepts that have seldom before been considered.

There is little doubt that a purchaser, an individual or company, at some point makes a judgment regarding the trade-off between performance or benefit and cost when considering the expenditure of funds. Moreover, the greater is the cost, the greater is the effort to justify it in terms of the benefits expected.

From the point of view of the manufacturer, these ruminations on the part of the customer come under the heading of marketability. Does the product offer a sufficient amount of benefit for the cost to cause the buyer to select one product over that offered by a competitor? Therefore, in the commercial world, affordability relates to competitiveness.

However, the important thing is this—that assessment on the part of the buyer of a commercial product is made at or near the point of sale. The product has already been designed and manufactured and, although the buyer may have certain optional equipment or other features that he or she can elect to purchase, the basic configuration of the product is set. Most importantly for our discussion, the IPPD team has made its judgments without the involvement of the customer. Not so in the defense world.

The design, development, and purchase of a new fighter plane by the military establishment proceeds in a substantially different way and so it follows that the functions of the IPPD teams in the military arena vary greatly from their counterparts in the civil sector. As a matter of fact, one finds IPPD teams at several points along the developmental path of a new weapon. In the case of our new fighter, judgments of the cost and affordability of advanced technologies will be made long before they are designed into it.

Today, we expect that a new military airplane will be operational for many years. Billions of taxpayer dollars are at stake here so the characteristics and cost elements of the new plane must be fully probed and understood. As a result, the road to the development and production of our new fighter is long and complex, far different, for example, from that of an auto company modifying an existing design, manufacturing a new model, and marketing it so that a customer can look at and test drive it and decide to purchase.

A determination to acquire by the Department of Defense will be made based on its evaluation of competing *designs* along with perhaps some data from tests of prototypes or demonstration models used to assess the efficacy of new technologies. Judgments of cost and affordability will have been made based on *estimates* or *predictions* by several IPPD teams involved along the way.

Therefore, the teams operating in the aerospace defense sector will need to have an array of tools for assessing the trade-offs among cost, affordability, and other design criteria. The techniques for evolving those tools cannot be addressed casually, rather they must be approached with great concern for their accuracy and validity. The company that wins the contract to build the airplane can ill afford to have its cost and affordability projections proved wrong. Therefore, the IPPD teams will carry a considerable burden.

There is yet another important aspect of integrated operations that is different in the defense sector from the commercial world and that relates to the manner in which new technologies are inserted into the design of a new product. We have already explained that manufacturers who offer commercial products for sale in the domestic or world marketplace will generally have evolved those new and hopefully attractive attributes within their own organizations and will have judged their impact on the competitiveness of the product. Will it attract buyers? The most important thing is this, the civil-sector IPPD team will generally have been associated with the maturation processes that the new technologies have and will be familiar with the characteristics and costs they bring with them. That familiarity will be a great advantage.

The system of research, technology development, and product/process design that brings new concepts and technologies to a fighter airplane is different. Many of the advances that are considered by competing aerospace defense organizations will have their roots in work carried out at universities and government laboratories. Especially in the latter arena, it is likely that cost and affordability judgments will have been made during the process of developing new technologies out of research advances. Clearly, the defense industry IPPD team must include those evaluations in its own deliberations.

Therefore, the framework in which an IPPD team functions will be heavily dependent on the ultimate destination of its product, the commercial or the defense sector. Let us now focus on the requirements that will be applied to cost and affordability tools and their availability for use in the design of our fighter aircraft.

10.3.1 Cost Models

There is a rather diverse array of cost estimation tools available to designers and to IPPD teams. Many of them are computer-based technologies offered for sale by commercial firms, but actually these are only the tip of the iceberg. There are likely hundreds of such models residing in the proprietary annals of industrial firms, tools based, in some situations, on many years of design and manufacturing experience. They are closely held and strongly protected. One can only conjecture about them.

Cost-estimating "rules of thumb" that might be used in the conceptual design phase are rather well established. For an airplane, cost per pound of airframe weight is a useful parameter, whereas, for a turbine engine, cost per pound thrust is well established. However, as the design becomes more detailed, different requirements must be placed on those models. In particular, they must be constructed around the same performance parameters that the IPPD team will be considering. To be effective, designers must be able to "see" the cost change as they grapple with alternative configurations and performance characteristics. It is impossible to state whether such models exist, given the proprietary interests of the industrial firms that use them. They seem not to have appeared in any publication.

To return to our main theme, differences in IPPD operations in the civil and military sectors, there will certainly be distinctions in the cost models used. In the commercial world, those costs that most directly affect market price of the product are of primary importance. Models for estimating operating and support costs would be of comparatively little interest in the civil sector but would be of great value to the military. Indeed, defense officials will want good estimates of life-cycle costs, the total cost of ownership of a weapons system from the day it is purchased until it is removed from operational status and retired or recycled.

Some of the cost increments would be rather well defined by the nature of the actions of DoD. That is, the initial contract for the design of the plane would cover all of the costs up to and including the submission of the final design in the competition. Following that, there would be the cost of the purchase of the first group of production airplanes. Then come the costs of operating, maintaining, upgrading, and the like—the list becomes rather long, depending on how one chooses to break down the larger elements.

As has been said many times and in many contexts, "The devil is in the details," and this matter of cost is no exception. Naturally, one of the elements of cost will be that associated with manufacturing; moreover, it should encompass both direct and indirect elements. In the civil world where the indirect cost is, say, 40 percent of the direct cost, methodologies exist for handling that item. In the aerospace defense arena, the indirect costs may be closer to 400 percent[7] and techniques for assigning it are by no means straightforward. This is an area where contemporary accounting methods appear to be unreliable.

10.3.2 Affordability Concerns

Now, what of the tools for making affordability assessments, for comparing performance or other benefits of a product to its cost? At the present time, those judgments, made either by an individual purchasing a car or by DoD acquiring a weapons system, are in the eye of the beholder. That does not mean they are not performed satisfactorily—obviously, the auto buyer believes his judgment is sound or he would not complete the purchase. The military makes use of an array of boards and evaluation teams to assist in making the decision. However, eventually, it usually comes down to one person.

It must be kept in mind that for the military this entire endeavor to make solid cost estimates and affordability judgments is a new ball game and the learning curve is steep. One notes that, in the absence of tools for making good evaluations of affordability, the ratio of performance or benefit over cost, there are efforts being made to *minimize* cost. Clearly, this is not the condition for maximizing affordability but it is, nonetheless, a worthwhile and understandable step, given the dearth of affordability tools. Until those tools are developed, decisions will hinge largely on cost.

In looking to the future, it seems almost inescapable that these "soft" methods of making affordability judgments are likely to be increasingly unacceptable. Budget pressures on the military will only increase and, as they do, there will result a search for better tools, leading to more deterministic evaluations of affordability, perhaps some way to quantify those assessments. This seems a natural result of what is now called "the new paradigm."

How might that be done? How would it be possible to develop tools that would allow the calculation of a *quantified* measure of affordability? It may be possible, for example, to nondimensionalize elements of performance and other benefits, as well as cost. The main difficulty here may be a restriction to the consideration of one element at a time. That is, if one had a group of nondimensionalized performance measures, one would need to scale each one in a way as to make its magnitude consistent with all the others. A 10 percent improvement in one parameter may be more important than a 30 percent change in another. However, that issue will confound any attempt to combine parameters and to ensure an even-handed judgment.

Some thought has been given to a different approach.[15] Let us first restate the definition of affordability described earlier but let us formalize it in the following way:

$$\text{Affordability} = \frac{\text{Performance or Benefit}}{\text{Cost}}$$

As has been said, if one wants to maximize this ratio, the condition of minimum cost is not appropriate. Therefore, when confronting the notion of a quantified index to affordability, it is necessary to assess the relationship between the numerator and the denominator. The obvious inference is that they are functions of some of the same variables and, in the largest sense, that must be true. One would naturally expect that higher performance or greater benefits must involve higher costs.

If the ratio is going to be quantified in a consistent way, the two elements must be dimensionally similar. Therefore, because the cost will be in dollars, the performance or benefit must also be in dollars and herein lies the challenge. It is by no means clear how to accomplish this—what follows is merely a suggested approach.

Although we will continue to use a fighter plane as our general framework for this discussion, this definition of affordability can also be applied to components or other capabilities. An airborne radar for our fighter will have some presumably measurable benefit and some cost; therefore, its affordability can be assessed independently. If the designers are considering a stealth capability and if they can estimate the cost of building that technology into the airplane and if its impact on survivability can be determined, its affordability can be calculated. These single-item estimates are, of course, easier to make than the total array of costs and attributes of something as complex as a fighter.

At this point, we shall assume that the denominator of the affordability equation, the cost element, can be handled with sufficient detail and accuracy, again, keeping in mind that we are addressing total life-cycle costs. Here we focus on the numerator and ways to evaluate its elements in terms of dollars.

What are some of the performance and other attributes of our fighter plane? The major design parameters include speed, maneuverability, radar cross section, range, takeoff and landing distances, payload, stability as a launch or gun platform, and night/all-weather capability.

One approach may be to group the attributes according to their overall role as follows:

Combat Capability   Those elements that relate to the ability to inflict damage on the enemy

Survivability   Those elements that relate to the ability to withstand the enemy's effort to inflict damage

Availability   Those factors involving component failure, maintenance, repair, and parts

Upgradability   Those factors that pertain to reequipping and modernizing a weapon so as to extend its useful life

Retirability   Those parameters affecting the recycling or disposal of components

Manufacturability might appear, at first glance, to be an attribute deserving of a place in this array, but most experts agree that its bottom-line metric is cost, specifically the minimization of the cost of manufacturing. Therefore, its rightful place is in the denominator of the affordability equation.

The first three of these groups have probabilities associated with them, which may lead to a way of expressing them in terms of dollars. Consider those items associated with Survivability. They all relate to the system's P_{bk}, the probability of its being killed by enemy action. Obviously, the lower the P_{bk}, the better. Looking at it in the affordability framework, the greater the prospects for survival, the more affordable the system will be.

Now, how does one translate that into dollars? Assume that the plane and its crew are assigned some value, some dollar amount, which reflects the investment that has been made in bringing the system to combat-ready status along with an evaluation of its contribution to combat capability. Clearly, if the plane is shot down and the crew killed, that investment will have been lost. Therefore, the survivability attribute has a value, which can be expressed as

$$\text{Survivability} = \frac{1}{P_{bk}} \quad \text{(Dollar value of the system)}$$

The issue of maintainability or availability can be handled in a similar fashion. There is a certain probability, says P_m, that the airplane and its crew will be unavailable because of a maintenance problem. Therefore, the investment made in the system will again have been lost. As was mentioned previously, maintainability can be enlarged to encompass the probability that a properly trained mechanic is available to repair the plane and that the necessary parts are on hand. Then the probability would be that of its availability, which we could designate P_a.

Other attributes such as range, loiter time, and upgradability can be couched in terms of an overall combat capability. Some related probabilities could be associated with them; however, this is not a trivial problem. A great amount of work is needed here.

Some complications are obvious. If the plane is designed to haul cargo or wounded soldiers, it generally would not be put in harm's way as would our fighter, so survivability takes on a different degree of importance. If an airplane is intended to be used as a launch platform for stand-off weapons, that attribute must receive higher marks in judging its affordability. All of this says that each element of the numerator in the affordability equation would need to be accompanied by some kind of weighting factor, an influence coefficient, which would be determined by the relative importance of the item being judged.

Clearly, much more needs to be written on this topic, but our intention here was to suggest that there may be ways of making more rigorous judgments of affordability than are currently available. Researchers will likely be pursuing them in the future.

Hopefully, this brief discussion of the issues of cost and affordability will serve to illustrate to the reader one of the newer and more challenging aspects of IPPD in the aerospace defense world. There are other factors, other forces at work, that are affecting the ways in which those integrated activities are evolving.

10.4   OTHER FACTORS

As we have seen, there are many differences between integrated operations in the aerospace defense sector and those in the world of commercial manufacturing. These differences result, in part, from the diverse character of the products of the two systems, from their recent history, and especially their objectives. We have noted that the departure from the world scene of the Soviet Union has had and is continuing to have an enormous effect on the U.S. defense establishment. As has been pointed out, that one event caused U.S. defense manufacturers to adopt as their own one of the enduring concerns of those who make commercial products, cost.

When the Soviet Union disappeared almost overnight, many of the leaders of those organizations at the top of the U.S. defense industrial base could see the handwriting on the wall. Some could not believe, in fact, refused to believe, the revolution that was at hand. Others perceived quickly that things would never again be as they had been and began to change. One of these changes was to seek ways to lower costs. In this endeavor, they looked to their counterparts in the civil sector, who were themselves attempting to cope with moves of the Japanese and other competitors on the world marketplace. A book, *The Machine That Changed The World,*[17] by researchers at MIT seemed to capture the mood. The term "lean manufacturing" became both fashionable and powerful.

10.4.1   Lean Aircraft Initiative

It seemed to some Air Force leaders that the "lean" framework might be a good way to restructure, indeed, to establish a new culture, for aerospace defense manufacturing. Therefore, in 1992, the Lean Aircraft Initiative (LAI), a $5.5 million program, was undertaken as a joint operation funded two-thirds by 18 defense contractors, one-third by the Air Force, with MIT providing coordination and research support.[16]

The objective was not to copy the auto industry—the defense establishment will always have a different mission—rather the intention was to see if the practices of car makers could show defense firms how to be more efficient, leaner but still meaner, as any military organization must be. In the view of the Air Force, "lean" means less of everything: less time, less inventory, fewer management layers, less capital, and fewer suppliers.

A special goal was the reduction or elimination, as shown in Figure 10.2, of nonvalue-adding elements associated with the manufacture and operation

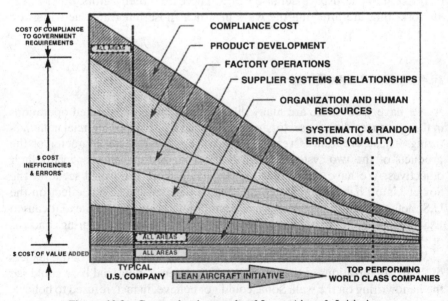

Figure 10.2   Cost reduction goals of Lean Aircraft Initiative.

of airframes, engines, avionics, with everything that comprises the life-cycle cost of weapons systems and components. A new culture? Indeed!

The Lean Aircraft Initiative is organized around focus groups, whose names serve to describe the program's primary thrusts. These include product development, where IPPD concerns are addressed; factory operations; supplier systems and relationships; organization and human resources and policy; and external environment. A general repository for the data and findings of these groups is the Lean Enterprise Model (LEM)—it also contains benchmarking data related to the practices of commercial firms.

The program has made several noteworthy advances. Research showed that if integrated operations were to be successful, databases for design, manufacturing, procurement, maintenance, and finance were required to be both common and relatable. In the words of the MIT researchers, "Such integration is fundamental for optimizing product and process flow, successful IPPD, design for manufacturing and design for cost, as well as for facilitating supplier involvement in development and production."

Evidence shows that when this is done reductions of about 60 percent are achieved in cost and schedule variances.

The program has also evolved several useful studies by MIT students. For example, one of these[4] presents the concept of a design structure matrix (DSM) and shows how it can be used to structure properly an integrated product team (IPT) and its subunits. As the author points out, "The DSM has several close relatives in the systems engineering world. In some ways, it is similar to the 'roof' of the House of Quality (in QFD) or an N^2 diagram."

New Air Force weaponry involves a steadily increasing amount of avionics, automated controls, and other subsystems requiring software, the cost of which has risen substantially. The LAI research has identified a "best practice," called the "software factory process," which utilizes systems for automated coding from requirements. Despite increases in the number of lines of code, errors decreased by more than 80 percent.

Studies of factory operations have led to methods for the development of plans for the flow of material, information, and processes, which dramatically increased efficiency. Flow optimization resulted in shorter flow times, reduced work-in-process inventories, lowered nontouch labor costs, and increased flexibility.

Investigation of supply chain dynamics revealed the benefits of long-term relationships between customers and suppliers. This has led to better integration of the subcontractors and improved utilization of their intellectual resources. As a result, the second-tier organizations participate in the IPTs, share risks, join in efforts to target costs, and seek to improve the efficiency of the entire system. Prime contractors are using certification programs, "best value" subcontracting, and electronic linkages to implement their new relationships with suppliers.

A major issue in the human resources arena is associated with downsizing and layoffs, especially getting workers to maintain a commitment to continuous improvement. Open sharing of company expectations and plans, together with active placement and training programs, appears to pay dividends. If the worst happens and layoffs are inevitable, early notice, coupled with active negotiations for other positions, serves to help maintain morale.

The MIT researchers found firm evidence that many of the barriers to the implementation of lean practices have their roots in government activities and they comment rather wryly, ". . . a lean aircraft industry must have a lean government customer." Further studies of different facets of these relationships are being conducted.

One of the more interesting aspects of the LAI program has been the rapid implementation of results by the Air Force in its sponsorship of projects involving a new fighter, the F-22, and a new transport, the C-17. The office in charge of the Joint Defense Attack Munitions (JDAM) program credits LAI findings and practices for a production phase bid that was less than half the cost estimated originally by the government.

No doubt the downsizing and restructuring of the aerospace defense industry will continue for some time into the future. The United States has yet to spell out clearly just what its role in the post–Cold War world will be and, as a result, the associated requirements for weapons, munitions, and manpower are not well established. Regardless of how that evolves, the Lean Aircraft Initiative has shown that defense manufacturers are willing to take from the civil sector whatever concepts, methodologies, and technologies can help it lower costs yet maintain a robust defense industrial base suited to the national security needs of the nation.

10.4.2 Boeing 777

In addition to the LAI program, there has been one other highly visible development in the aviation industry that seems to have secured forever the place of IPPD as a vital part of any advanced manufacturing enterprise. In 1990, Boeing made the commitment to implement the so-called digital design process for its new commercial airliner, the 777. The company invested literally billions of dollars in an effort to change its previously conservative ways and to embrace the concepts of concurrent engineering on a truly grand scale. The results have electrified the aviation industry and surprised even the staff involved in the effort.[2]

Boeing representatives describe the project as the single largest trial of CAD technology mounted to date. The 777 was designed totally using three-dimensional modeling technology implemented with CATIA (computer-aided, three-dimensional interactive application) software at 2200 workstations networked to an eight-mainframe computing cluster storing 3 terabytes of data. In addition to using CATIA as a three-dimensional design tool, it was also utilized as a digital preassembly tool. This eliminated the need for the usual full-scale, nonflying mockup in order to check fits and to identify either useful available space or interference problems, that is, where two parts try to occupy the same space. The digital model allowed engineers to check the accessibility of parts for maintenance by maneuvering a digital "mechanic" in three-dimensional space.

IPPD? In spades! There were some 238 teams all linked together by the CATIA system, all contributing to the heavier emphasis on planning during the early stages. On average, each drawing in the system is changed approximately four times before final release but with the concurrent operations, the design iterations did not have to go from designer to planner to manufacturer to assembler and back again. Everyone contributed to finding imperfections in the original design so it did not leave a team without having been reviewed by everyone.

Boeing's customers were also heavily involved in the teams so the resulting airplanes were tailored to their desires. An unexpected benefit was the education of the customers regarding the complexity of designing and building something like a 777. An official said that the many of them came to understand that "some of their previous demands were difficult to accomplish so they compromised the product."

The 777 is a very efficient and affordable airplane, as evidenced by its having captured over 75 percent of its market. However, its effect on the Boeing organization is interesting to observe—hard-core traditionalists have been won over to the world of IPPD. One engineer said, "For the first time we put leading edges on the wings without shims. For the first time we filled the fuel system and had no leaks. For the first time we powered up the avionics system and everything worked!" The precision achieved in construction was noteworthy—when the left wing was installed on the first airplane, the wing tip reference point was off by 0.001 inch while the same point on the right wing was within the accuracy of the measuring system. The 777 fuselage is over 200 feet long—past experience led the Boeing engineers to expect a front-to-back misalignment of at least half an inch—the 777 fuselage was just 0.023 inch off.

These experiences have made a host of believers in the aerospace engineering fraternity both inside and outside Boeing. Company management has decided to apply the same procedures to reengineering a new model of the 737 and will likely plan to use it in Boeing's entry in the Air Force's Joint Strike Fighter (JSF) competition.

10.4.3   Other Approaches

Aerospace defense firms have embraced IPPD in various ways using approaches that best suited their particular organizational and operational frameworks. Litton's Guidance and Control Systems Division (GCSD) reported[9] that it had begun teaming manufacturing and product design personnel in the early 1980s but there was little concurrence between the groups unless the individual project engineer took the initiative. Gradually, through a series of small steps, they moved ahead.

First, they adopted DoD's Best Practice templates into their internal policies, procedures, and processes, which were then reviewed, modified, and implemented by working level total quality management teams. More teams were set up to examine and streamline long-established practices. One of these groups reworked a manual used to guide all software engineering activities in the company. Another overhauled the engineering change system, which resulted in an impressive reduction of 25 to 35 percent in processing costs.

A computer-integrated manufacturing (CIM) team was given the responsibility for developing a comprehensive plan and architecture for the incorporation of computer-aided engineering into the generation of digital data, its subsequent control, and its eventual use in downstream processes. That team was so successful that ultimately it was accorded authority to review and approve or disapprove all budgetary requests for design automation equipment and tools.

Officials state that, "On a program specific basis, Litton GCSD now forms multi discipline teams which generate the total proposal and implement the program when and if funded. These proposal teams draw on a group of experts throughout Litton GCSD to ensure optimum consideration of program aspects such as customer requirements, technical approaches, manufacturing processes, quality control, customer support, and warranties."

Honeywell's Air Transport Systems Division implemented IPPD in a different way using SDRC's I-DEAS Master Series tool set to effect joint activities among product designers, manufacturing engineers, and methods people.[11] Designers use the software for development work in solid modeling and mechanical engineers employ it to support analysis. Importantly, they have developed a common base of data-linking designers and analysts, a vital step, as has been reported in the Lean Aircraft Initiative. Workers report that they find it easier to get high-quality products to market. There is nothing like success to invigorate the adoption of a new idea although the notion of integrated operations in design and manufacturing is scarcely new. After all, some 20 years ago the Air Force initiated the ICAM program, Integrated Computer-Aided Manufacturing, and one sees now the seeds that were planted as a result of that effort coming to flower. Wisely, that program was dedicated to the extension of those new ideas beyond the defense arena—organizations such as Caterpillar and Deere and Co. are implementing the fruits of that initiative.

Clearly, there are forces at work which are causing those firms that make military weapons to examine, test, and adopt concepts like IPPD. Without doubt, the most significant of these is cost/affordability and it is certain to continue into the future. Although there are practices in the world of commercial manufacturing that programs like the Lean Aircraft Initiative can bring to the attention of organizations in the defense arena, practices that will help lower cost and increase affordability, the business of making military aircraft and their associated components is sufficiently different as to merit its own set of new and dedicated methods.

10.5 CONCLUDING REMARKS

As the aerospace defense establishment, including both its industrial and government components, continues to come to grips with "the new paradigm," it seems clear that IPPD will be one of the fundamental management structures that will serve to anchor that effort. However, it is not possible to predict exactly how that will come about, owing to the uncertainty occasioned by the special characteristics of defense manufacturing.

Indeed, this is one facet of a larger issue dealing with basic research in manufacturing and with the array of fundamental data and knowledge that results therefrom. Recently, the Air Force Manufacturing Technology Directorate joined with the Design and Manufacturing Division of the National Science Foundation to sponsor basic research dedicated to issues of interest to defense manufacturers. This program will, over time, alter the structure of the base of fundamental information that undergirds the manufacturing discipline. The research literature of the past is dominated by papers that have come out of investigations focused on commercial manufacturing problems. With the advent of this new program, the research spectrum will be broadened and those special concerns of the defense manufacturing world will be addressed. Moreover, it is likely that the advances that come out of defense manufacturing research will lead to spinoffs that will be useful in the civil sector.

Although it is beyond the scope of this presentation to attempt to describe the entire spectrum of this defense-related research, it is certainly appropriate to mention briefly some of those associated with integrated operations.

10.5.1 Cost Estimation Tools

First and foremost is the need for investigations that can lead to improved methods and tools for estimating the cost and affordability of new weapons, especially in an environment involving the highly complex items manufactured at low volumes and rates. The cost-incurring activities obviously take place in industry so that is where the research should focus, but very quickly one encounters the wall of proprietary concerns. Insurmountable? Not at all.

There are high-level issues, questions that transcend the kinds of tools currently used by individual companies. A major question concerns the propriety of the cost data being gathered, an issue that concerns industrial accounting systems. Cost estimation models clearly reflect the base of data from which they spring and if the data assembled are not selected with the view of producing models of use to engineers, then, in fact, the models will not help in IPPD operations.

The literature is replete with examples of the engineering shortcomings of accounting systems[14] and the major problem, mentioned earlier, is that those systems are designed to satisfy externalities, entities such as corporate offices and the Internal Revenue Service. They can present data on return on investment but have difficulty illuminating product development costs.

Accounting systems serve a purpose so they must continue to function. Suppose, however, that a research group of *engineers* designed a financial data gathering and analysis system. It would likely look rather different. For one thing, engineers are willing to make estimates—they are not fixated on balancing the books. It is also a fair bet that the models that emerge from such a system would better serve activities such as those conducted by IPPD teams.

Proprietary concerns? The real problem here is not that companies want to safeguard the sources of their competitiveness—that is proper and understandable. The real problem is that academicians want to publish the results in the open literature, an activity they consider essential in order to obtain pay raises, promotion, and tenure. Although, in recent years, some institutions have taken a more benign view of proprietary and classified research and are committed to letting engineering professors meet the needs of the industrial segment of their professional community, many universities still cling to their old ways. Is there an answer to this? Yes!

Publications are generally of interest to those lower-ranking academicians seeking to advance up the ladder. Because tenured full professors generally have their reputations and credentials established, what is to prevent them from doing proprietary research if they so desire? Yes, they may have to sign confidentiality agreements with the industrial firms with whom they are working but, given a proper concern for the welfare of the company, ways can usually be found to disseminate research results.

The bottom line is this. Research into cost and affordability issues is essential to the U.S. defense community. Engineering professors can and should help.

10.5.2 Implementing IPPD

There are many other related matters to be examined. What must be done to make them function with maximum effectiveness? Here the published research reflects the usual disciplinary biases—behavioralists will cite the need for strong leadership, whereas information systems specialists believe that if the proper data are supplied, success is assured. The essential thing here is for investigations that transcend these narrow interests, for studies that are as "integrated" and as high level as the teams they seek to empower.

There are also concerns for the proper organization and structuring of an array of teams operating within a single setting. Boeing reports that over 200 groups worked on the 777—surely they all did not function at the same level. There must have been a hierarchy of sorts with some groups reporting to other groups. What are the guidelines for establishing that framework?

What of technology development efforts, the different levels of design and the need to scale IPPD operations appropriately? One does not expect IPPD activities to be pursued with equal fervor and intensity at all levels of engineering design. What is proper for a preliminary design effort compared to a detailed design undertaking?

In the earlier discussion of differences between the defense manufacturing sector and the commercial world, much was made of the issue of product complexity, the best example being a modern fighter plane. There an enormous number of components and disparate subsystems are crammed into an incredibly small package. It would seem logical to expect that the greater the complexity of the product, the more IPPD teams that would be required to design both the product and the manufacturing system needed to produce it. How does one configure and operate such a system?

There are related technology issues, notably that of dense assembly. In addition, how does one probe and develop data representing the variability that will inevitably be encountered in the manufacturing process? What experiments and tests are needed to produce that information? What IPPD issues relate to the use of the special materials found in most advanced military systems?

The list is not endless but neither is it short. The defense community can use the products of bright and agile minds dedicated to the study of its special set of problems.

In a larger sense, it is clear that integrated operations can do some marvelous things for the U.S. industry, military or civil. It is our hope that this brief presentation has served as not only a kind of status report for a rapidly changing defense establishment, but that it has exposed the reader to some of the special challenges residing therein.

REFERENCES

1. Black, J. E., 1994, AI Assistants for Requirements Management, *Concurrent Eng. Res. Appl.,* no. 2, pp. 254–264.
2. Boeing 777 Home Page, 1997, http://www.boeing.com/777.html.
3. Brown, D. C., S. E. Lander, and C. J. Petrie, 1996, The Application of Multiagent Systems to Concurrent Engineering, *Concurrent Eng. Res. Appl.,* Vol. 4, No. 1.
4. Browning, T. R., 1996, Systematic IPT Integration in Lean Development Programs, M.S. Thesis, MIT.
5. Cooper, R., 1996, Activity-Based Costing and the Lean Enterprise, *J. Cost Management,* Winter.
6. Deitz, D., 1996, Next Generation CAD Systems, *Mech. Eng.* August, p. 68.
7. Gutowski, T., D. Hoult, G. Dillon, S. Muter, E. Kim, M. Tse, and E. Neoh, 1994, Development of a Theoretical Cost Model for Advanced Composites Fabrication, *Composites Manufacturing,* May.
8. Klein, M., 1995, iDCSS: Integrating Workflow, Conflict, and Rationale-Based Concurrent Engineering Coordination Technologies, *Concurrent Eng. Res. Appl.,* Vol. 3, No. 1.

9. Litton Home Page, 1997, http://www.bmpcoe.org/surveys/LITG2__C/LITG2__C__bp.html.

10. Perry, W. J., 1995, Memorandum to the Secretaries of the Military Departments et al, Office of the Secretary of Defense, Department of Defense, Washington, DC.

11. SDRC Home Page, http://www.sdrc.com/ideas/case/honeywell/.

12. Sheilds, M. D., and M. A. McEwen, 1996, Implementing Activity-Based Costing Systems Successfully, *J. Cost Management,* Winter.

13. Sobelewski, M., 1996, Multiagent Knowledge-Based Environment for Concurrent Engineering Applications, *Concurrent Eng. Res. Appl.,* Vol. 4, No. 1.

14. Thomas, R. E., 1994, Manufacturing Cost Estimation, Modeling, and Control: A Review of Research Literature, StraTech, Inc., Beavercreek, OH.

15. Thomas, R. E., 1996, Foundations of Affordability, StraTech, Inc., Beavercreek, OH.

16. Weiss, S. I., E. M. Murman, and D. Roos, 1996, The Air Force and Industry Think Lean, *Aerospace America,* May.

17. Womack, J. P., D. Jones, and D. Roos, 1990, *The Machine That Changed the World,* Harper Collins.

11

ANTICIPATING MANUFACTURING CONSTRAINTS AND OPPORTUNITIES IN THE CONCEPT GENERATION AND PRODUCT PLANNING PHASES

ROBERTO VERGANTI

Politecnico di Milano

11.1 INTRODUCTION

Most practitioners and researchers agree that the early phases of product development projects are the most critical ones. These phases are generally known by different names, such as preproject activities, concept generation, product planning, idea generation, and product definition. Although empirical research demonstrates that decisions taken in those early phases have a major impact on the outcome of the innovation, the mechanisms that allow those phases to be properly managed are still largely unexplored. Indeed, anticipating decisions and problem solving in the early phases is a complex task. It requires predicting the behavior of the product throughout its life cycle and, in particular, the early analysis of the product and process interactions. The

Integrated Product and Process Development, Edited by John Usher, Utpal Roy, and Hamid Parsaei
ISBN 0-471-15597-7 © 1998 John Wiley & Sons, Inc.

major issue is that this information does not become available until one gets into detailed design and therefore it is usually unknown at the outset of a project. To overcome this issue and support the management of the early development phases, some general principles and tools have been proposed in the literature, such as teamwork and quality function deployment. However, there is a lack of insight on their complex mutual interactions and empirical validation of their actual effectiveness is often overlooked. This chapter identifies an articulated and coherent set of methods, organizational mechanisms, and behavioral patterns that successful companies adopt in the concept generation and product planning phases to anticipate manufacturing constraints and opportunities. In particular, three mechanisms are investigated: (1) how information about product–process interactions is gathered, (2) how this information becomes actual knowledge about future constraints and opportunities, and (3) how this knowledge is integrated into detailed design without affecting both the degree of innovation and the capability to react to late unexpected opportunities. The discussion is based on field research concerning 12 in-depth case studies of companies operating in the automobile, helicopter, and white-goods industries. It emerges that teamwork is necessary, but not sufficient. Two other driving forces appear as the basic engine of concept generation and product planning: (1) the systemic capability of learning from past innovation projects and (2) the balance between early decisions and late reactive capabilities.

11.2 NEED TO ANTICIPATE CONSTRAINTS AND OPPORTUNITIES

The life cycle of a product may be seen as a process with different phases and several internal and external interactions (see Fig. 11.1). The main phases are product development (which is articulated in concept generation and product planning, product design, and process design), production, consumption, disposal, and evolution into a next-generation product. Most authors agree that the success of product innovation has its roots in the early phases of the product life cycle.[5, 9, 25, 27, 36, 46] These phases are known in the literature by different names, such as concept generation and product planning, preproject activities, idea generation, product definition, and so on (in this chapter, these names will be indifferently used as synonymous). The importance of preproject activities has also been empirically validated in well-known research studies based on extensive surveys.[11–13, 15]

The reason for the relevance of the early product development phases may be traced back to the structural nature of the product life cycle. In fact, the different stages of the life cycle are linked by reciprocal interdependencies.[17, 41]

For example, in order to develop a new product and launch it into the market, customer expectations have to be anticipated and constraints due to

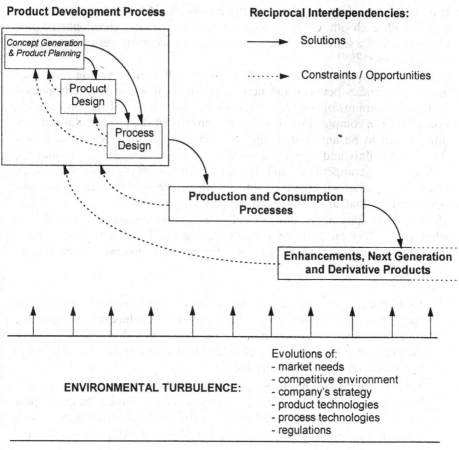

Figure 11.1 Structure of the information flows in the product life cycle.

manufacturability have to be accounted for; in these regards, Clark and Fuji-moto[8] represent the product development process as a simulation of consumption and production: Developing a new solution means forecasting the future constraints and opportunities that will arise during the life cycle of the product. Reciprocal interdependencies also occur with next-generation products. In fact, new ideas and solutions may be carried over, especially when the current project develops a new platform to be used in other derivative products.[14, 16, 45, 47] Constraints and opportunities coming from future production and consumption of the latter should therefore be anticipated in the former. Finally, reciprocal interdependencies characterize the internal activities of the product development process. A new concept must fit with the product technologies, whereas a product design must be feasible and should account for the constraints coming from process design. In other words, the concept, the product package, and the process should be continuously integrated with each other.[37] As a consequence, the product development process consists of a series of problem-solving cycles occurring between its different activities.[20, 21, 39]

Managing product innovation entails effectively handling these reciprocal interdependencies between the new solutions designed in the early phases and the constraints/opportunities arising in later phases of the product life cycle. This is a complex task, because the novelty of the product implies new information to be anticipated; that is, each new solution generates a new set of constraints and opportunities to be forecast. Moreover, the market, technological, competitive, and regulatory environments are continuously changing, together with the company's strategy, creating new constraints and opportunities that are unlikely to be predicted at the outset of a project. Hence, uncertainty about downstream information creates severe problems when integrating product and process innovation. This eventually compromises the performance of the product development process for the following reasons:

- Early product designs have to be changed later to embody unexpected constraints and opportunities. This is typical of engineering changes due to unfeasibility or poor manufacturability, issues that usually emerge during pilot production. Late changes entail costly reworks and increase the time to market of the product.
- Process engineering has to put additional work into the project in order to handle incompatible solutions (e.g., special fixtures have to be developed, because product drawings did not contemplate the constraints of production technologies). This again implies higher development costs and delays.
- Quality is compromised; for example, the actual customer needs are not satisfied or new technological opportunities are not incorporated into the product and process.

In order to deal with reciprocal interdependencies in the product life cycle, companies have two alternatives:

1. *Feedback Planning (or Reactive Approach).* The product is developed without intensively anticipating information in the early phases. The reasons behind this approach can be ascribed to the high uncertainty of the downstream information when examined in the upstream phases of the product development project. Because this uncertainty decreases as the production and consumption processes become closer, changes to design solutions may be introduced late in the project to handle unexpected constraints or opportunities. This approach is effective provided these late corrective actions (i.e., engineering changes or

additional work wasted in process engineering) have a minor impact on product development performance. However, this is seldom true: the later the downstream constraints emerge, the higher the time and cost due to reworks and engineering changes (see also Hayes et al.,[27] Meredith and Mantel,[35] and Trygg[42]). In fact, late changes entail a larger number of already developed solutions to be redesigned, including also those most costly and time consuming such as detailed product design and process design. In some cases, unexpected technological opportunities or flaws in interpreting the market needs occur when the product is already on the market and it is therefore too late to introduce any change.

2. *Feedforward Planning (or Proactive Approach, or Front Loading).*[22] Information is anticipated as early as possible in the product development project, so that the new solutions generated in the early phases already account for future constraints and opportunities. The basic principle of this approach is that the impact of design choices on market requirements, on design feasibility, on product manufacturability, usability, reliability, maintainability, recyclability, and so on, should be analyzed as early as possible in the development project, that is, in concept generation and product planning. Here, in fact, there is the highest ability to influence the outcome of the project. Hence, great efforts and managerial focus are devoted to properly define the product concept and specifications in order to reduce possible late problems and reworks. The drawback of feedforward planning is the uncertainty that affects the early phases. Indeed, often downstream information does not become available until the team becomes involved in detailed design; moreover, future changes in the market, technological, competitive, and regulatory environments are hardly known at the outset of the project. Therefore, this approach is effective provided a company has outstanding capabilities to anticipate highly uncertain information. Note that uncertainty could be reduced by shortening the cycle time of the product development process. In fact, this would also shorten the future time horizon and therefore improve the accuracy of early predictions. However, shortening the cycle time requires avoiding late reworks and corrective actions. This again leads to feedforward planning; that is, the cycle time may be shortened provided a company has great capabilities to anticipate downstream constraints and opportunities. Hence, a lack of proactive capabilities generates a vicious circle between late corrective actions, cycle time, time horizon, and uncertainty in the early phases (Hayes et al.[27] use the metaphor "following a moving target" to describe this circle). Companies with poor mechanisms to anticipate downstream information simply make feedforward planning a sterile exercise: Time at the beginning of the project is wasted, whereas reworks in later stages are not avoided.

11.3   AIM AND METHODOLOGY OF THE STUDY

Balancing between feedback and feedforward planning is a dilemma. As mentioned previously, a number of studies demonstrate the importance of preproject activities and the effectiveness of feedforward planning. Indeed, this approach is also a major principle of the new product development paradigms. Trygg,[42] for example, considers "early involvement of manufacturing" as one of the main principles of concurrent engineering. Clark and Fujimoto,[8] in their study of product innovation throughout the auto industry, describe the preliminary exchange of information with the downstream phases as the more complete, albeit complex, coordinating mechanism in new product development. However, although there is a wide agreement on the effectiveness of the feedforward approach, there is still a lack of research on its inherent mechanisms. Hence, the dilemma still remains: How can this approach be properly implemented? What are the organizational mechanisms that make feedforward effective? How does one avoid feedforward becoming a sterile exercise and a useless waste of time?

In these regards, most authors agree that the preliminary analysis of constraints and opportunities may be successful only if carried out at a cross-functional level, that is, through multifunctional and multidisciplinary teams (see, e.g., Bacon et al.,[2] Clark and Fujimoto,[8] Gupta and Wilemon,[25] and Houser and Clousing[29]). New managerial techniques have also been proposed to support the early planning of product development such as quality function deployment[29] or life-cycle costing.[4] As a matter of fact, most companies still face substantial problems in implementing the feedforward approach, notwithstanding they resort to multifunctional teams and dedicate great efforts to anticipate manufacturing constraints. Hence, the reasons for failures in early involvement cannot be simply traced back to a lack of communication between different departments or to time pressure that induces one to skip the product planning activities. There is therefore a need to further understand the complex, articulated, and coherent set of managerial criteria, organizational mechanisms, and behavioral patterns that allow uncertain information to be properly anticipated and that transform early manufacturing involvement into a successful planning principle.

This chapter investigates the factors and guidelines for proper implementation of feedforward planning and identifies the major reasons that undermine its effectiveness. It reports the results of a research study based on 12 in-depth case studies of Italian and Swedish companies operating in three different industries (for the sake of confidentiality, each company has been denoted by a letter): vehicles (cars, buses, trucks, tractors; companies A, B, C, and D), white goods (cooling, washing, and cooking devices; companies E, F, G, H, I, J, and K), and helicopters (both for defense and civilian use; company L). The research method entailed collecting both qualitative and quantitative information at the company level and at a single-project level (consisting of the most recent and finished innovation project of a product line or platform).

Qualitative data were gathered by interviewing four managers in each company (marketing, research and development or product engineering, process engineering or manufacturing, project manager or platform manager). A questionnaire with more than 200 parameters was used to collect quantitative data and structured subjective information from managers. These parameters were integrated into a few variables in order to synthetically represent the drivers and mechanisms of feedforward planning. Because these variables are complex, multidimensional and heterogeneous, fuzzy-set functions and operators have been used when necessary to integrate the parameters. These variables, functions, operators, and parameters are reported in the Appendix.

Later sections in this chapter discuss the lessons drawn from observations of successes and failures in the case studies. In particular, a general framework is illustrated first in order to show the major drivers of feedforward planning and their mutual interactions. Then the single drivers are briefly analyzed.

11.4 MECHANISMS AND DRIVERS OF FEEDFORWARD PLANNING: A GENERAL FRAMEWORK

The mechanisms that allow manufacturing constraints and opportunities to be anticipated in the concept generation and product planning phases are fairly complex. The main reason is that feedforward planning is not bounded in a set of well-structured techniques and methods. Of course, managerial techniques such as quality function deployment or life-cycle costing are relevant; checklists indicating the classes of information to be anticipated may also provide significant support. However, as this study will show, the basic principles of feedforward planning mainly consist of "soft" rules and behavioral patterns.

Consider the following example. A white-goods company has decided to develop a smaller refrigerator. New technological solutions are proposed and, from a technical point of view, the new product seems to be feasible. However, the concept development team wants to detect early any problem that could emerge in the assembly line and that could make the product unmanufacturable (unless at very high costs). To deal with this issue, a feedback or a feedforward approach could be implemented. A purely reactive approach would entail that product engineering starts developing the new refrigerator and provides information and drawings to process engineering as soon as possible. Then process engineering begins to develop dies, tools, and fixtures and verifies any assembly problems. Any feedback concerning manufacturing constraints would reflect the actual practice of process engineering in attempting to implement the upstream design. In other words, feedback is based on a *learning-by-development* mechanism.

Feedforward planning is based on a completely different and far more complex procedure. This entails that during the product planning phase product engineering and process engineering, both belonging to the preproject team, exchange uncertain information on future constraints and opportunities. Teamwork is essential in this process of early analysis. However, early detection of possible issues is unlikely to occur on the basis of unregulated discussions between product and process engineering, unless every design detail is defined early and investigated. On the contrary, both product and process engineering lead the discussion toward what they believe are possible sources of late issues. In particular, among all the information that could be anticipated, product engineering selects only those parts of the new technological solution that could generate problems in process design and manufacturing. Note that these critical areas are not necessarily concerned with the most complex part of the product; in other words, the product engineer does not have to anticipate issues on his or her own future tasks (usually these issues are easily known). The difficult thing is that product engineering should identify the parts of its new solution that could create problems in other phases, without entering into detailed design. Process engineering would then analyze this preliminary information in order to detect possible downstream constraints and opportunities linked to those prospective solutions. Hence, the capability of process engineering to anticipate downstream constraints strongly depends on the information provided by the product engineer and therefore on his or her understanding of the process technologies.

This example shows that feedforward planning is made effective by the *systemic knowledge* held by each member of the preproject team, that is, by his or her capability to early detect which elements in his or her own decisions may have significant consequences on the other phases of the product life cycle. Systemic knowledge is therefore the engine of feedforward planning. Figure 11.2 proposes a general framework that illustrates how systemic knowledge is built in the early phases and integrated into the product development process. In particular, three basic mechanisms are illustrated:

1. *Making Early Information Be Systemic.* This means bringing and integrating into the preproject phase the necessary information concerning the product life cycle (such as customer needs, evolution of product and process technologies, dynamics on the competitive environment, etc.). The factors that make early information be systemic are teamwork and communication, harmonized objectives, and encouraged and supported proactive thinking.

2. *Making Systemic Information Become Systemic Knowledge.* This means gaining actual knowledge of critical areas by selecting relevant information and combining possible patterns. The factor that allows systemic knowledge to be nurtured is systemic learning.

3. *Integrating Systemic Knowledge into Detailed Design.* After being anticipated, early information and decisions have to be transferred and exploited in the subsequent product design and process design phases. This implies acting on two factors: integration with detailed design and planned flexibility.

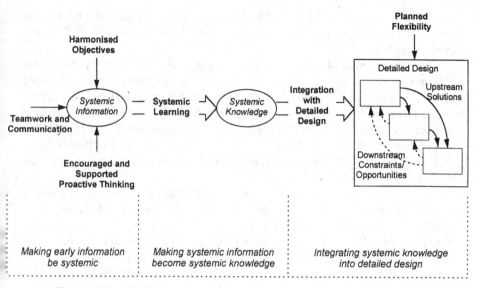

Figure 11.2 Mechanisms and driving forces of feedforward planning.

In the following sections, the previously articulated and coherent set of mechanisms and drivers is analyzed in detail.

11.5 MAKING EARLY INFORMATION BE SYSTEMIC

11.5.1 Teamwork and Communication

Anticipation of constraints and opportunities entails collecting a large amount of information. Bacon et al.,[2] for example, consider as information to be anticipated: the customer and user needs, the competitive product offerings, the technological risks and opportunities, and the regulatory environment into which the product will be delivered. Recently, concurrent engineering also stressed the importance of anticipating in the concept generation phase a thorough analysis of the manufacturing constraints. In some companies, this information is collected and analyzed by a specialized department (e.g., the product planning department); in other companies, it is the product manager or the platform manager who specifies the product concept. Sometimes a single functional department is involved in the early phases of product development, such as marketing (in market-driven companies) or research and development (in high-technology industries). However, most studies demonstrate that the most successful criterion of collecting and analyzing such an articulated set of information is the early involvement of all major departments having a direct contact with future constraints and opportunities. As far as production constraints are concerned, research studies have investigated the benefits of early involvement of process engineering and manufacturing[19, 23, 26, 29] and major suppliers.[7, 18, 33, 34]

The effectiveness of teamwork and early manufacturing involvement also emerges in our study. First, all companies (except company E) allocated concept generation and product planning to a multifunctional team. In particular, as to production constraints, companies A, B, K, and I involved process engineering in their preproject team, whereas companies F and H involved manufacturing. Both departments belonged to the team in companies C, D, G, J, and L. Also, companies A, H, K, and L involved selected major suppliers when specifically needed.

The contribution of teamwork and communication to the anticipation of information is also shown in Figure 11.3, where "Feedforward Effectiveness" denotes the capability of a company to anticipate constraints and opportunities, avoiding late reworks and unexpected problems. Measures are drawn by data in the structured questionnaires, which are aggregated by means of fuzzy functions and fuzzy operators. Details on measurements for the feedforward effectiveness and the degree of teamwork and communication are provided in the Appendix. Here, two major points are highlighted. First, all measures are relative, thus making it possible to compare companies operating in different industries. Second, feedforward effectiveness does not measure the overall performance of a product development project (e.g., product success in terms

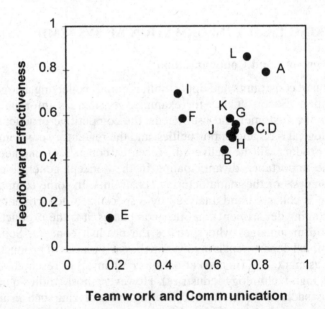

Figure 11.3 Relation between teamwork in the early development phases and feedforward planning.

of sales), because this overall performance is affected by several parameters and phenomena (company E, for example, ranks third in terms of product functional performance; however, because of a lack of feedforward capabilities, it achieved this performance to the detriment of extremely high product costs and time to market). On the contrary, the analysis is restricted to the sole parameters that reveal the capability to anticipate uncertain downstream information in the early phases of product development. Feedforward effectiveness is therefore measured through a fuzzy function depending on the occurrence of the typical drawbacks of ineffective preproject planning (such as reworks, engineering changes, unexpected events that call for additional work in later stages of product development, unanticipated product costs, and a longer time to market than originally foreseen). This metric, being directly linked to the detrimental impacts of reciprocal interdependencies, allows the benefits and drawbacks of the proactive approach to be separated from other performances.

As the diagram shows, teamwork seems to positively influence the capability to anticipate information. Indeed, projects with a higher degree of teamwork experience four major benefits due to this cross-functional early exchange of information: First, it enlarges the knowledge base available in the early phases of product development and therefore reduces the uncertainty on future constraints and opportunities; second, it ensures alignment of the product concept with the company strategy and with the functional strategies (in particular, with the process technology strategy); third, it allows process engineering to start earlier with some preliminary activities, thus encouraging the parallelization of design phases; fourth, because the owners of downstream constraints are involved in concept generation, their commitment to upstream decisions is fostered.

11.5.2 Harmonized Objectives

Gathering different departments in a preproject cross-functional team, however, does not ensure communication and anticipation of information from the product life cycle. If team members have different functional objectives (e.g., if the product engineering department is measured only according to the time needed to provide detailed drawings), integration in the preproject team is unlikely to occur. Without a proper system of objectives, early involvement may result in a fruitless exercise: Upstream departments, exposed to time pressure, do not see any advantage in anticipating and communicating their likely product design to downstream phases, whereas process engineering and manufacturing bring to preproject planning only their constraints and feasibility problems, restraining the creativity and innovation typical of concept development. Of course, a new culture and distinctive capabilities are necessary when applying the feedforward approach: Downstream phases should learn to be creative, whereas upstream phases have to master the development of new solutions in the presence of a higher number of constraints. However, such a culture and capabilities are ineffective (and, more-over, are unlikely to be nourished) if the preproject team does not have common objectives. This means, for example, all team members being mea-sured on the overall time to market of the project or on the break-even time of the product.[28] A harmonized performance measurement system fosters anticipation of information and communication, because all the preproject team members are measured on their ability to avoid the detrimental impacts of late and unexpected constraints and opportunities.

This was also confirmed by our sample. For instance, companies I and L (highly effective) used such incentives for the product engineering department as the overall time to market and the in-house defectiveness (i.e., the percent-age of defects occurring in production). On the contrary, company E (lowest effectiveness) adopted as a performance measurement index for product engi-neering only the functional features of the product.

11.5.3   Encouraged and Supported Proactive Thinking

Anticipating information in the early phases of a product development project is a difficult task, because a large amount of uncertain and unstructured information should be analyzed. Because of a lack of guidance and support, most preproject teams eventually make only a superficial anticipation and quickly proceed straight into detailed design. Proactive thinking should, on the contrary, be stimulated, nurtured, and supported, as much as possible in the early phases. This means that detailed design should be pushed further into subsequent phases in order to keep the preproject team on its feedforward track. The problem is how to encourage uncertain information to be gathered in the early phases, without entering into detailed design. To this purpose, three methods have been found to be useful:

1. *Use of a Formal Preproject Stage.* A typical solution is to define a formal procedure that delineates the activities and documents to be produced in the concept generation and product planning phases, which usually end with top management's approval. Albeit useful and often necessary, this method is not the keystone to support feedforward planning. In fact, all companies in our sample have a formal preproject phase, and most of them also have highly structured preproject reports and formal approval gates, including those companies that are not effective in feed-forward planning, such as companies E, B, and H.

2. *Use of Managerial Techniques and Tools.* Several methods have been proposed to support the feedforward effort of the preproject team. Some of them are traditional project management tools, such as the work breakdown structure or network diagrams (i.e., PERT and Gantt). Oth-ers have been proposed in recent years specifically for concept generation and product planning. Examples are quality function deployment (QFD),

target costing, life-cycle costing, value engineering, variety reduction program, and failure mode/failure effect analysis. Network diagrams and target costing were the most widely applied techniques in our sample (see Fig. 11.4). Quality function deployment was used in six cases (B, C, D, G, J, and K), not necessarily successfully (it was not implemented in the most effective projects: A, L, and I). Indeed, as other studies demonstrate,[1, 24] QFD appears as a useful technique to stimulate proactive thinking, but it also bounds the preproject team to thorough and long early analyses, which may be useless and counterproductive in turbulent environments.

3. *Early Prototyping.* One of the most effective mechanisms for fostering early discussion and enhancing the feedforward exercise is to simulate as early as possible the physical characteristics of the new product. To this purpose, a useful method is early prototyping, that is, the early and rapid development of a prototype that reflects the expected features, functions, and characteristics of the product in a rough and approximated manner. Early prototyping should only aim at stimulating proactive thinking and not at testing detailed solutions; hence, it may be easily and quickly realized by the preproject team. A white-goods company, for instance, simply modifies a product of the previous generation in order to simulate the functions and characteristics of the new product.

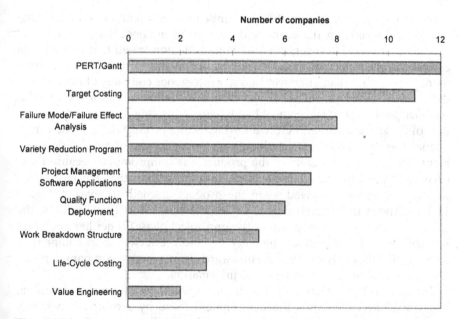

Figure 11.4 Frequency of adoption of managerial techniques and tools in the early phases of the product development process.

A car manufacturer has decided to anticipate the development of prototypes in the first phase of product development, although these prototypes may not be an accurate and detailed reproduction of the new car idea. The helicopter company uses electronic mockup systems to detect any geometric interference of parts during concept generation. For further information on current technologies that are accelerating the realization of early prototypes, see Bullinger et al.[6]

11.6. MAKING SYSTEMIC INFORMATION BECOME SYSTEMIC KNOWLEDGE

Figure 11.3 demonstrated the effectiveness of teamwork in the preproject phases and underlined how communication in the early phases is necessary in order to anticipate constraints and opportunities. However, teamwork does not appear as a sufficient condition for a successful feedforward approach. Companies C, G, J, K, and D have very high degrees of teamwork; however, this does not directly reflect a high effectiveness of feedforward, especially if compared to company I. The effectiveness of early involvement therefore also depends on other mechanisms that should be added to those creating systemic information.

11.6.1   Feedforward Effort and Selective Anticipation

One may argue that the capability to anticipate information and avoid late problems depends on the effort and time spent in concept generation and product planning. However, our case studies demonstrated that detail is not essential in preproject planning. Moreover, too much information is sometimes more detrimental than effective. In one white-goods company of our sample, the concept development team was forced to define in detail the product specifications. A strategic council had to certify the coherence and completeness of these specifications before allowing further product design. The consequence was that the preproject phase took twice the time of product development. Moreover, the quality of the product was compromised because these early analyses consumed such a large amount of time that the customer's needs were already changed when the product reached the market. Figure 11.5 illustrates that a high degree of feedforward effort (measured by the number of decisions anticipated in the preproject phase) is neither necessary nor sufficient. For example, company I is more effective than company J, although I makes only a limited feedforward effort, whereas company J makes a detailed and forced anticipation of information.

Indeed, a high degree of detail in product specifications is seldom necessary, especially when the environment faced by a company is highly turbulent (see also Iansiti[30] in these regards). In the white-goods industry, for example, the market is continuously changing as to functions and

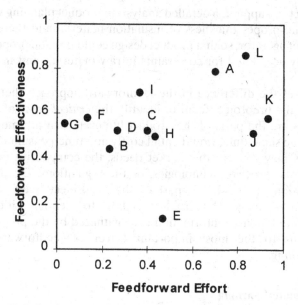

Figure 11.5 Relation between feedforward effort and feedforward effectiveness.

aesthetics of the products. In the helicopter industry, the product technology evolves rapidly. In both these cases, detailed specifications are extremely difficult to define and, at the same time, are very unreliable, so they are often changed during the downstream development stages. Detailed feedforward planning in these contexts therefore does not provide significant insight on future constraints/opportunities and eventually results in a waste of precious time. Moreover, overspecification may impair creativity in the concept development team.[31]

A successful early involvement may therefore consist of anticipating only a limited amount of information that allows one to verify the coherence between the product concept and the future constraints, leaving, at the same time, a maximum degree of freedom on the downstream phases to seize any unexpected opportunity. We call this approach *selective anticipation*: The early phases define only the overall concept and structure of the product and of the process, dedicating most attention and anticipation effort only on a few critical areas. For example, in helicopter development, the distribution of weights has to be defined and thoroughly specified as early as possible, because any subsequent change could imply a modification of several components and of the architecture of the product. A company in the white-goods industry involved in the development of a new refrigerator recognized as a critical area the high variability of the quality of the polyurethane insulation provided by its prospective supplier; a detailed analysis in product planning was devoted to defining the proper thickness of insulation material and therefore of the body of the refrigerator, so that product design could provide proper solutions, which already accounted for constraints in raw materials and in the supplying process.

What makes the difference in the feedforward approach therefore is the capability of the preproject team to identify the *critical areas* of the project. Critical areas are the parts of the product life cycle that are most uncertain and that, at the same time, would entail costly and time-consuming corrective actions. They may concern the market needs, the competitive environment, the product and process technologies, or the regulations. Detailed analysis and anticipation effort on the part of the preproject team should focus mostly on those areas. The problem is how to identify critical areas in advance, within the large information base gathered by the preproject team. This leads us to the most important driver of feedforward planning: *systemic learning*.

11.6.2   Systemic Learning

Selective anticipation is the most difficult exercise in feedforward planning. In fact, it cannot only be based on structured and well-defined rules and managerial techniques. For instance, consider the initial example that explained the feedforward behavior of product engineering and process engineering when developing a new refrigerator. This example clearly shows how critical areas were not identified by attempting to develop a new solution, that is, by means of *learning by development*. Being in the early phases, feedforward can only rely on knowledge drawn from past development projects, that is, on *learning from experience*.

The capability of a company to learn from past projects and to incorporate this experience into the preproject phase appears therefore as the basic rule for effective feedforward planning. This has been confirmed by the 12 case studies in our sample. To this purpose, the impact of the learning mechanisms on the effectiveness of feedforward planning has been analyzed by means of the structured data in the questionnaires. Figures 11.6 and 11.7 show the results of this analysis. Note that the degree of systemic learning in the diagrams does not account for the specialized and individual experiences of the team members in their own departments. Because it represents the ability to develop *systemic* knowledge, systemic learning measures how companies learn from past mismatching between upstream solutions and downstream constraints. The systemic learning index therefore consists of a fuzzy function that depends on: the existence of a formal project termination or project audit phase; the relative importance, in that phase, of the analysis of mismatching between initial solutions and resulting constraints/opportunities; and the presence in the project termination team of the same members who belonged to the preproject team. Figure 11.6, in particular, illustrates the impact of systemic

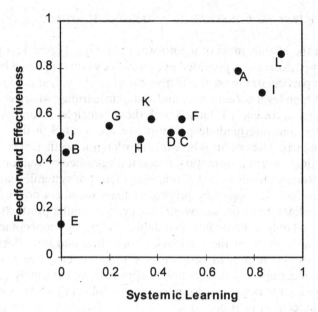

Figure 11.6 Relation between systemic learning and feedforward effectiveness.

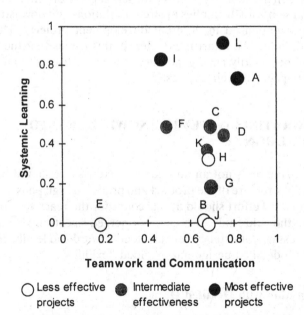

Figure 11.7 Benefits of the use of systemic learning jointly with teamwork.

learning on the effectiveness of feedforward planning. Figure 11.7 points out the superior performances provided by teamwork when supported by systemic learning. In particular, the most effective companies (A, I, and L) are characterized by a high level of teamwork and systemic learning, whereas the unsuccessful project in company E lacked both these principles. Moreover, companies with low and intermediate performance are located in the lower-right area of the matrix. This confirms how teamwork is not a sufficient condition for effective anticipation of information, unless it is sustained by systemic learning.

The keystone of feedforward planning is therefore systemic learning, that is, the experience developed by preproject team members not only in their specific disciplines, but from the overall life cycle of similar products. Systemic learning occurs only if the people who define the product concept and specifications are also aware of the feedback information and the effects of their initial choices. Therefore, in order to stimulate systemic learning, project termination meetings and project audits appear to be extremely useful. Preproject team members should also play a major role in these ex-post analyses of project outcomes, in order to learn from variations between prospected early solutions and actual final results (further details on the methods for learning from past projects may be found in Bartezzaghi et al.[3] and Clark and Wheelwright[9]). In companies A, L, and I, for example, the preproject team also participates in the project termination phase; in these project termination meetings, priority is given to discussing any unexpected problems and to understanding the reasons for variations in project results. Job rotation of team members, on the contrary, seems to be less capable of generating systemic learning, because it usually implies upstream solutions and downstream information being experienced in different development projects. The learning cycle would therefore be interrupted. Hence, this means that the preproject team does not necessarily need generalists, but specialists with knowledge of the systemic impacts of their choices.

11.7.   INTEGRATING SYSTEMIC KNOWLEDGE INTO DETAILED DESIGN

Feedforward planning is not an autonomous exercise. It should be coherent with the internal structure of the product and process design phases. In particular, the feedforward effort should also account for the reactive capabilities of a company. In the field study, two factors emerged as extremely important to ensure the necessary consistency between feedforward and feedback planning: integration with detailed design and planned flexibility.

11.7.1   Integration with Detailed Design

Efforts aimed at anticipating information in preproject activities are fruitless if this information is not transferred and exploited in the following phases of product development.[21] A white-goods company, for example, designated its R&D, marketing, and manufacturing directors as responsible for developing the concept and specifications of a new strategic refrigerator. These highly experienced people were able to identify and anticipate information on a number of critical areas of the new product. However, the refrigerator faced many problems and modifications during production ramp-up because product engineers had only a minor knowledge of the issues that were identified by this high-level preproject team.

We observed several ways of integrating preproject planning with the downstream detailed design. Most companies rely on reports that define product specifications. Albeit necessary, these reports are rarely able to transfer the richness of the unstructured information discussed in preproject team meetings. The most effective solution is to provide continuity between the people involved in the early phases and the design teams: first of all, by ensuring that the project manager already belongs to the preproject team; second, by keeping the preproject team together also during product and process design (e.g., as a product planning council); and third, by involving working-level members early in the preproject phase, at least in some meetings or in the preliminary collection of downstream information.

11.7.2 Planned Flexibility

The anticipation of constraints and opportunities aims at reducing the probability of late corrective actions due to unexpected problems. However, not all the information may be anticipated at the beginning of a project, and the risk of implementing corrective actions cannot be avoided for all the critical areas. In the helicopter industry, for example, defining detailed specifications for the avionics systems was useless, because their technology is continuously changing. In an oven development project, the reliability of a new temperature control system was not certifiable in the product planning phase. In these cases, where the information needed for defining product specifications does not become available until the team is involved in detailed design, a feedback approach seems to be more effective.[38, 40, 43, 44]). However, feedforward planning again plays a central role in this reactive policy. In fact, the cost and time of late corrective actions may be limited if preventive measures are taken at the concept generation phase. For example, the product may be conceived with a modular structure so that any change in a single part will not affect the overall system. Protective and alternative solutions may also be previously arranged to overcome any unfeasibility problem due to a new technology. Finally, additional resources may be allocated to the most uncertain activities. These preventive measures confer *flexibility* to the product development process (on the concept of development flexibility, see Thomke[40]). Without this flexibility, the feedback approach would provide poor performance. Note that preventive measures are effective provided they are arranged in the early phases of the project. For this reason, an essential driving force of feedforward planning is what we call *planned flexibility*. This entails that the flexibility of the product development process be planned early so that late adjustments and innovations may be accepted with marginal consequences. In other words, feedforward planning should aim at both:

- Reducing uncertainty about constraints and opportunities by anticipating downstream information, thus decreasing the probability of implementing late corrective actions.
- Reducing the cost and time of possible corrective actions through planned flexibility.

In order to plan for flexibility, the preproject team should also account for the internal structure of the subsequent phases of the product development process. In particular, there are two major alternatives to articulate the development process: a stage–gate system and overlapping activities.[31] The former means that product development is split into sequential phases (e.g., concept generation, product planning, product design, and process design), separated by checkpoints that verify the completeness of upstream solutions, which thereafter are not allowed to change.[10] The overlapped approach, on the contrary, consists of simultaneously carrying out different phases. Clark and Fujimoto[8] analyzed the superior performance provided by overlapping product design and process design in car development, while Iansiti[30] analyzed the benefits of overlapping preproject and detailed design in the mainframe and multimedia industries. Overlapping activities is a feedback approach. In fact, it is based on the feedback provided by actual practices in attempting to implement the upstream design.[20] Moreover, as there are two or more phases still in progress, overlapping activities entails that late changes may be accepted with marginal consequences. For example, insofar as the product specifications are not frozen, the product concept may be adjusted to handle sudden changes in customers' needs or in the competitive environment and product design may be updated according to new technological developments, especially those concerning the manufacturing process. In other words, the overlapped approach enlarges the window of reacting to late constraints and opportunities.

Figure 11.8 depicts the degree of overlap between product design and process design in the 12 projects and its correlation with the time-to-market performance of the companies (considered as the total development effectiveness). The figure shows that the overlapped approach in itself does not provide high performance (see companies B, C, and E). Indeed, as discussed previously, reactive approaches may be adopted, provided the development flexibility they need is properly planned in the early phases. Hence, one may expect that overlapping activities are effective, provided they are supported by the planned flexibility principle. For this reason, we further investigated the mutual interactions between feedforward and feedback planning in the 12 companies.

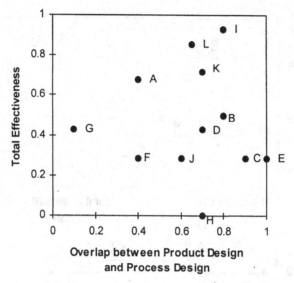

Figure 11.8 Impact of overlapping activities on product development performance.

Figure 11.9 shows the results of this analysis. The companies have been classified according to two dimensions: the degree of detailed anticipation (number of detailed decisions taken in the preproject phase) and the degree of selective anticipation (number of critical areas investigated in the preproject phase). The figure shows five classes of companies that may be identified as:

1. *Low Feedforward and Low Feedback.* Exemplified by companies F and G, these companies place low integration efforts in the product development process. They also have a poor time-to-market performance.
2. *Detailed Feedforward and Low Feedback.* Exemplified by companies A, J, and L, these companies adopt a stage–gate system. Because they do not resort to the overlapped structure, they must devote a large and detailed effort to feedforward planning, in order to be sure to avoid any subsequent modification (which would be extremely costly and time consuming). This approach is suitable when the environment is not highly turbulent and when a company has high feedforward capabilities (note that company J, which has poor systemic learning mechanisms, is not able to make this approach fruitful).
3. *Low Feedforward and High Feedback.* Exemplified by companies B and D, these companies extensively overlap the development activities, but do not plan for the necessary flexibility in the early phases. This approach

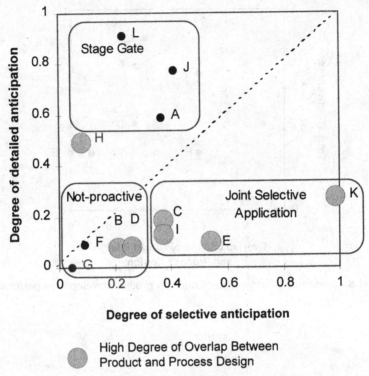

Figure 11.9 Interactions between feedforward and feedback planning.

does not provide high performance, confirming that overlapping is not effective if not assisted by feedforward planning.

4. *Detailed Feedforward and High Feedback.* Exemplified by company H, these companies try to be proactive, but make improper use of feedforward mechanisms (company H has poor systemic learning). Hence, they are also compelled to adopt a reactive approach to deal with unexpected problems. Note that the performance of company H is poor when compared with companies B and D. This means that feedforward may result in a waste of time and may become a negative exercise if not properly applied, and that feedback is not able to compensate for errors of an improperly detailed preproject stage. Under these conditions, company H would probably achieve better performance by accelerating feedforward and going rapidly into detailed design.

5. *Selective Feedforward and High Feedback.* Exemplified by companies C, E, I, and K, these companies jointly apply both early involvement and overlapping. This approach may be extremely successful (see com-

pany I). In fact, on the one hand, planned flexibility makes overlapping activities effective. On the other hand, overlapped activities support the selective anticipation principle. This approach avoids early phases that go into deep details, because upstream solutions (concept and product drawings) are unfrozen and may be changed later on at low cost and time. However, it is also the most complex to be performed, because it strongly depends on the systemic learning mechanisms that allow critical areas to be selected (see the opposite behaviors of companies I and E).

These considerations highlight that feedforward and feedback planning are not reciprocally incompatible. On the contrary, they may be jointly applied with outstanding results and mutual support, through a proper combination of proactive capabilities, planned flexibility, and reactive mechanisms.

11.8   CONCLUSIONS

Anticipating information in concept generation and product planning is a complex task, because of the high uncertainty induced by innovation and environmental turbulence. In this chapter, the basic feedforward mechanisms that make early project activities successful have been investigated, namely, teamwork and communication, harmonized objectives, encouraged and supported proactive thinking, systemic learning, integration with detailed design, and planned flexibility. Most of these principles rely on the capabilities of the preproject team to balance between two extremes: on the one side, the need to provide freedom to downstream phases in order to incorporate new technological advances or market requirements that emerge in late development; on the other side, the necessity to establish fixed targets and specifications in order to focus the development activities on feasible and strategically consistent products, thus avoiding the detrimental impacts of late corrective actions. This art of balancing between feedback and feedforward planning is not encoded in managerial techniques. It is incorporated in the behavioral and learning patterns of managers, designers, and engineers. Indeed, systemic knowledge appears as the basic engine for the anticipation of manufacturing constraints and opportunities in the early project phases. This means that product designers and process engineers involved in concept generation and product planning should deeply understand the reciprocal impacts of their innovative solutions before entering into detailed design. This knowledge is nurtured by a continuous and systemic development of capabilities, that is, by systemic learning. Appointing the preproject team for investigating, during project termination and project audits, the reasons for downstream problems, accumulating systemic knowledge into structured databases, continuously improving procedures and organizational systems, and transferring knowledge from past projects, are necessary keys toward the proper early anticipation of constraints and opportunities. Learning from the past is not marginal, even in highly innovative product and process developments.

Findings from this chapter need to be supplemented with further investigation. An important issue is to explore the effect of context factors, such as industry or company size, on the preceding mechanisms. A preliminary look at the data collected in the 12 companies, however, seems to maintain that the six mechanisms are effective whatever the industry and context. For example, the three most effective companies, A, I, and L, which apply all those mechanisms, belong to three different industries (i.e., vehicles, white goods, and helicopter, respectively). Also, the choice of a detailed or selective anticipation approach does not seem to be influenced by the type of industry. Probably other factors, such as the innovation strategy of a company or its behavioral and learning patterns, are more relevant in leading toward a principally proactive or reactive approach. Understanding these patterns would require enlarging the sample of investigated companies. This will be a topic of further research.

APPENDIX

In order to compare the behavior and performance of the 12 companies, structural data where collected by means of a questionnaire with more than 200 parameters. By aggregating some of these parameters, we obtained measures for the eight synthetic variables reported in Figures 11.3 and 11.5 to 11.9 (feedforward effectiveness, teamwork and communication, feedforward effort, systemic learning, total effectiveness, overlap between product and process design, degree of selective anticipation, and degree of detailed anticipation). In some cases, aggregation is implemented using fuzzy-set theory.[48] This is necessary when a synthetic variable has two major characteristics: (1) It is *multidimensional,* being dependent on several constituting parameters through a hierarchical structure (e.g., systemic learning depends on the main focus of project termination, on the main roles involved in project termination, and on the main roles involved in the preproject phase); and (2) its constituting parameters are measured in the questionnaire by *heterogeneous* indexes (e.g., the focus of project termination is a Likert-like scale, whereas the involvement of specific roles, such as manufacturing or product engineering, in project termination is a logical variable). In this case, "crisp" fuzzy logic is adopted. That is, heterogeneous parameters are transformed into homogeneous fuzzy functions whose values range between 0 and 1; then these constituting fuzzy functions are aggregated through fuzzy operators. In particular, two compensatory operators are used in the study:

$$X.Fuzzyand.Y = \alpha \min (X, Y) + (1 - \alpha) (X + Y)/2$$
$$X.Fuzzyor.Y = \alpha \max (X, Y) + (1 - \alpha) (X + Y)/2 \quad \text{where } \alpha = 0.2$$

Table 11.1 reports the aggregation operators (either simple or fuzzy) for the eight synthetic variables analyzed in the study. A brief explanation of the constituting parameters from which the synthetic measures have been derived is also provided.

For example, consider the variable "feedforward effectiveness" for company B. This variable was measured by aggregating information from the structured questionnaire of the company.

First, we analyzed the flaws that occurred in the company's project. To this end, several typical shortcomings and problems that may occur during a project were enumerated in the questionnaire. Some of them were related to feedforward capabilities (such as the occurrence of reworks and of unexpected events). Other flaws were dependent on other factors (such as a lack of top management support or problems in integrating CAD/CAM systems). The company indicated the occurrence of each flaw on a Likert-like scale (with 1 denoting that the flaw did not occur and 5 denoting that the flaw caused major problems in the project). The reason for this approach is that managers may provide robust judgments only if they make relative evaluations and cannot directly identify what is commonly considered as a "good" answer in absolute terms. Company B assigned a 3 for the parameter "Reworks" and a 4 for "Unexpectedevents." These parameters are transformed into values ranging between 0 and 1 by means of standard fuzzy functions. In particular, "Reworks" becomes 0.6 and "Unexpectedevents" becomes 0.76. Then these functions are aggregated by means of a *Fuzzyor* operator, which provides a result of 0.70.

The second type of information collected concerned the variance between initial target performance and actual project outcome. The cycle time of the project was 25 percent longer than the time to market defined in the concept generation phase. Similarly, the product cost was 15 percent higher than expected. These variances are transformed into a number ranging between 0 and 1 by means of fuzzy functions whose shape depends on the competitive priority of time and cost for the company (see Fig. 11.10). In fact, company B also enumerated its competitive priorities (ranging among several parameters such as time to market, product cost, product functional performance, delivery dependability) on a Likert-like scale (with 1 indicating that the performance is not relevant and 5 denoting that the performance is crucial for competition). If the competitive priority is low, then the variance is transformed into a small value, whereas if the competitive priority is high, the variance results in a high value. As to company B, time to market is considered as a top priority (5 on the Likert-like scale) and product costs follow just behind (4). Then the resulting values are 0.5 and 0.15. These values are then aggregated through a *Fuzzyor* operator. The result (0.36) is again combined with the fuzzy function, indicating the occurrence of flaws (0.70), by means of a *Fuzzyor* operator, which provides 0.56. Finally, $1 - 0.56$ equals 0.44, that is the feedforward effectiveness of company B.

TABLE 11.1 Measures for the Synthetic Variables Analyzed in Figures 11.3 to 11.9

Synthetic Variable	Aggregation Operator	Notes on Constituting Parameters
Feedforward effectiveness	$1 - [(Reworks.Fuzzyor.Unexpectedevents).Fuzzyor.(Costvariance.Fuzzyor.Timevariance))] -$	Reworks = occurrence of reworks during development, relative to other flaws in the project not depending on feedforward capabilities (e.g., a lack of top management support); Unexpectedevents = occurrence of events initially unforeseen, relative to other flaws in the project; Costvariance = variance of final product costs with respect to the target cost established in the early phases. This variance is considered to be extremely critical if product cost is a competitive priority for the company; Timevariance = variance of the actual time to market with respect to the expected time to market established in the initial plan. This variance is considered to be extremely critical if time is a competitive priority for the company.
Teamwork and communication	Departments.Fuzzyand.Attitude.Fuzzyand.Communication	Departments = number of roles involved in the preproject team (platform manager or product planning, marketing, research or product engineering, process engineering, manufacturing, purchasing, lead customers, major suppliers); Attitude = importance of attitude toward integration when selecting the preproject team members relative to other skill parameters (e.g., technical competencies); Communication = percentage of colocated team members.
Feedforward effort	Σ Anticipated decisions/30	Thirty-five typical development decisions were analyzed (concerning the product concept, the product package, the manufacturing process, and the organization of the development project). "Anticipated decisions" are the decisions among these 35 that have been taken in the concept generation or product planning phases of the project. 30 is simply a scaling factor.

Systemic learning	Team.*Fuzzyand*.Focus	Team = percentage of preproject team members involved in project termination of previous similar projects. Focus = importance of analysis of variances during project termination of past projects, relative to other purposes of project termination (execution of production ramp-up, late problem fighting, resource evaluation.
Total effectiveness	TTMcompetitive.*Fuzzyand*.TTMstrategy	TTMcompetitive = time to market of the project compared to the time to market of competitors for similar products. TTMstrategy = ranking of strategic relevance of time to market, relative to other strategic performances of the company.
Overlap between product and process design	$\dfrac{\text{Product design time} + \text{Process design time}}{\text{Total detailed design time}} - 1$	The ratio is the simultaneity.ratio proposed by Clark and Fujimoto.[8] 1 is a scaling factor.
Degree of detailed anticipation	Σ Detailed anticipated decisions/22	Number of decisions (among the 35 considered in the questionnaire) that have been analyzed and established with a high level of detail in concept generation or product planning. 22 is a scaling factor.
Degree of selective anticipation	Σ Selectively anticipated decisions/22	Number of decisions that have been analyzed and established only by focusing on critical areas in concept generation or product planning, while detailed decisions have been forwarded to detailed design. 22 is a scaling factor.

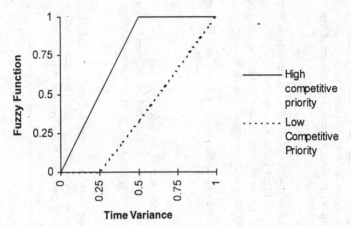

Figure 11.10 Transformation of a constituting parameter into a fuzzy function.

REFERENCES

1. Akao, Y. (ed.), 1990, *Quality Function Deployment. Integrating Customer Requirements into Product Design,* Productivity Press, Cambridge, MA.
2. Bacon, G., S. Beckman, D. Mowery, and E. Wilson, 1994, Managing Product Definition in High-Technology Industries: A Pilot Study, *California Management Rev.,* Vol. 36, pp. 32–57.
3. Bartezzaghi, E., M. Corso, and R. Verganti, 1997, Continuous Improvement and Inter-Project Learning in New Product Development, *Internat. J. Technol. Management,* Vol. 14, No. 1, pp. 116–138.
4. Blanchard, B. S., 1979, *Design and Manage to Life Cycle Cost,* M/A Press, Portland.
5. Brown, S. L., and K. M. Eisenhardt, 1995, Product Development: Past Research, Present Findings, and Future Directions, *Academy of Management Rev.,* Vol. 20, No. 2, pp. 343–378.
6. Bullinger, H. J., D. Fischer, and K. Worner, 1995, Rapid Prototyping—A Method for the Development of Innovative Products, *13th International Conference on Production Research,* Jerusalem.
7. Black, K. B., 1989, Project Scope and Project Performance: The Effect of Parts Strategy and Supplier Involvement on Product Development," *Management Sci.,* Vol. 35, pp. 1247–263.
8. Clark, K. B., and T. Fujimoto, 1991, *Product Development Performance,* Harvard Business School Press, Cambridge, MA.
9. Clark, K. B., and S. C. Wheelwright, 1993, *Managing New Product and Process Development,* Free Press, New York.
10. Cooper, R. G., Stage–Gate Systems: A New Tool for Managing New Products, *Business Horizons,* Vol. 33, No. 3.
11. Cooper, R. G., 1994, Third-Generation New Product Process, *J. Product Innovation Management,* Vol. 11, pp. 3–14.

12. Cooper, R. G., and E. J. Kleinschmidt, 1994, Determinants of Timeliness in Product Development, *J. Product Innovation Management*, Vol. 11, pp. 381–396.

13. Cooper, R. G., and E. J. Kleinschmidt, 1995, Benchmarking the Firm's Critical Success Factors in New Product Development, *J. Product Innovation Management*, Vol. 12, pp. 374–391.

14. Corso, M., M. Muffatto, and R. Verganti, 1996, Multi-Product Innovation: Emerging Policies in Automotive, Motorcycle, and Earthmoving Machinery Industries, *Third International Product Development Management Conference, EIASM*, Fontainebleau, France.

15. De Brentani, U., 1991, Success Factors in Developing New Business Services, *European J. Marketing*, Vol. 15, No. 2, pp. 33–59.

16. De Maio, A., R. Verganti, and M. Corso, 1994, A Multi-Project Management Framework for New Product Development, *European J. Oper. Res.* Vol. 78, pp. 178–191.

17. De Weerd-Nederhof, P. C., I. C. Kerssens-Van Drongelen, and R. Verganti (eds.), 1994, *Managing the R&D Process*, TQC, Enschede, The Netherlands.

18. Dyer, J. H., and W. G. Ouchi, 1993, Japanese-style Partnerships: Giving Companies the Competitive Edge, *Sloan Management Rev.*, Fall, pp. 51–63.

19. Ettlie, J. E., 1996, Global Comparisons of Early Manufacturing Involvement (EMI) in New Product Development, *Third Product Development Management Conference, EIASM*, Fontainebleau, France, pp. 275–298.

20. Fujimoto, T., 1989, Organization for Effective Product Development: The Case of the Global Automobile Industry, Ph.D. Dissertation, Graduate School of Business Administration, Harvard University, Boston.

21. Fujimoto, T., 1993, Information Asset Map and Cumulative Concept Translation in Product Development, *Design Management J.*, Fall, pp. 34–42.

22. Fujimoto, T., 1997, Shortening Lead Time through Early Problem Solving—A New Round of Capability Building Competition in the Auto Industry. *International Conference on New Product Development and Production Networks, WZB*, Berlin.

23. Gerwin, D., 1993, Integrating Manufacturing into the Strategic Phases of New Product Development, *California Management Rev.* Vol. 35, pp. 123–138.

24. Govers, C. P. M., 1996, QFD as a way of Management, *R&D Management Conference on Quality and R&D*, Enschede, The Netherlands, pp. 144–153.

25. Gupta, A. K., and D. L. Wilemon, 1990, Accelerating the Development of Technology-Based New Products, *California Management Rev.*, Vol. 32, No. 2, pp. 24–44.

26. Haddad, C., 1997, Involving Manufacturing in the Early Stages of Product Development: A Case Study from the North American Automotive Industy, International Conference on New Product Development and Production Networks, WZB, Berlin.

27. Hayes, R., S. C. Wheelwright, and K. B. Clark, 1988, *Dynamic Manufacturing. Creating the Learning Organization*, Free Press, New York.

28. House, C. H., and R. L. Price, 1991, The Return Map: Tracking Product Teams, *Harvard Business Rev.*, January–February, pp. 92–101.

29. Hauser, J. R., and D. Clausing, 1988, The House of Quality, *Harvard Business Rev.*, May–June, pp. 63–73.

30. Lansiti, M., 1995, Shooting the Rapids: System Focused Product Development in the Computer and Multimedia Environment, Working Paper 95-026, Harvard Business School, Boston.

31. Imai, K., I. Nonaka, and H. Takeuchi, 1988, Managing the New Product Development Process: How Japanese Companies Learn and Unlearn, in *The Uneasy Alliance: Managing the Productivity–Technology Dilemma*, K. B. Clark, R. H. Hayes, and C. Lorenz (eds.), Harvard Business School Press, Boston, pp. 337–375.

32. Jürgens, U., 1995, Anticipating Problems with Assembly in the Early Stages of Product Development, *Third International Workshop on Assembly Automation*, Venice.

33. Kamath, R. R., and J. K. Liker, 1994, A Second Look at Japanese Product Development, *Harvard Business Rev.*, November–December, pp. 154–173.

34. Liker, J. K., R. R. Kamath, S. N. Wasti, and M. Nagamachi, 1996, Supplier Involvement in Automotive Component Design: Are There Really Large US Japan Differences? *Res. Policy*, Vol. 25, pp. 59–89.

35. Meredith, J. R., and S. J. Mantel, 1989, Project Management. A Managerial Approach, Wiley, New York.

36. Rosenthal, S. R., 1992, *Effective Product Design and Development: How to Cut the Lead-Time and Increase Customer Satisfaction*, Business One Irwin, Homewood, IL.

37. Slack, N., S. Chambers, C. Harland, A. Harrison, and R. Johnston, 1995, *Operations Management*, Pitman, London.

38. Smith, P. G., and D. G. Reinertsen, 1991, *Developing Products in Half the Time*, Van Nostrand Reinhold, New York.

39. Souder, W. E., and R. K. Monaert, 1992, Integrating Marketing and R&D Project Personnel within Innovation Projects: An Information Uncertainty Model, *J. Management Studies*, Vol. 29, No. 4, pp. 485–512.

40. Thomke, S. H., The Role of Flexibility in the Development of New Products: An Empirical Study, Working Paper 96-066, Harvard Business School, Boston, MA.

41. Thompson, J. D., 1967, *Organisations in Action*, McGraw-Hill, New York.

42. Trygg, L., 1991, The Need for Simultaneous Engineering in Time-Based Competition, in *Technology Management: Technology Strategies and Integrated Information Systems in Manufacturing*, J. Ranta, and S. Å. Hörte (eds.), VTT, Helsinki.

43. Von Hippel, E., and M. J. Tyre, 1995, How Learning by Doing is Done: Problem Identification in Novel Process Equipment, *Res. Policy*, Vol. 24, pp. 1–12.

44. Ward, A., J. K. Liker, J. J. Cristiano, and K. S. Durward II, The second Toyota Paradox: How Delaying Decisions Can Make Better Cars Faster, *Sloan Management Rev.*, Spring, pp. 43–61.

45. Wheelwright, S. C., and K. B. Clark, 1992, Creating Project Plans to Focus Product Development, *Harvard Business Rev.*, March–April, pp. 70–92.

46. Wheelwright, S. C., and K. B. Clark, 1992, *Revolutionizing Product Development: Quantum Leaps in Speed, Efficiency and Quality*, Free Press, New York.

47. Wheelwright, S. C., and W. E. Sasser, 1989, The New Product Development Map, *Harvard Business Rev.*, May–July, pp. 112–123.

48. Zimmerman, H. J., 1993, *Fuzzy Set Theory and Its Application*, 6th ed., Kluwer Academic, London.

12

EFFECTIVE PRODUCT AND PROCESS DEVELOPMENT USING QUALITY FUNCTION DEPLOYMENT

ROHIT VERMA and TODD MAHER
DePaul University

MADELEINE PULLMAN
Southern Methodist University

12.1 INTRODUCTION

Designing new products and/or modifying the attributes of existing products to satisfy the needs of customers in the marketplace has captured the attention of engineering, business, and product development researchers for the last several decades. Therefore, a number of books and journals continue to publish research articles highlighting different steps of the product development process (see e.g., Green and Krieger,[5] Ulrich and Eppinger,[10] Urban and Hauser,[11] and Wheelwright and Clark[12]). The national and international conferences of several professional organizations (e.g., Product Development and Management Association, American Production and Inventory Control Society, Decision Sciences Institute, Production and Operations Management Society) also conduct multiple sessions on designing and developing products based on customers preferences.

During recent years, quality function deployment (QFD) has been recognized as an effective method for product and process development.[1] Quality

Integrated Product and Process Development, Edited by John Usher, Utpal Roy, and Hamid Parsaei
ISBN 0-471-15597-7 © 1998 John Wiley & Sons, Inc.

function deployment is a structured approach for integrating the *voice of the customer* into the product design/development process.[6,7] The purpose of QFD is to ensure that customer requirements are factored into every aspect of product development from planning to production floor. Quality function deployment uses a series of matrices, which look like houses, to deploy customer input throughout design, manufacturing, and delivery of products. The premise is that cooperation and communication among marketing, manufacturing, engineering, and R&D leads to greater new-product success.

As mentioned earlier, QFD connects the *voice of the customer* into the design, development, and production process. The *voice of the customer* is a generic term representing a hierarchical set of customer needs where each need (or set of needs) has assigned to it a priority, which indicates its importance to the customer. The first QFD matrix, called the *house of quality,* links the voice of the customer to the product design attributes (*voice of the engineer*). The second matrix (*design matrix*) of QFD links the design attributes to the product components or features. The *operating matrix* further links the product components to process decisions. And, finally, the *control matrix* links the operating processes to production planning and control decisions.

Since the publication of Hauser and Clausing's[7] article, the customer preferences aspect of QFD has received a lot of attention in the literature. A number of published articles have documented the benefits of using the *house of quality* and/or shown how to collect data for constructing the house of quality. For example, Griffin and Hauser[6] present a comparison of different approaches for collecting customer preferences in QFD. Kim et al.[8] developed a decision support system for QFD using fuzzy multicriteria methodologies. Their models allow the product designer to consider trade-offs among various customer attributes, as well as to simultaneously consider the inherent fuzziness in the associated relationships. Chakravarty and Ghose[3] demonstrate the need and use of system theory–related paradigms for developing quantitative and qualitative models for tracking product/process interactions in QFD.

Even though a number of publications stress the usefulness of the house of quality, none of the well-cited articles shows an example of the complete QFD process. To successfully integrate the voice of the customer into the product design and development process, it is critical that the house of quality information is translated downstream to the other three matrices. Both new and existing products are made of several components, which are manufactured by multiple processes. Completing all four matrices will allow managers to identify and control the critical process parameters and will therefore lead to effective product and process development. Therefore, this chapter explains the main ideas behind the QFD process and presents an extended example of the use of QFD in manufacturing electrical transformers. This example will show how customer preferences can be deployed throughout product design and process development using quality function deployment.

12.2 QUALITY FUNCTION DEPLOYMENT PROCESS

Mitsubishi's Kobe Shipyard is credited for developing and using QFD for the first time as a product/process design tool in 1972. Shortly thereafter, building on earlier efforts at Mitsubishi, Toyota developed advanced QFD concepts and has used the technique since 1977 with very impressive results. According to Evans and Lindsay,[4] between 1977 and 1979 Toyota realized a 20 percent reduction in startup costs on the launch of a new van. The startup costs were down 38 percent in 1982 and were down 61 percent in 1984, with respect to 1977 costs. Additionally, the new product development lead time for Toyota was reduced by one-third and the product quality improved dramatically. Since then, a number of companies in Japan, the United States, Europe, and the rest of the world have implemented QFD with good results. Xerox and Ford initiated the use of QFD in the United States in 1986. Since then, a number of leading companies, including General Motors, Motorola, Kodak, IBM, Procter & Gamble, AT&T, and Hewlett-Packard, have successfully used QFD for product/process design and development.[4] An overview of the QFD process in presented in Table 12.1.

Figure 12.1 shows the relationship between the four QFD matrices. The QFD concepts presented in Table 12.1 and Figure 12.1 can be implemented for almost any manufacturing product and/or process. Here we present a QFD example for the design and manufacture of electrical transformers. This manufacturing company is located in the midwestern United States and will be refered to as Electric Equipment Company throughout the chapter.

Electric Equipment Company (EEC) designs and manufactures custom ferroresonant (regulating) transformers, linear/isolation transformers, inductors/chokes, and mercury vapor ballasts. It has been in business for over 25 years, primarily in magnetic technology. Electric Equipment Company uses just-in-time (JIT)/continuous-flow production concepts in a cellular manufacturing environment, stressing the Kaizen philosophy of continuous improve-

TABLE 12.1   Quality Function Deployment Process

Step	Activity
1	Identification of customer needs and preferences
2	Relationship between customer needs and engineering design characteristics
3	Interrelationships between the engineering design characteristics
4	Competitive evaluation of competing products and targets for design attributes
5	Linking engineering design characteristics and component characteristics
6	Linking component characteristics and the process operations
7	Linking process operations and control parameters
8	Implementation and continuous improvement

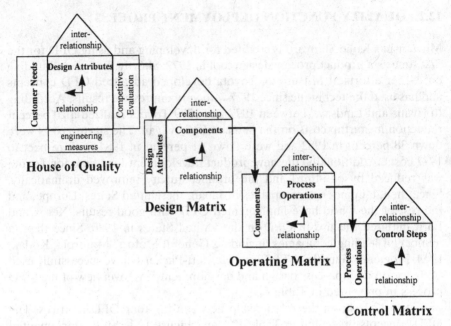

Figure 12.1   Quality function deployment process.

ment. The following sections show how QFD concepts were implemented at EEC.

12.2.1   Customer Needs and Preferences

The QFD process starts with the identification of the needs of the customers. A customer need is a description, in the customer's own words, of the benefit to be fulfilled by a product. Identifying and prioritizing customer needs are extremely important for effective product development because, generally, consumers evaluate product(s) on more than one criterion.[2, 9] Therefore, the QFD process starts with the collection of qualitative and/or quantitative information from the customer about his or her needs and preferences.

Companies can use a variety of methods or "listening posts" to collect information from customers. Griffin and Hauser[6] consider the gathering of customer information to be a qualitative task (including personal interviews, group interviews, and focus groups). In a typical study, between 10 and 30 customers are interviewed for approximately one hour each in a one-to-one setting. The interviewer probes the customer, searching for a better description of his or her needs. The interview ends when the interviewer feels that no new needs can be articulated from the customer. Group interviews and/or focus groups can also be used to identify the needs of customers. A focus group is a randomly selected panel of individuals who answer questions about a product(s) and discuss the attributes of the products that satisfy the customer needs. In a recent study, however, Griffin and Hauser[6] found both person-to-person and focus groups to be equally effective in articulating customer needs.

The customer needs for EEC were determined jointly by staff members in the marketing, sales, and engineering divisions, as all three groups interact with the customer. The information was obtained through interviews, surveys, and brainstorming over several months. One production manager at EEC was responsible for information collection and for developing the four QFD matrices, while other members of the organization provided feedback (in individual or group settings) whenever necessary. The customers of EEC identified cost and product reliability to be the two most important attributes of the transformers. They emphasized that the transformers should be reliable and should conform to government regulations at the lowest possible cost. They also regarded on-time delivery to be very important. The low rate of temperature rise in transformers was identified as another important product attribute. Conforming to industrial and professional standards (UL/CSA/VDE/CE and others) was also considered to be very important. Because several types of transformers manufactured by EEC are installed in other electronic products (e.g., the power supply for medical diagnostic equipment), a low noise level was very important to customers. Additionally, the customers identified efficiency, small size, and aesthetics as other important attributes.

After the identification of product attributes, it is important to prioritize them based on their relative importance to consumers. A number of quantitative techniques can be used in combination with the qualitative methods described previously to prioritize the customer needs. For example, marketing professionals have used rating and/or ranking methods to identify the relative importance of product attributes for a long time. Refer to the text by Urban and Hauser[11] for detailed discussions on various marketing research methods. Other techniques for prioritizing customer requirements include conjoint analysis and discrete-choice analysis, which are based on factorial experimental design procedures and econometric models.[2, 9] Electric Equipment Company used traditional customer surveys in addition to qualitative information collected during interviews to prioritize customer requirements on a scale of 1 to 10 (1 = least important, 10 = extremely important). The project team chose not to use other sophisticated techniques (conjoint and/or discrete-choice analyses) because of time and resource constraints. The objective of this QFD project was to understand the relationships among customer preferences, design attributes, process operations, and control parameters. Therefore, the project team decided to first develop the four QFD matrices and understand the interrelationships between various product/process parameters before undertaking considerable quantitative data collection efforts.

Figure 12.2 shows a completed house of quality for transformer manufacturing. The prioritized customer information is presented on the left-hand side of the house on a scale of 1 to 10 (10 = extremely important, 1 = least

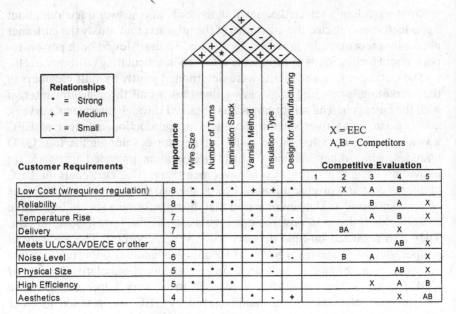

Figure 12.2 House of quality.

Customer Requirements	Importance	Wire Size	Number of Turns	Lamination Stack	Varnish Method	Insulation Type	Design for Manufacturing	Competitive Evaluation 1	2	3	4	5
Low Cost (w/required regulation)	8	*	*	*	+	+	*			X	A	B
Reliability	8	*	*	*		*				B	A	X
Temperature Rise	7			*	*	-				A	B	X
Delivery	7			*			*			BA	X	
Meets UL/CSA/VDE/CE or other	6				*	*					AB	X
Noise Level	6				*	*	-			B	A	X
Physical Size	5	*	*	*		-					AB	X
High Efficiency	5	*	*	*						X	A	B
Aesthetics	4				*	-	+				X	AB

Relationships: * = Strong, + = Medium, - = Small

X = EEC
A,B = Competitors

important). For the sake of clarity, only the more important attributes are presented in Figure 12.2. Cost and reliability were identified as the two most important attributes, followed by temperature and delivery performance. The next sections explain how these customer preferences were linked to the engineering design attributes of the transformer.

12.2.2 Engineering Design Characteristics

The information collected in the previous QFD step presents the prioritized needs of the customer. For effective product design and development, it is necessary to translate those customer requirements into the relevant engineering design attributes of the product. A careful analysis by members from the engineering, product development, manufacturing, and marketing departments is often necessary to identify all relevant product attributes that affect customer preferences.

An electrical transformer consists of metallic coil(s) (mostly copper or copper alloys) tightly wound over a stack of metallic lamination sheets. The finished transformer is used as a device for controlling electrical current and/or voltage in the power module for various products (e.g., audio systems, medical equipment, and power supply lines). Even though the exact engineering specifications are different for different types of transformers, the following six product elements were identified (by the engineering staff at EEC) to be the most important for all types of transformers: wire size, number of wire turns, lamination stack size, varnish method, insulation type, and design for manufacturability. The wire size, the number of turns, and the lamination stack directly affect the properties of the electromagnetic field generated in the transformer. The varnish and insulation processes are required for controlling the quality of the transformer properties. Finally, it is necessary for the transformer design to be manufacturable.

The engineering design attributes are presented on the roof of the house of quality, as shown in Figure 12.2. Such an arrangement of customer preferences and engineering characteristics makes it very easy to graphically represent the relationship between the two sets of variables. The middle part of the house of quality (Fig. 12.2) shows the relationship between the customer preferences and engineering design attributes for EEC. For example, the wire size, number of turns, and lamination stack are strongly related to the cost, reliability, size, and efficiency of the transformer. Similarly, the varnish method strongly affects the temperature rise, aesthetics, and noise level of the transformer. Specific varnish methods are necessary to meet various industrial standards. Varnishing is a time-consuming process and therefore affects the delivery performance. The type of insulation used in the transformers strongly affects reliability, temperature rise, noise level, and conformance to industrial standards. The insulation is also related to cost and transformer size. Finally, design for manufacturability affects the cost and delivery performance and is also related to temperature rise, noise level, and aesthetics.

12.2.2.1 Interrelationships among Engineering Design Characteristics

For a majority of products, it is often difficult to change one engineering design attribute without affecting others. For example, transformer insulation is related to the varnish method used and vice versa. The roof of the house of quality presents such interrelationships among the engineering design attributes. The construction of the roof involves a careful engineering study of the design variables and an understanding of how one attribute affects the other. Companies can use a variety of methods to determine such interrelationships. Brainstorming sessions might be enough for identifying the relationships among a few variables, whereas others might require design experiments or analysis of production data. However, the exact choice of the method depends on the specific product and the processes used to manufacture it. At EEC the interrelationships were identified jointly by the engineering and manufacturing staff. Formal experiments were not conducted but production data for the last few months were analyzed. For example, the relationship between the number of turns and the lamination stack is strongest among the engineering attributes. Figure 12.2 shows other medium and small relationships among the engineering attributes.

12.2.3 Competitive Evaluation

This step includes identifying existing competing products and evaluating them for each of the customer preferences. Such evaluation helps in highlighting the relative strengths and weaknesses of the current product offerings and provides directions for improvements to the product development personnel. It also gives an opportunity to identify the "selling points." For example, the right-hand side of Figure 12.2 shows the competitive evaluation of transformers manufactured by EEC (labeled X) and two of its major competitors (A and B) on a scale of 1 to 5 (5 = best, 1 = worst). The competitive evaluation of the customer requirements was conducted by feedback received from customers by sales, engineering, and quality control staff. The evaluation of the technical requirements was completed by performing a benchmarking study, as well as by customer feedback. The transformer manufactured by EEC is better than its competitors for almost all customer-based attributes except cost and efficiency. The delivery performance and low noise level for EEC transformers are much better than its competitors and therefore are "selling points" for the company. Because EEC's transformers have better quality than that of its competitors in more than one dimension, it can target its products to the high end of the market.

The "basement" of the house of quality presents the engineering targets for process improvement. This space can also be utilized to develop a competitive evaluation of the competing products on the basis of engineering design attributes. Often this step involves benchmarking and/or "reverse engineering" the competitors' products. The target levels and competitive engineering evaluation further provide guidelines for translating customer information to the rest of the product/process development procedures. The EEC manufacturing staff evaluated the engineering design attributes by performing a benchmarking study. (*Note:* Because of the proprietary nature of the engineering process, Figure 12.2 does not show the completed "basement" for transformer manufacturing at EEC.)

The proceding four steps complete the first QFD matrix: the house of quality. Next we present a description of completing the other three QFD matrices.

12.2.4 Design Matrix

The second QFD matrix, the *design matrix,* links the engineering design attributes to the individual components of the product. This matrix can be constructed either for all engineering design attributes (from the house of quality) or for a selected few important attributes. Similar to the house of quality, the design matrix requires a careful analysis of the product, its components, and manufacturing processes. The roof of the design matrix presents the interrelationships among the component characteristics. Similar to the house of quality, the basement of the design matrix can be used for engineering targets for the components.

The design matrix for EEC was constructed by determining the component characteristics of the transformer and comparing those characteristics with the engineering design requirements of the house of quality. The engineering department was primarily responsible for determining the component characteristics. The next step was to determine the relationships between design requirements and the component characteristics, followed by the interrelationships among the component characteristics. Designing a transformer always involves trade-offs and, by constructing this matrix, engineering can make design decisions based on how they will affect the attributes that are important to the customer.

Figure 12.3 shows a completed design matrix for transformers. The engineering team at EEC identified the following seven components, which are related to the design characteristics: coils, insulation, terminal, lamination type, shunts, hardware components, and brackets. The type of coils used in the transformer affects all of the engineering design characteristics. For example, the coil type determines the size of the wire and how many turns it should be wound on a lamination stack of a given size; the insulation material used depends on the size of the lamination stack; the type of electrical terminal attached to the transformer is associated with wire size and insulation type; the lamination type used determines the varnish method and insulation type; and so on. The roof of the design matrix presents the interrelationships among the transformer components. As shown in Figure 12.3, coils, lamination type, and brackets are related to most of the other components.

The design matrix shows how the components affect and are affected by each other and by the design requirements of a product. By understanding

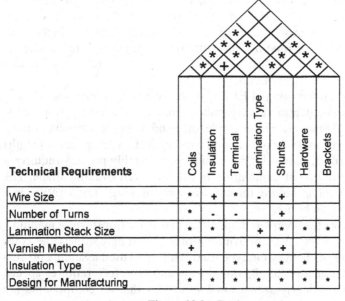

Technical Requirements	Coils	Insulation	Terminal	Lamination Type	Shunts	Hardware	Brackets
Wire Size	*	+	*	-	+		
Number of Turns	*	-	-		+		
Lamination Stack Size	*	*		+	*	*	*
Varnish Method	+			*	+		
Insulation Type	*		*		*	*	
Design for Manufacturing	*	*	*	*	*	*	*

Relationships
* = Strong
+ = Medium
- = Small

Figure 12.3 Design matrix.

these relationships, management can effectively trade off the less important attributes for the more relevant ones. For example, Figure 12.3 shows that the type of coils used is related to most of the engineering design attributes and also to the other components. Therefore, the selection of appropriate coils should precede the selection of other components. Similar analyses can also be conducted for other components. In other words, the design matrix effectively breaks down each product into its components. Because individual components are fabricated and assembled at manufacturing stations, the design matrix eliminates the need for directly linking each product with each production process. The next two sections show how components can be linked to the process operations and to the control parameters.

12.2.5 Operating Matrix

The third QFD matrix, the *operating matrix,* connects the components to the key process operations. Usually, the manufacturing staff (managers, production supervisors, line workers) are actively involved in the process of developing this matrix. It is interesting to note that for a large variety of industries the same basic process operations manufacture all the components. The same processes are mixed and matched in various configurations to generate the product line for a company. Therefore, identifying the critical processes and linking them to the components (from the design matrix) allows managers to connect the customer requirements (from the house of quality) directly to the manufacturing operations. In other words, the operating matrix provides a road map from plant level operations to the customer needs.

Electric Equipment Company developed the third matrix by determining the critical operating processes used to manufacture the components of the transformers. Once key process operations were identified, their interrelationships (the roof) and their relationships with the components were developed. These relationships show how specific processes are related to each other and to various transformer components.

Transformer manufacturing at EEC involves seven key processes: winding process, coil finishing, lamination assembly, lamination test, varnish process, final assembly, and final test. For example, the winding process involves taking the appropriate type and size of wire and winding it (to a prespecified length) on a lamination stack. Similarly, the lamination assembly process requires a specified number of precut lamination sheets of a given size and shape. The varnish process involves baking the coil assembly at a high temperature in a large oven for several hours.

Figure 12.4 shows the completed operating matrix for transformer manufacturing at EEC. It shows that the winding process strongly affects the coils, the insulation, and the lamination type used. The lamination assembly process is the only process related to all components and therefore requires special attention. It can be clearly seen from Figure 12.4 that all components are related to more than one process. We believe that the development of the

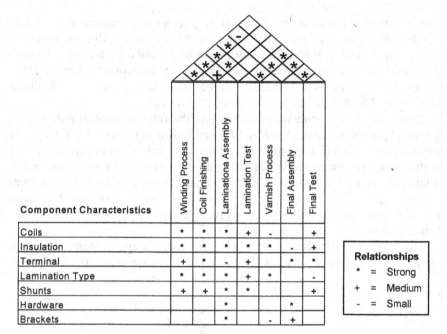

Component Characteristics	Winding Process	Coil Finishing	Laminationa Assembly	Lamination Test	Varnish Process	Final Assembly	Final Test
Coils	*	*	*	+	-		+
Insulation	*	*	*	*	*	-	+
Terminal	+	*	-	+		*	*
Lamination Type	*	*	*	+	*		-
Shunts	+	+	*	*			+
Hardware			*			*	
Brackets			*			-	+

Relationships
* = Strong
+ = Medium
- = Small

Figure 12.4 Operating matrix.

operating matrix is the key to successful product/process development because it translates the requirements into the actual processes used to manufacture components. Additionally, because a few key processes are used to manufacture a multitude of components, the control of these operations will lead to effective operations management. The last QFD matrix (the controls matrix) develops the control parameters for the key processes.

12.2.6 Control Matrix

The role of the fourth QFD matrix, the *control matrix,* is to develop specific quality control plans for the key operating processes (from the operating matrix). Developing control parameters for the key process operations is necessary for effective operations management because these processes manufacture the components for the products that satisfy the customer needs. A careful analysis of the processes and the identification of critical operating parameters are necessary to develop this matrix. A good control matrix is more than just a small part of a large method—it is the first step toward the process of continuous improvement.

Seven critical control points were determined for transformer manufacturing at EEC: coil dimensions, terminals check, winding turns test, mechanical requirements check, electrical testing, varnish curve plot, and final testing.

These control parameters are universal to all types of transformers manufactured by EEC. For example, the winding turns test ensures that the length of the coil equals the specifications. This is very important because the coil length determines the electromagnetic properties of the transformer. Similarly, the critical baking process is monitored by the varnish curve plot, which shows how the coil temperature changes inside the oven.

Based on the key process operations and the relevant control parameters, a control matrix was developed by EEC, as shown in Figure 12.5. Except the varnish process, which is monitored by the curve plot, all other processes have multiple control parameters. For example, the critical winding turns test is related to the winding process, the coil finishing process, the lamination assembly, and the lamination test. In other words, a "pass" in the winding turns test confirms that the previous four processes were operating within established parameters.

The control matrix is very valuable for process improvement because it identifies the parameters that can and should be monitored for effective opera-

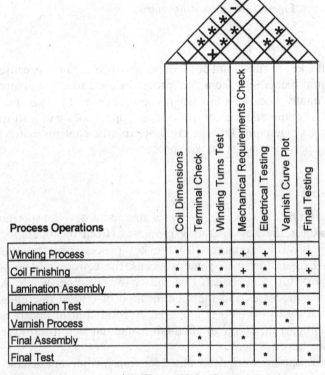

Process Operations	Coil Dimensions	Terminal Check	Winding Turns Test	Mechanical Requirements Check	Electrical Testing	Varnish Curve Plot	Final Testing
Winding Process	*	*	*	+	+		+
Coil Finishing	*	*	*	+	*		+
Lamination Assembly	*		*	*	*		*
Lamination Test	-	-	*	*	*		*
Varnish Process						*	
Final Assembly		*		*			
Final Test		*			*		*

Relationships
* = Strong
+ = Medium
- = Small

Figure 12.5 Control matrix.

tions management. For example, from the completed control matrix, EEC determined the most important areas to work on were the winding turns test, electrical testing, and final testing. The need to test the turns of the windings in the winding process caused EEC to look at the existing equipment and determine a method to check turns in the coil-winding process. This led to the development of a programmable handwinding machine that not only checks turns in the windings, but also automatically slows down and stops the machine after the required number of turns. Ensuring that the number of turns is correct (as the coil is being wound) eliminates the need to check the coils after they have been wound. This improvement drastically improved the quality of the winding process and virtually eliminated all defects associated with it.

Recognizing the need for further improvement, the autowinding process was also enhanced. The house of quality pointed out that delivery was important to the customer and that EEC was ahead of its competitors in that area. The previous method of autowinding required an average of 10 minutes to change over from one coil to another. By improving the equipment being used and by modifying the autowinding machine, the changeover currently takes approximately 90 seconds (one-piece flow). Hence, by implementing one-piece flow, the lead time has been reduced, improving delivery performance to the customer. The quality has improved by eliminating batch processing and the work in process. In addition, the process is now more visual and it is easy to determine were there is a problem in the process.

12.2.7 Continuous Improvement

Development of the four QFD matrices is only the beginning of an effective product development process. To continuously meet and exceed a customer's needs at a reasonable price, a company must constantly monitor its operations and strive for improvement. A completed QFD process provides a clear snapshot of the current state of operations from customer needs to product design to process operations. It provides directions for further improvement. However, in order to be the market leader in the future, a company needs to focus on identifying ways to improve its processes. Over the last 10 years or so, a number of process improvement methods and techniques have been identified, which can be easily implemented in combination with, QFD.[4] These methods include management and planning tools (e.g., affinity diagram, interrelationship digraph, tree diagram, prioritization matrices, matrix diagram, process decision program chart, and activity network diagram), continuous improvement tools (e.g., cause-and-effect diagram, run chart, scatter diagram, flowchart, Pareto diagram, histogram, and control chart), and creative thinking tools (e.g., problem definition, brainstorming, brainwriting, mindmapping, word-and-picture association, advanced analogies, and morphological chart).

12.3   CONCLUSIONS

This chapter presented an overview of the quality function deployment process and showed how it can be implemented for product/process development of a relatively complex electronic product. We reviewed the basic QFD concepts and provided a step-by-step guideline from identifying customer needs to developing critical control points.

As shown in this chapter, QFD provides valuable information with respect to the design and development of products and processes in a very systematic manner. The first QFD matrix, the house of quality, not only shows hierarchical customer needs and their relation to design attributes but also presents an evaluation of competing products and identifies the selling points for a given product. The second QFD matrix extends the design attributes to the components of the products. As discussed earlier in the chapter, the third and fourth QFD matrices provide very useful information relating to the development/ identification of key processes and control parameters.

A number of companies in Japan, the United States, and Europe have used the QFD process and have witnessed improved results. As a final note, we would like to point out that QFD should not be considered a "tactical" method but a "framework" for effective product/process development. The minor details about the individual matrices are not as important as identifying the key relationships, which link the customer needs to various aspects of the product development process.

ACKNOWLEDGMENTS

The authors would like to thank the electrical transformer manufacturing company (referred to as EEC in this chapter) for allowing them to conduct and publish the research described in this chapter. Partial funding for this research was provided jointly by APICS E&R Foundation and the Marketing Science Institute.

REFERENCES

1. Akao, Y., 1994, *Quality Function Deployment: Integrating Customer Requirement into Product Design,* Productivity Press, Portland.

2. Ben-Akiva, M., and S. R. Lerman, 1991, *Discrete Choice Analysis, MIT Press,* Boston.

3. Chakravarty, A. K., and S. Ghose, 1993, Tracking Product–Process Interactions: A Research Paradigm, *Production and Operations Management,* Vol. 2, pp. 72–93.

4. Evans, J. R., and W. M. Lindsay, 1996, *The Management and Control of Quality,* 3rd ed., West Publishing, New York.

5. Green, P. E., and A. M. Krieger, 1989, Recent Contributions to Optimal Product Positioning and Buyer Segmentation, *European J. Oper. Res.,* Vol. 41, pp. 127–141.

6. Griffin, A., and J. R. Hauser, 1993, The Voice of the Customer, *Marketing Sci.,* Vol. 12, pp. 1–27.

7. Hauser, J. R., and D. Clausing, 1988, The House of Quality, *Harvard Business Rev.,* Vol. 66, pp. 63–73.

8. Kim, K., H. Moskowitz, A. Dhingra, and G. Evans, 1993, Fuzzy Multicriteria Methodologies and Decision Support System for Quality Function Deployment, Working Paper, Purdue University.

9. Louviere, J. J., 1988, *Analyzing Decision Making: Metric Conjoint Analysis,* Sage, Newbury Park, CA.

10. Ulrich, K. Y., and S. D. Eppinger, 1995, *Product Design and Development,* McGraw-Hill, New York.

11. Urban, G. L., and J. R. Hauser, 1993, *Design and Marketing of New Products,* 2nd ed., Prentice-Hall, Englewood Cliffs, NJ.

12. Wheelwright, S. C., and K. B. Clark, 1992, *Revolutionizing Product Development: Quantum Leaps in Speed, Efficiency and Quality,* Free Press, New York.

13

DESIGN FLOW MANAGEMENT AND MULTIDISCIPLINARY DESIGN OPTIMIZATION IN APPLICATION TO AIRCRAFT CONCEPT SIZING

BRETT A. WUJEK
Engineous Software, Inc.

ERIC W. JOHNSON
Valparaiso University

JOHN E. RENAUD, J. B. BROCKMAN,
and STEPHEN M. BATILL
University of Notre Dame

13.1 INTRODUCTION

This chapter reports on two ongoing efforts, one centered in the more traditional multidisciplinary design optimization (MDO) arena of optimization algorithm development and the other introducing a technique for workflow management in a multidisciplinary design environment. A concurrent subspace optimization (CSSO) approach for MDO is successfully implemented

Integrated Product and Process Development, Edited by John Usher, Utpal Roy, and Hamid Parsaei
ISBN 0-471-15597-7 © 1998 John Wiley & Sons, Inc.

in application to an aircraft concept sizing test problem. The CSSO approach implemented in this study provides for design variable sharing and distributed computing across disciplines. Significant time savings are observed when using distributed computing for concurrent design. More importantly, a significant savings in the number of analysis calls required for optimization is observed using the CSSO approach as opposed to an all-at-once approach. An approach for workflow management of complex multidisciplinary problems is introduced and applied in an analysis of the aircraft concept sizing test problem. The approach implemented in this chapter is based on a novel information model, called the task schema, that classifies the various tool and data objects used in a design process and defines the rules by which tools and data may be used to form tasks.

13.2 BACKGROUND

The research presented in this chapter is part of an ongoing effort to develop a framework and systematic methodology to facilitate the application of multidisciplinary design optimization (MDO) to a diverse class of system design problems. Multidisciplinary design optimization is based on the philosophy of identifying the appropriate combination of parameters under the control of a designer (or design team) that will result in the most effective product or system. For all practical aerospace systems, the design of the system is a complex sequence of events which integrates the activities of a variety of discipline "experts" and their associated "tools." The development, archiving, and exchange of information between these individual experts is central to the design task, and it is this information that provides the basis for these experts to make their design decisions—resulting in the final product.

Two types of tools for multidisciplinary design are implemented in this investigation. One is a traditional "hardwired" algorithm approach to MDO, whereas the other makes use of an interactive multidisciplinary design environment, which facilitates the development, archiving, and exchange of information between these individual experts. *Future work is targeted toward the integration of these two approaches, combining the best aspects of each.*

Research in MDO has traditionally been centered on the development of optimization strategies and advanced analysis methods for aerospace systems design. One such strategy is the CSSO–MDO approach of Renaud and Gabriele.[13-15] The CSSO–MDO algorithm implemented in this study makes use of a modular software environment NDOPT (Notre Dame Optimization Tool)[19] which incorporates a well-developed graphical user interface for menu-driven execution and results display. This enhanced programming environment highlights the modularity of the CSSO algorithm while providing for distributed processing of analyses.

Widespread acceptance and application of MDO practices in existing multidisciplinary design organizations has been hindered by the perception that most MDO algorithms have been designed to work in a generic fashion on analysis codes with little or no human interaction. The aforementioned CSSO approach using NDOPT is one such example. Industry recognizes that automated design tools are limited by the expertise and experience of the automated design tool users. One approach to multidisciplinary design that provides for a high degree of human interaction in an automated design environment is found in application of the Hercules[5] workflow management system. Workflow management encompasses a set of tools and techniques for organizing, monitoring, and automating complex processes. The Hercules system utilizes a schema-based approach to workflow management and has been previously implemented in application to integrated circuit design. In this research, the Hercules system is successfully adapted for use in an aircraft concept sizing problem. These two different approaches to multidisciplinary design are reviewed and contrasted in the remainder of this chapter. Implementation results for the CSSO–MDO algorithm are presented first, followed by a discussion of the workflow management approach to multidisciplinary design.

13.3 CONCURRENT SUBSPACE OPTIMIZATION

The goal of concurrent engineering is to avoid costly problems that occur in the downstream activities of a product development cycle. This is accomplished by forming product design teams made up of experts from various functional groups to work concurrently instead of sequentially on the design issues. The promise of concurrent engineering is reduced product development time and greater understanding of the interdisciplinary couplings that complicate designs, leading to higher-quality, lower-cost designs. The reality of the practice of concurrent engineering is that, to reap these benefits, design teams must be able to master the interdisciplinary couplings that exist in complex engineering systems and to develop an approach to managing trade-offs between disciplines that lead to more optimal designs. In recent years, several researchers have reported on the development of MDO strategies that accommodate concurrent engineering practice.

The evolution of these strategies can be traced to Sobieszczanski-Sobieski,[18] in which the concept of a CSSO-based MDO algorithm is first introduced. The successful implementation of the Sobieszczanski-Sobieski CSSO–MDO algorithm in a distributed computing environment is reported in Bloebaum et al.[2] The Bloebaum study reports that modifications to the coordination optimization procedure (COP) of the Sobieszczanski-Sobieski CSSO algorithm were required for successful convergence. Additional modifications to the COP are reported in Bloebaum and Hajela.[1] These modifications make use of an embedded expert system (EES) for heuristic coordination in an approach labeled CSSO–EES. In Pan and Diaz,[12] the existence of "pseudo-optimal" designs, which present convergence problems for subspace optimization strategies, is introduced. It is the existence of "pseudo-optimal" designs that create problems in the COP of the Sobieszczanski-Sobieski CSSO algorithm. A study by Shankar et al.[17] indicates that the COP, as proposed in Sobieszczanski-Sobieski,[18] fails on a simple two-dimensional quadratic programming problem. The Shankar study suggests a number of possible modifications. It should be noted that the original Sobieszczanski-Sobieski CSSO algorithm has been implemented as a commercial multidisciplinary design optimization package SYSOPT, as described in Eason et al.[8]

A CSSO-based MDO algorithm, which introduced a coordination procedure of system approximation (CP–SA) to replace the COP of Sobieszczanski-Sobieski, was introduced in Renaud and Gabriele[13] and modified in Renaud and Gabriele.[14, 15] The Renaud–Gabriele approach is fundamentally different from the Sobieszczanski-Sobieski approach in that the design variable update is not based on the simple combination of subspace solutions. The Renaud–Gabriele coordination approach is based on archiving design information generated during nonlinear subspace optimizations and then using that data to form an approximate representation of the full system design problem. The optimal solution of this approximate system design problem, subject to move limits, is used as the next design iterate in the sequential algorithm. The Renaud–Gabriele system approximation–based coordination strategy inherently avoids converging to "pseudo-optimal" designs.

A flowchart of the Renaud–Gabriele CSSO–MDO algorithm is depicted in Figure 13.1, illustrating the different modules. Beginning with some initial design x^0, a system analysis is performed to obtain the values of the states for the initial design. The next module calculates the sensitivities at this initial design point. This information is used in the subspace optimizers to approximate nonlocal states, allowing the optimizers to be executed concurrently. Data obtained during these subspace optimizations are retained in a database, from which a response surface is constructed to approximate the design space at the system level. Finally, an optimization of this system approximation is performed, subject to move limits on the design variables. The result of this restricted system level approximate optimization is used as the next design iterate, and the process is repeated until convergence is achieved. For complete algorithm implementation details, the reader is referred to Wujek et al.[19]

Wujek et al.[19] focus on recent implementation advances and modifications in the continued development of the Renaud–Gabriele CSSO–MDO algorithm based on the CP–SA. In this research, a similar implementation study is conducted in application to an aircraft concept sizing problem. The implementation includes design variable sharing across disciplines (i.e., subspaces) in the subspace design regime. A modular software environment NDOPT, which incorporates a well-developed graphical user interface for menu-driven execution and results display, is used in implementation. Timing results are reported for the CSSO–MDO algorithm operating in a distributed computing environment.

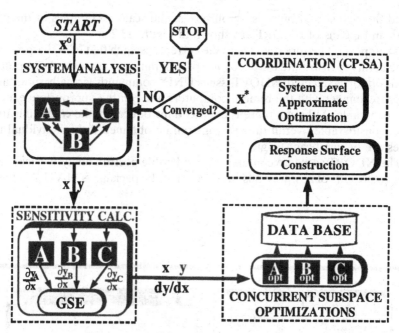

Figure 13.1 CSSO algorithm flowchart.

13.3.1 Modular Algorithms and GUI Development

Algorithms are typically composed of individual modules, which serve separate purposes and are somehow interconnected. The CSSO algorithm described previously contains four main modules, and these modules can be further decomposed into submodules, as shown in Figure 13.1. Whether they be FORTRAN programs or industrial software packages, these submodules typically represent analysis or optimization tools and are the building blocks of the entire algorithm.

The CSSO algorithm studied here utilizes several individual FORTRAN programs as its building-block submodules. To automate the process of launching the submodules of the algorithm and to coordinate the communication among them, a graphical user interface (GUI) called NDOPT (Notre Dame Optimization Tool) was developed using Tcl (tool command language) and Tk (toolkit).[11] These two packages together provide a programming system for the development of GUI applications and are the basis for the code written to develop NDOPT. The NDOPT package provides a generic framework through which many tasks associated with studying test problems, running algorithms, and reviewing results can be accomplished. All options are executed through sets of buttons, scroll boxes, and scales. Examples of the windows and menus of NDOPT are shown in Figure 13.2.

An important feature relevant to the current research that NDOPT provides is the ability to distribute the execution of programs among several computers. If this option is selected, NDOPT issues UNIX commands virtually simultaneously to connect with the remote computers and run the appropriate executable programs. Although there is overhead introduced in the communication time, considerable overall time savings can be obtained if the individual run times are of significant length.

The NDOPT software contains a considerable postrun analysis package to view the results of any previous runs. Much of the package is MATLAB based

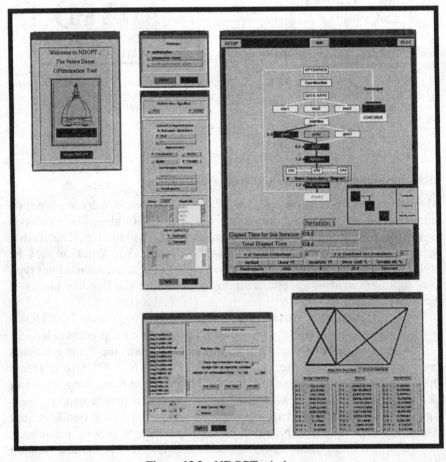

Figure 13.2   NDOPT windows.

for plotting purposes, with several options added through the NDOPT script. The capabilities of this package provide a quick and efficient method to study results because everything is available at the push of a button rather than through command line execution. Such useful information as convergence histories, approximation plots, and time results are immediately available to the designer.

13.3.2 Test Problem: Aircraft Concept Sizing

To test the CSSO algorithm described previously, a new MDO test problem was developed by the MDO research group at the University of Notre Dame. The problem is directed toward the preliminary sizing of a general aviation aircraft concept subject to certain performance requirements. The system analysis for this test problem is based on empirical relations and databases for this specific class of aircraft. The problem formulation and system decomposition have since been modified and updated, but this study reports on results using the original formulation described as follows.

The design vector in this problem is comprised of variables relating to the geometry of the aircraft, propulsion and aerodynamic characteristics, and flight regime. In all, 10 design variables exist for this problem:

x_1 = Aspect ratio

x_2 = Wing area

x_3 = Maximum lift coefficient (CL_{max})

x_4 = Propeller efficiency

x_5 = Specific fuel consumption

x_6 = Fuselage length

x_7 = Fuselage diameter

x_8 = Density at cruise altitude

x_9 = Cruise speed

x_{10} = Fuel weight

Appropriate bounds are placed on all of the design variables. The problem also includes a number of parameters that are fixed during the design process to represent constraints on mission requirements, available technologies, and aircraft class regulations. These include the number of passengers (2), payload weight (398 lb), number of engines (2), engine weight (197 lb), and ultimate load factor (5.7). An alternate formulation could consider these parameters as design variables.

The design of the system is decomposed into six contributing analyses (CAs), as shown in Figure 13.3. Decomposition into six contributing analysis modules, which produce one state each, is obviously extreme and used here for illustrative purposes only. An alternate form of decomposition for this

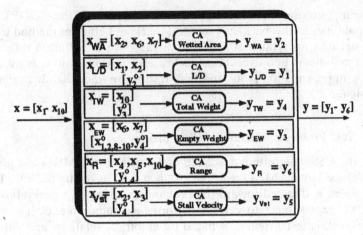

Figure 13.3 System analysis with design variable and state.

problem is given in Sellar et al.[16] The six states calculated for this design problem are as follows:

y_1 = Maximum lift-to-drag ratio
y_2 = Total aircraft wetted area
y_3 = Empty weight
y_4 = Gross takeoff weight
y_5 = Stall speed
y_6 = Aircraft range

As illustrated in the dependency diagram of Figure 13.4, five feedforward couplings and one feedback coupling exist among the contributing analyses. The feedback coupling is due to the dependence of empty weight on total weight, and iteration among these contributing analyses is necessary to converge to a consistent set of system states. In implementation, the system analysis is carried out through execution of FORTRAN codes, applying the empiricisms for this class of aircraft as discussed previously. A tolerance of 0.01 percent was set on the weight iteration between the empty- and total-weight CAs.

The objective in this problem is to determine the least gross-takeoff weight design within the bounded design space, subject to two performance constraints. The first constraint is that the aircraft range must be greater than a specified requirement, and the second constraint is that the stall speed must be less than or equal to some prescribed maximum stall speed.

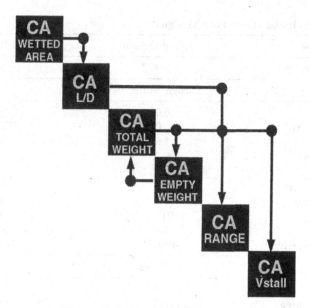

Figure 13.4 Dependency diagram.

In standard form, the system optimization problem is

$$\text{Minimize} \quad f(x) = \text{Weight} = y_4$$
$$\text{Subject to} \quad 1 - \frac{y_5}{V_{\text{Stall}_{\text{req}}}} \geq 0$$
$$1 - \frac{\text{Range}_{\text{req}}}{y_6} \geq 0$$
$$V_{\text{Stall}_{\text{req}}} = 70 \text{ ft/sec} \qquad (13.1)$$
$$\text{Range}_{\text{req}} = 565 \text{ miles}$$

13.3.3 CSSO Results

The CSSO algorithm was run in NDOPT, starting from an initial design (see Table 13.1) that is slightly infeasible with respect to the range constraint. At this initial design point, the gross takeoff weight of the aircraft (the objective function) is 2290.2 lb. As shown in Figure 13.5, the algorithm converges after eight iterations to a final weight of 1762.2 lb.

The final values of the design variables and states are also listed in Table 13.1. The trends of the design variables follow what is expected in aircraft design. As can be seen from the table, eight of the design variables end up at their bounds (L = lower, U = upper). In fact, seven of these, x_4 (propeller efficiency) being the lone exception, converge to the bound immediately during the first iteration, which is a feasible search, as shown in Figure 13.6a, given

TABLE 13.1 Initial versus Final Designs

Objective	Initial Design	CSSO Final Design
Function	2290.2	1762.2
Design Variables		
x_1 = AR	6.0	5.0L
x_2 = Wing area	200.0	177.9
x_3 = CL_{max}	1.6	1.7U
x_4 = Prop. eff.	0.75	0.85U
x_5 = c	0.45	0.45L
x_6 = Fuse. length	22.0	20.0L
x_7 = Fuse. diam.	4.2	4.0L
x_8 = Density	0.002378	0.0019L
x_9 = Speed	200.0	200.0L
x_{10} = Fuel Weight	400.0	145.2
States		
y_1 = L/D	11.9	11.0
y_2 = Wetted area	810.3	714.0
y_3 = Empty weight	1492.2	1219.0
y_4 = Total weight	2290.2	1762.2
y_5 = V stall	77.6	70.0
y_6 = Range	1197.6	565.0

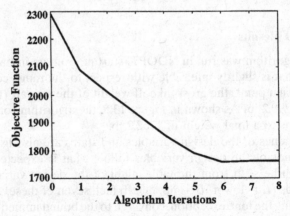

Figure 13.5 CSSO convergence history of total weight.

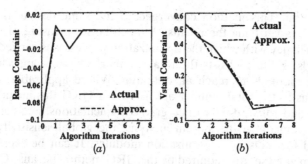

Figure 13.6 Convergence history of (a) g_2 and (b) g_1.

the range constraint. Figure 13.7a shows the history of the wing area, a major driver in the design of an aircraft. It can be seen that increases in the wing area during iterations 1 and 3 are used to meet the range constraint. Gradual decreases in the amount of fuel (Fig. 13.7c) drive the weight of the aircraft down. Another interesting observation is that during iteration 5, when the design becomes infeasible with respect to the stall speed (Fig. 13.6b), the propeller efficiency is increased from its minimum to its maximum value (Fig. 13.7b) to account for this and take full advantage of the propeller technology.

As a means of comparison, this test problem was optimized all at once using a conventional optimizer (OPT3.2, developed by Gabriele and Beltracchi), which uses the generalized reduced gradient (GRG) method applied to

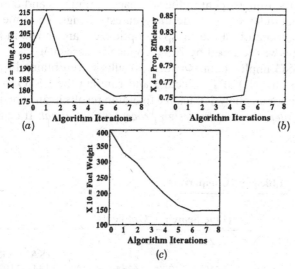

Figure 13.7 Convergence history of (a) x_2, (b) x_4, and (c) x_{10}.

the full system. This method converged to the same design as the CSSO algorithm, thus verifying the results. A much better measure of comparison is in the efficiency with which the optimization process was carried out by the two methods. Table 13.2 lists the total number of system analyses required by the two methods to reach an optimum. Where applicable, the various contributions to the total number are listed; for GRG this includes calls for line search and forward-difference gradient calculations, and for CSSO the CA calls are made in the system analysis, the global sensitivity equations (GSE), and the subspace optimization modules. It can be seen that many more system analyses are required by the GRG method because CSSO avoids these costly analyses by utilizing approximations to the system. Perhaps a better indication of the relative cost of each method is the total number of contributing analyses required, also listed in Table 13.2. There is a great benefit (91 percent reduction in CA calls) in using the CSSO algorithm due mainly to the fact that complex coupling exists in this problem (between empty- and total-weight CAs). In fact, iteration between these two CAs accounts for 4788 of the 6384 CA calls (75 percent) in the GRG run. The decoupled nature of the subspace optimizers in the CSSO algorithm avoids this expensive iteration. Only 96 extra CA calls (16 percent of the total) are required in the system analysis of the CSSO algorithm to account for the coupling in this problem.

As previously mentioned, distributed computing was utilized to implement true concurrency in the design process. In this study, only the execution of the subspace optimizers was carried out in parallel. Although the sensitivity submodules could easily be distributed, little or no benefit would be seen, considering the small size of the problem with overhead of the distribution process. Figure 13.8a shows the amount of time spent in each module of the CSSO algorithm for sequentially executed subspace optimizers. It can be seen that most of the time is spent in the subspace optimizers and that distribution of these submodules will provide the greatest savings. The time results from a run with distribution of the subspace optimizers are shown in Figure 13.8b. The total time was reduced by 200 seconds (35 percent).

In the CSSO implementation study, results were obtained by loading the test problem into NDOPT, selecting and running the CSSO algorithm, and utilizing the postrun analysis package to obtain results in an expedient manner. *The human interaction in the design process was minimal.* It could be argued

TABLE 13.2   Efficiency Comparison

	GRG (Line Search + Gradient)	CSSO
System analyses	216 + 50 = **266**	**8**
CA calls	5184 + 1200 = **6384**	(SA + GSE + SSO) 144 + 192 + 256 = **592**

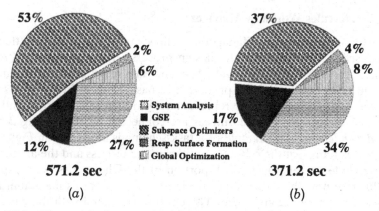

Figure 13.8 Time results for runs with (*a*) sequential and (*b*) distributed subspace optimizers.

that NDOPT does not accurately model the design process due to the fact that there is no human interaction in the actual execution. Instead, NDOPT is a tool that provides for automated *testing* of new MDO algorithms and methodologies. The human interaction is limited to a selection of algorithm options and a push of a start button so that the testing can be carried out in a timely and orderly fashion. The postrun analysis package developed serves these same purposes. *In contrast, the workflow management approach to multidisciplinary design provides a high degree of human interaction at the cost of increased execution time.* This alternative approach to multidisciplinary design, which provides for a high degree of designer interaction, is introduced and discussed in the remainder of this chapter.

13.4 DESIGN FLOW MANAGEMENT SYSTEMS

Design flow management systems are systems that provide designers with services that aid in the development and execution of the design process. The central element in these systems is a *design flow*. A design flow represents the design process by defining all of the tasks or activities that will be performed and how they are connected together to form the process. There are several different graph-based representations that have been used to model the design flow ranging from hierarchical networks[3, 7, 9] to Petri nets.[4, 6] Depending on the model representation, flow management systems can offer a variety of services. For example, these systems provide a formal methodology for designers to follow, the ability to trace the history of individual pieces of design data, and the ability to track the state of the design process. Kleinfeldt et al.[10] present a summary of services for existing flow management systems.

13.4.1   Hercules Workflow Manager

The Hercules Workflow Manager is a flow management system that uses a task schema to represent the design process.[5] The task schema provides construction rules that define how tool and data resources can be combined to form tasks that model the process. The basic element in the task schema is a *task entity*. Task entities represent each design object (tools and data) found in the process. These entities are connected together using *data* and *tool dependencies*. An example of a task schema is illustrated in Figure 13.9. The boxes represent the different entities in the process and the arrows show the dependency relationships. A portion of the schema has been expanded to illustrate how the conceptual calculation of an aircraft's maximum lift-to-drag ratio can be accomplished. The entities associated with the calculation include the wetted area, the aspect ratio, the wing area, and the equations used to calculate the ratio. The dependencies describe how the entities are applied to create the value. The entity associated with the functional dependency is applied to the entities connected to the data dependencies to create the result. For our example, the lift-to-drag ratio is generated by applying the lift-to-drag equations to the wetted area, aspect ratio, and wing area.

13.4.2   Design Data versus Design Metadata

When using Hercules, it is important to distinguish between design data and metadata. Design data are the raw data objects used or created in the design process. The raw data can be either a data artifact, represented as a value, set of values, or specific representation, or a tool, represented as an executable that transforms the artifact from one state to another. Design metadata, however, *characterize* the actual design data and include information such as who created the data, when they were created, or some distinguishing comments. An example of the relationship between design data and design metadata is

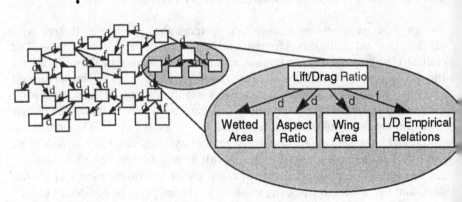

Figure 13.9   Example of task schema diagram.

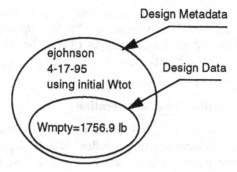

Figure 13.10 Design data versus metadata.

illustrated in Figure 13.10. In this example, a piece of the design data is the value of an aircraft's empty weight (1756.9 lb). The design metadata describe when the empty weight was created (4-17-95), by whom (ejohnson), and how (using the initial total weight).

As a flow management system, Hercules is concerned with managing design metadata rather than design data. Design metadata management is accomplished through a *task database*. The task database stores instances of metadata objects that are created during process execution along with the relationships between instances. The advantage of collecting this information is that it provides continuous documentation of the design process.

A portion of a populated task database for our lift-to-drag ratio example is shown in Figure 13.11. The boxes in the figure represent the entity classes that were defined in the task schema, whereas the spheres are instantiations of the classes that contain the design metadata and links to the actual data.

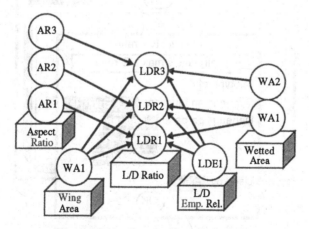

Figure 13.11 Populated task database.

The arcs between spheres provide the ability to determine how the design data were actually created. For example, the lift-to-drag ratio instance LDR2 was created using the lift-to-drag equations LDE1, the wing area WA1, the aspect ratio AR2, and the wetted area WA1.

13.4.3   Example Hercules Process Execution

In this section, we will go through a simple example to demonstrate how the Hercules graphical user interface supports flow-based design. We will illustrate designer interaction using the task, "create a lift-to-drag ratio," which was shown in Figure 13.9.

The first step in preparing for execution is to select a portion of the task schema, called a *task tree,* which covers the scope of the task to be executed. This is accomplished by selecting any entity involved in the task and then either expanding forward toward a goal or backward toward the primary inputs of the process. We will illustrate how this is accomplished by selecting the entity *L/D ratio* and then expanding backward to create the task that is used to generate the ratio.

Figure 13.12 shows the selection of the entity through the Hercules help facility. The help facility is implemented in a hypertext format, which allows the user to traverse through the task schema while deciding on a starting point.

Once the *L/D ratio* entity is selected as the starting point, an icon appears in the task window. A variety of options are then available through a pulldown menu that appears when clicking on the icon. The menu options for the *L/D*

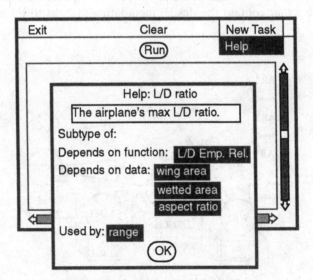

Figure 13.12   Hercules help facility.

ratio entity are shown in Figure 13.13. These options include entering the help facility, selecting a browser to view instances of the *L/D ratio* entity that are stored in the task database, and different expanding and unexpanding operations.

The backward and forward expansion operations are created from the construction rules in the task schema and will be unique to the different entities. Because we are interested in creating a task to calculate the lift-to-drag ratio, we use the **Create** option to generate a task tree that will perform that specific task. This backward expansion is created from the following construction rule:

L/D-Ratio ← *L/D-Equations* (*Wetted Area, Aspect Ratio, Wing Area*)   (13.2)

Figure 13.14 illustrates the expanded tree after the **Create** option was chosen.

Once the scope of the task has been selected, the next step in preparing for execution is to bind instances to the leaf nodes, or primary inputs, of the task. In our example, an instance must be bound to each of the four inputs. Because the binding order is arbitrary, we select the *wetted area* entity first. Again using the pulldown menu, we select the **Browse** option, which allows us to browse the instances associated with the *wetted area* entity. This selection is presented in Figure 13.15.

Instance browsers are used to view a set of instances found in the task database linked to a specific entity. The instance browser associated with the *wetted area* entity is shown in Figure 13.16. Each instance browser has several options that allow the user to filter the instances found in the database. In Figure 13.16, two different filters were used, date limits and user limits, to find all of the instances that were created by the user "jbb" between July 25

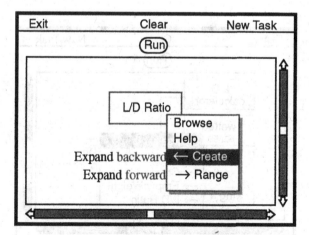

Figure 13.13 Entity popup menu.

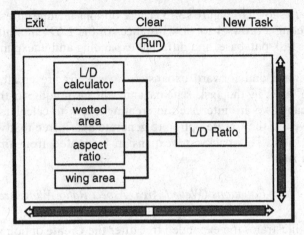

Figure 13.14 Task tree describing "create *L/D* ratio" task.

and 26. The query is performed when the **Select** button is depressed. In the figure, the selected instance, "circular fuselage," was the only instance that satisfied the query. Because entities model both tools and data, the binding of the remaining leaf nodes associated with the task is accomplished in the same manner.

Once instances have been assigned to all of the leaf nodes, the task is ready to be executed. Execution is done by simply pressing the **Run** button. During execution, instances of design metadata are created and stored in the task database at each step in the tree. Although we have selected only one task for execution, the task tree can be expanded to include more tasks that may

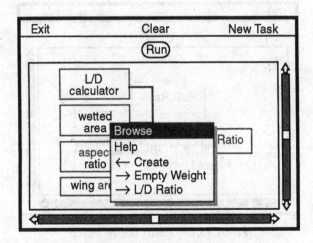

Figure 13.15 Selecting instance browser.

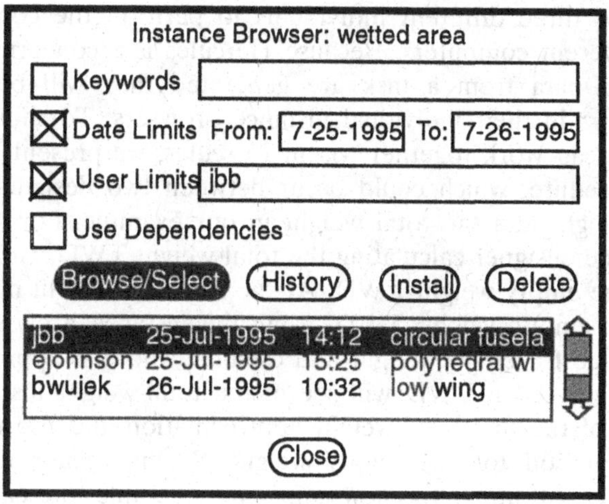

Figure 13.16 Wetted-area instance browser.

be executed via a postorder traversal. Although this example uses only simple closed-form expressions, Hercules is also capable of utilizing highly specialized CAD tools because all tools are encapsulated into the Hercules environment. Encapsulation involves placing a *wrapper* around the executable tool. The wrapper is used to translate the input data into a format the tool can use, execute the tool, and then, if necessary, translate the output data into a specified format.

After execution, Hercules provides two different mechanisms for retrieving design history from the database. *Backward-chaining* queries determine how a piece of data was created. These queries are performed using the **History** button shown in Figure 13.16. When pressed, a query is sent to the database to find the specific tool and data instances that were applied in creating the selected instance. In contrast, *forward-chaining* queries are utilized to view the design derivation history. For example, suppose a user wanted to find a set of all lift-to-drag ratio instances that were created using the wetted-area instance "polyhedral wing." This would be accomplished by expanding a task tree surrounding the lift-to-drag task, binding the "polyhedral wing" instance to the wetted area, and depressing the **Select** button with the *use dependencies* option in the lift-to-drag instance browser. With that filter, the lift-to-drag instances returned would be limited to those that had the "polyhedral wing" as part of their history.

13.4.4 Aircraft Concept Sizing Example

To show how Hercules could be used in an MDO environment, we have implemented the aircraft sizing example described previously. The objective was to show how Hercules could be used to perform a full system analysis.

We employed three different individuals to perform the contributing analyses on different computers. Because Hercules is a concurrent execution system, when data from a task are generated they will become visible to all designers in their individual instance browsers. To show how multiple designers can work together within Hercules, we present the following iteration procedure, which could occur between two designers calculating the empty weight and the total weight in our example. For simplification, we will call the designer calculating the total weight TWD, and the designer calculating the empty weight EWD. At the start, EWD will make an initial guess for the empty weight. TWD, using the instance browser, will select the initial value for empty weight and calculate the first approximation for the total weight. Then EWD will see in the total-weight instance browser the creation of the first total-weight approximation and use it to create a new approximation for the empty weight. Seeing a new update in the browser, TWD will utilize the new empty-weight value to create the second total-weight approximation. This cycle will continue until one or both of the values converge to a specified limit.

13.5   CONCLUSIONS

Two approaches for multidisciplinary design are demonstrated in application to an aircraft concept sizing problem composed of six subdisciplines. One approach is a "traditional" CSSO–MDO algorithm implemented in a generic framework (NDOPT) through which many tasks associated with studying test problems, running algorithms, and reviewing results can be accomplished. The other approach utilizes the Hercules Workflow Manager to provide an interactive multidisciplinary design environment that facilitates the development, archiving, and exchange of information between individual design experts. *Future work is targeted toward the integration of these two approaches, combining the best aspects of each.*

Significant time savings are observed when using distributed computing for concurrent design across disciplines in the CSSO–MDO implementation of the aircraft concept sizing test problem. More importantly, a significant savings in the number of analysis calls required for optimization is observed when comparing the CSSO approach to a traditional single-level optimization strategy. The number of analysis calls required for optimization is reduced by a factor of 10. This reduction is made possible through the use of linear approximations in the concurrent subspace optimizations and the use of response surface approximations for coordination. Use of these approximations provides for the temporary decoupling of the system design problem, resulting in significant analysis savings.

Widespread acceptance and application of MDO practices in existing multidisciplinary design organizations have been hindered by the perception that most MDO algorithms have been designed to work in a generic fashion on analysis codes with little or no human interaction. The aforementioned CSSO approach using NDOPT is one such example. Industry recognizes that automated design tools are limited by the expertise and experience of the automated design tool users. Workflow management systems can provide an interactive multidisciplinary design environment that will facilitate the development, archiving, and exchange of information between individual design experts. In this chapter the Hercules Workflow Management system is successfully adapted for use in analysis of the aircraft concept sizing problem. The Hercules approach is based on a novel information model, called the task schema, that classifies the various tool and data objects used in a design process and defines the rules by which tools and data may be used to form tasks.

ACKNOWLEDGMENTS

This multidisciplinary research effort was supported in part by the following grants and contracts: NASA Research Grant Award NAG-1-1561, NSF Grant DMI93-08083, and NSF Grant DMI94-57179.

REFERENCES

1. Bloebaum, C. L., P. Hajela, 1991, Heuristic Decomposition for Non-Hierarchic Systems, *Proceedings of the 32nd AIAA/ASME/ASCE/ AHS/ASC Structures, Structural Dynamics and Materials Conference,* Baltimore, pp. 344–353.

2. Bloebaum, C. L., P. Hajela, J. Sobieszczanski-Sobieski, 1992, Non-Hierarchic System Decomposition in Structural Optimization, *Eng. Optim.,* Vol. 19, pp. 171–186.

3. Ten Bosch, K. O., P. Bingley, and P. Van Der Wolf, 1991, Design Flow Management in the NELSIS CAD Framework. *Proceedings of the 28th ACM/IEEE Design Automation Conference,* pp. 711–716.

4. Bretschneider, F., C. Kopf, H. Lagger, A. Hsu, and E. Wei, 1990, Knowledge Based Design Flow Management. *Proceedings of the IEEE International Conference on Computer-Aided Design,* IEEE, New York, pp. 350–353.

5. Brockman, J. B., and S. W. Director, 1991, The Hercules CAD Task Management System, *Proceedings of the IEEE International Conference of Computer-Aided Design,* pp. 254–257.

6. Casotto, A., and A. Sangiovanni-Vincentelli, 1993, Automated Design Management Using Traces, *IEEE Transactions on Computer-Aided Design of Integrated Circuits and Systems,* Vol. 12, pp. 1077–1095.

7. Chiueh, T., and R. Katz, 1990, A History Model for Managing the VLSI Design Process, *Proceedings of the IEEE International Conference on Computer-Aided Design,* IEEE, New York, pp. 50–52.

8. Eason, E. D., G. A. Nystrom, A. Burlingham, E. E. Nelson, 1994, Robustness Testing of Non-Hierarchic System Software, *Proceedings of the Fifth AIAA/USAF/ NASA/ISSMO Symposium,* Panama City, FL, pp. 817–824.

9. Van Den Hamer, P., and M. A. Treffers, 1990, A Data Flow Based Architecture for CAD Frameworks. *Proceedings of the International Conference on Computer-Aided Design,* IEEE, New York, pp. 482–485.

10. Kleinfeldt, S., M. Guiney, J. K. Miller, and M. Barnes, 1994, Design Methodology Management, *Proceedings of the IEEE,* Vol. 82, No. 2, pp. 231–250.

11. Ousterhout, J. K., 1994, *Tcl and the Tk Toolkit,* Addison-Wesley, Reading, MA.

12. Pan, J., and A. R. Diaz, 1989, Some Results in Optimization of Non-Hierarchic Systems, *Advances in Design Automation,* B. Ravani (ed.), ASME, New York, pp. 15–20.

13. Renaud, J. E., and G. A. Gabriele, 1991, Sequential Global Approximation in Non-Hierarchic System Decomposition and Optimization, *Advances in Design Automation, Design Automation and Design Optimization,* G. Gabriele (ed.), ASME, New York, pp. 191–200.

14. Renaud, J. E., and G. A. Gabriele, 1993, Improved Coordination in Non-Hierarchic System Optimization, *AIAA J.,* Vol. 31, No. 12, pp. 2367–2373.

15. Renaud, J. E., and G. A. Gabriele, 1994, Approximation in Nonhierarchic System Optimization, *AIAA J.,* Vol. 32, No. 1, pp. 198–205.

16. Sellar, R. S., S. M. Batill, and J. E. Renaud, 1996, A Neural Network-Based, Concurrent Subspace Optimization Approach to Multidisciplinary Design Optimization, Conference Paper AIAA 96-0714, Presented at the 34th AIAA Aerospace Sciences Meeting, Reno.

17. Shankar, J., C. J. Ribbens, R. T. Haftka, and L. T. Watson, 1993, Computational Study of a Nonhierarchical Decomposition Algorithm, *Comput. Optim. Appl.,* Vol. 2, pp. 273–293.

18. Sobieszczanski-Sobieski, J., 1988, Optimization by Decomposition: A Step from Hierarchic to Non-Hierarchic Systems, in *NASA Conference Publication 3031, Part 1,* Second NASA/Air Force Symposium on Recent Advances in Multidisciplinary Analysis and Optimization.

19. Wujek, B. A., J. E. Renaud, S. M. Batill, and J. B. Brockman, 1995, Concurrent Subspace Optimization Using Design Variable Sharing in a Distributed Computing Environment, *Proceedings of the 21st ASME Design Automation Conference.*

14

INTEGRATED DESIGN AND PROCESS PLANNING FOR MICROWAVE MODULES

JEFFREY W. HERRMANN, IOANNIS MINIS, and DANA S. NAU
University of Maryland

KIRAN HEBBAR
Bentley Systems

STEPHEN J. J. SMITH
Hood College

14.1 INTRODUCTION

The standard product development process includes the conversion of functional requirements to design specifications, conceptual design, detailed design, process planning, production planning, and, finally, production. However, decisions made during the early phases of the process commit a large percentage of the total product cost. Thus, designers need tools that support concurrent engineering at all stages of product development, from conceptual and preliminary design through detailed design and manufacturing planning. In general, existing CAD/CAM tools are useful only during or after the detailed design stage. Moreover, existing preliminary and conceptual design tools support only the capture of design specifications.

Integrated Product and Process Development, Edited by John Usher, Utpal Roy, and Hamid Parsaei
ISBN 0-471-15597-7 © 1998 John Wiley & Sons, Inc.

This chapter identifies the important issues in integrating design and planning of microwave modules and discusses our research efforts related to these issues. Although achieving complete design and planning integration is necessarily a long-range goal, this research explores the relevant issues, provides insight into the design and planning process, and develops sophisticated methods that can integrate the design and planning of microwave modules and other complex electromechanical systems.

14.1.1   Microwave Modules

Most commercial electronic products operate in the 10-kHz-to-1-GHz radio frequency spectrum. However, in the telecommunications arena, the range of operation frequency has been increasing at a tremendous pace. For scientific and commercial long-range defense applications—such as radar, satellite communications, and long-distance television and telephone signal transmissions—radio frequencies prove unsuitable, primarily due to the high noise-to-signal ratio associated with radio frequencies. Moreover, the lower-frequency bands have become overcrowded due to the overuse of these bands for commercial communications applications.[33]

Consequently, in contrast to other commercial electronic products, most modern telecommunications systems operate in the 1–20-GHz microwave range, and modules of such systems are termed microwave modules (see Fig. 14.1).

In earlier microwave circuit assemblies, different parts of the circuit were built separately using coaxial cables or waveguides and later assembled by fastening the parts together. Due to the size and configuration of the coaxial cables and waveguides, these were large and heavy assemblies, and the assembly procedure was a time-consuming and costly process. These earlier assemblies were replaced by microwave integrated circuits (MICs), in which all functional components of the circuit are fabricated as artwork on the same planar board, using the same fabrication technology. The artwork lies on the dielectric substrate, which lies on the metallic ground plane that also serves as a heat sink. Functional components such as transistors, resistors, and capacitors can be classified as either "integrated" or "hybrid." Integrated components are fabricated as a geometric manifestation of the artwork. Hybrid components are assembled separately using techniques such as soldering, wire bonding, and ultrasonic bonding. If all functional elements of the device are integrated, such devices are known as monolithic microwave integrated circuits (MMICs).

The production method depends on several factors, some of which are the choice of dielectric material and the degree of integration of functional elements in the design. If all elements are assembled as hybrids, then lamination, photomask deposition, etching, plating, adhesive deposition, application of flux, reflow soldering, trimming, cleaning, testing, tuning, drilling, milling, and casting form a superset of the operations used.[3, 7] If, however, some compo-

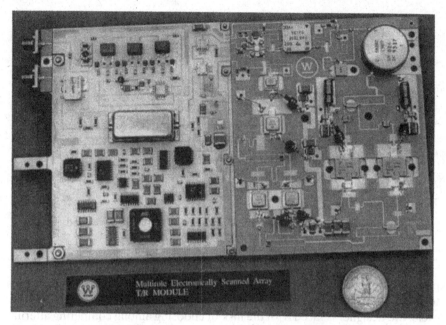

Figure 14.1 Typical microwave module.

nents are fabricated as integrated elements, then the product requires both thin-film and thick-film deposition.[19]

14.1.2 Motivation

The design and manufacturing cycle for microwave modules is shown in Figure 14.2. Electronics designers develop the detailed circuitry; mechanical designers design the device to resist shock and vibrational loadings and they also develop the assemblies, the heat removal systems, and the housing of the device; and manufacturing engineers plan the electronics-related manufacturing processes (such as lithography, soldering, cleaning, and testing) and the mechanical processes (such as drilling and milling) to manufacture the end product. These are not independent decisions: For microwave modules, mechanical properties such as component placement and artwork dimensions affect electrical behavior. This interrelationship further complicates the design and manufacturing cycle.

The task of communicating design and manufacturing requirements and design changes across disciplines could be greatly aided by tools that integrate both electronic and mechanical computer-aided design and provide access to process planning and design evaluation capabilities, as shown in Figure 14.3. A designer could use such tools for both the electronic and the mechanical aspects of a product, analyzing various aspects of the design's performance,

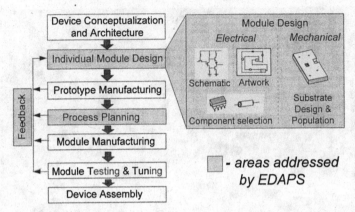

Figure 14.2 Design and manufacturing cycle for microwave modules.

planning how to manufacture the proposed design, and evaluating the plans to obtain feedback about the design. Throughout the design and manufacturing cycle, the designer is faced with the task of choosing among competing alternatives.

Consider first the typical case in which the manufacturer both designs and fabricates the microwave module. In this case, a number of choices are available for a given schematic, including alternate components, vendors, and processes. For example, a resistor of given specifications could be available as both leaded and surface mount types, and offered by a number of vendors with differing cost and quality ratings. These differences could, in turn, require different processes for assembly (board placement) and electrical connection

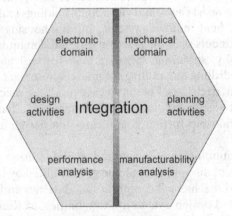

Figure 14.3 Integration of disciplines for design and manufacture of complex electro-mechanical devices.

(soldering). Also, the designer may need to evaluate both manual as well as automated options to carry out processes such as assembly and soldering. Additionally, there may exist quantity discounts and other intangible benefits associated with placing orders with a small number of suppliers—a fact that should be taken into account when choosing the components. The preceding factors therefore indicate that designers are typically faced with a large number of options in terms of component–process configurations and furthermore, there are cost and quality trade-offs between the various choices. Consequently, along with the manufacturability tools reported in the literature, there is a distinct need for models that efficiently explore the search space to identify "good" design options in terms of cost, quality, and other metrics.

Consider now the manufacturing firm's need to respond quickly to a market opportunity. The firm may wish to form a partnership with other manufacturers who may realize a portion of the product design and who cooperate to lower the product cost, improve its quality, and reduce the time span necessary to bring the product to market. Such a partnership may be a virtual enterprise: The partners electronically exchange the necessary information for design, process planning, production planning, inventory management, testing, distribution, and billing. Therefore, in addition to the design and manufacturing process described previously, the manufacturing firm must select the partners that can best realize the product. This goes beyond the classic make-or-buy decision. In addition, partner selection has design implications, because the designer should consider, during the early design phases, the partner-specific strengths that are related to the product's manufacturing requirements.

At this point, one can identify some required capabilities for integrated design and planning tools that support designers of complex electromechanical systems:

1. To manage alternative design and planning options throughout the design process.
2. To identify feasible options that designers might otherwise ignore and to provide information that they need to choose the best option.
3. To provide seamless access to external information sources such as CAD systems, design evaluation modules, parts catalogs, and supplier databases.

These requirements exceed the features of existing design support tools. Existing CAD/CAM tools are useful only during or after the detailed design stage. Designers need support during preliminary and conceptual design as well. Existing tools for preliminary and conceptual design only capture design specifications. In contrast, designers and manufacturing engineers need to develop and evaluate alternative designs and plans.

Thus, integrating design and planning raises numerous issues that need investigation: integrating electrical and mechanical design; representing design and process options that occur at different levels; generating feasible design and process options; evaluating feasible alternatives; comparing feasible alternatives on multiple criteria; and providing seamless access to external data sources. Our efforts to integrate the design and planning of microwave modules addresses many of these issues. In this chapter, we describe three major research efforts.

The first research effort is a detailed process planning procedure for microwave modules. The procedure integrates electrical and mechanical computer-aided design (CAD). It uses knowledge about the relevant manufacturing processes and information from the CAD models to generate a detailed process plan and evaluate the product's manufacturability.

The second effort is a trade-off analysis model that represents the design and process options associated with a microwave module and supports the designer's need to select options and balance multiple criteria such as cost, yield, and time.

The third research effort is a generative high-level process planning approach for partner selection and synthesis of virtual enterprises. The designer uses an object-oriented group technology scheme to represent the product design. Manufacturing resource models describe the manufacturing process capabilities and performance of potential partners. The generative high-level process planning methodology identifies feasible process planning and alternatives; represents them using a structured decision tree; estimates each alternative's total cost, quality, and cycle time; and allows the designer to select the most suitable one.

The remainder of the chapter is structured as follows: Section 14.2 describes the detailed process planning approach. Section 14.3 describes the electromechanical assembly model. Section 14.4 summarizes the high-level process planning approach. Section 14.5 discusses the issues that the previous research addresses and considers future research directions.

14.2 CAD INTEGRATION AND DETAILED PROCESS PLANNING

The detailed process planning approach forms the Electromechanical Design and Planning System (EDAPS), a toolkit for microwave module manufacture that integrates electronic and mechanical computer-aided design, electronic and mechanical process planning, and plan-based design evaluation.[16] The system generates process plans concurrently with the design and assists the designer in performing plan-based critiquing of microwave module designs. Process planning occurs both in the mechanical domain, including such processes as drilling and milling, and in the electronic domain, including such processes as through-hole plating, artwork deposition, placing components, and soldering. This provides feedback about manufacturability, cost, and cycle time to the designers, based on process plans for the manufacture of the device.

This research explores many issues related to integrated design and planning: integrating electrical and mechanical design, representing process options at different levels, generating feasible process options, evaluating feasible alternatives using multiple metrics, and providing seamless access to different modules and multiple data sources.

The detailed process planning approach includes CAD tools for electronic and mechanical design and an integrated process planner for mechanical and electronic manufacturing processes. The architecture of the corresponding system is shown in Figure 14.4 and contains three related modules:

- In the circuit schematic and circuit layout module, the designer generates electronic circuitry. An integrated set of commercial software supplied by EEsof's Series IV system[13] forms the core of this module. On top of this software, we have built routines that provide application-specific information. We address the circuit layout module in more detail in Section 14.2.1.

- In the substrate design module, the designer performs mechanical feature-based design. Bentley Systems' Microstation CAD software[24] supplies the set of tools required to achieve this functionality. Custom routines in C++ and the Microstation Development Language build the appropriate features, integrate Microstation with the rest of the system, and extract and supply relevant manufacturing information to individual modules. We address the substrate design module in more detail in Section 14.2.2.

- In the process planning and plan evaluation module, the AI-based process planner creates a process plan for the design and reports to the designer the cost and cycle time for the design. We describe the process planning and plan evaluation module in more detail in Section 14.2.3.

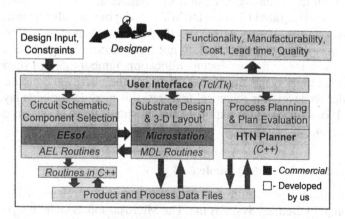

Figure 14.4   EDAPS system architecture.

The coordination of these modules and the exchange of data among them take place through a user interface written in the Tcl/Tk language.[25] This user interface allows the designer to smoothly interact with the heterogeneous modules that constitute the system.

14.2.1   Circuit Schematic and Circuit Layout Module

The microwave circuit design and layout module uses a powerful set of tools included in the EEsof electronic CAD tool. In particular, the module uses EEsof's Libra tool for linear and nonlinear schematic circuit design and EEsof's ACADEMY tool for layout generation.

Using Libra, the designer designs the "schematic circuit," choosing components from predefined and user-defined device libraries. In schematic circuits, the components and transmission lines are represented as symbols. The actual artwork shapes corresponding to the circuit elements are not represented in the schematic. The designer subjects this circuit to time and frequency domain response analyses to achieve the desirable functionality. The designer does several design iterations, and Libra evaluates each design until the designer obtains a functionally satisfactory circuit.

Libra incorporates some design-for-manufacturing principles. Based on the required circuit functionality, the limiting tolerances on each component's electrical parameters can be calculated and thus manufacturing yield can be predicted. Yield information calculated this way gives an idea of the required investment in postproduction. This yield metric is the maximum yield that can be expected out of the design. It is useful in performing sensitivity analysis of the design. However, manufacturing yields are not only a function of electrical parameter tolerances. Some of the other influences can be the defects that result from the soldering processes that are directly related to the package shape, dimensions, and materials.

Once the schematic circuit is complete, the artwork shapes necessary to realize circuit interconnections and other metallizations on the substrate are automatically generated by ACADEMY. The layout can also be interactively laid down to fit the artwork within specified size constraints and to incorporate those artwork layer elements that do not have electronic significance. Examples of such elements are product identification numbers, design version numbers, fiducial marks, and the global origin for the microwave module.

In order to develop mechanical features, this module converts layout data into the IGES format[20] for export to the mechanical CAD system described in Section 14.2.2.

14.2.2   Substrate Design Module

The substrate design module uses Microstation, a comprehensive CAD package supplied by Bentley Systems Inc. The Microstation modeler is a parametric feature-based design system. According to Salomons,[29] features are information sets that refer to aspects of form and other attributes of a part, such that these sets can be used in reasoning about the design, performance, or manufacture of the part or assemblies they constitute. The ACIS solid modeler[1] is used internally to represent and provide methods to generate and modify features defined in Microstation. In this approach, the following manufacturing features are most relevant to process planning and plan evaluation:

- *Dielectric.* The dielectric substrate is assumed to have prismatic geometry with designer-specified corner radii, thereby directly corresponding to the material removal shape volumes of end-milling features. The feature information set contains dimensions, corner radii, location, orientation, and electronic parameters such as the dielectric constant and dielectric material.

- *Heat Sink.* The initial geometry of the heat sink (or ground plane) is also assumed to be prismatic with corner radii. Related information describes its material, length, width, height, and corner radius. An additional constraint specifies that the widths and lengths of the heat sink and dielectric be equal, because the dielectric is fabricated on the heat sink.

- *Component Mounting Pockets.* For packaged components that require recesses in the substrate and heat sink for mounting and grounding, component mounting pocket features whose geometry corresponds to an end-milling feature have been provided. By default, the dimensions of such a feature are a function of the dimensions of the packaged component, and its location is the same as that of the packaged component. This generic end-milling feature can be used to construct all other cutouts, pockets, and grooves in the dielectric and heat sink.

- *Vias.* Conductive through-holes (vias) are represented as manufacturing features because they directly correspond to the material removal volumes of drilling features. In addition to the diameter, location, orientation, and length of the holes, the via feature stores useful manufacturing information such as electroplating thickness, if electroplated, and, if tapped, a reference to the pitch, nominal diameter, and the owner screw.

14.2.3 Process Planning and Plan Evaluation Module

To perform detailed process planning for microwave module designs, we use an approach from artificial intelligence called *hierarchical task network* (HTN) planning.[11, 28, 32, 34] We have also used this approach in some of our other work.[31]

Hierarchical task network planning proceeds by taking a complex task to be performed and considering alternate methods for accomplishing the task. Each method provides a way to decompose the task into a set of smaller tasks. By applying other methods to decompose these tasks into even smaller tasks, the planner will eventually produce a set of primitive tasks that it can perform directly.

As an example, one method for making the artwork is to perform the following series of tasks: precleaning for the artwork, followed by application of photoresist, followed by photolithography for the artwork, followed by etching. There are several alternate methods for applying photoresist: spindling the photoresist, spraying on the photoresist, painting on the photoresist, and spreading out the photoresist from a spinner. The relationships between tasks and methods form a task network, part of which is shown in Figure 14.5.

This decomposition of tasks into various subtasks is important for process planning for the manufacture of microwave modules for two reasons. First, the decomposition in an HTN naturally corresponds to the decomposition of a design into the parts and processes required to manufacture it. Second, the ability to include the complex tasks "make drilling and milling features," "make artwork," "assembly and soldering," and "testing and inspection" in sequence provides a uniform framework that can naturally accommodate all the processes in mechanical and electronic manufacturing.

This decomposition requires manufacturing knowledge. Sometimes a particular method can always be used to perform a particular task. For example, because spreading out the photoresist from a spinner is so accurate, this method can always be used to perform the task of applying the photoresist. Sometimes a particular method can only occasionally be used to perform a particular task. For example, because spraying on the photoresist is only somewhat accurate, this method cannot be used to apply the photoresist if a coupler in the artwork has a gap less than or equal to 10 mils.

Certain tasks are primitive, meaning that they do not break down into any other tasks. We consider a task to be primitive if it is considered to be a single small step in the manufacturing process. For example, precleaning for the artwork is a primitive task. Once the complex task of making the entire product has been broken down into a series of primitive tasks, a process plan has been created; carrying out the steps of the process plan will manufacture the product.

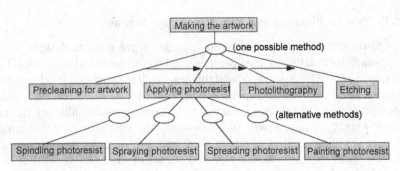

Figure 14.5 Part of the task network for microwave module manufacture.

Consider the substrate shown in Figure 14.6. "Make board" decomposes into "Make plated through-holes and features," "Make artwork," "Assembly," and "Testing and inspection." "Make plated through-holes and features" decomposes into "Drill plated through-holes," "Plate plated through-holes," and "Make features." "Drill plated through-holes" and "Plate plated through-holes" decompose into primitive tasks, which we do not discuss here.

"Make features" is the next task, and because there are features left to be made, it decomposes into "Make a single feature" and "Make features." This "loop" in the task network allows us to decompose a task, such as "Make features," into zero or more subtasks, such as "Make a single feature."

"Make a single feature" decomposes into "Setup and end-mill (the top cutout on the left-hand side of the substrate)," because, in our planner, we always do all the milling before we do any drilling. "Setup and end-mill (the top cutout on the left-hand side of the substrate)" decomposes into "Setup," "Setup end-milling tool," and "End-mill." Because the part is not currently set up on the machining center, "Setup" decomposes into "Orient the part," "Clamp the part," and "Establish a datum point." All three of these tasks are primitive.

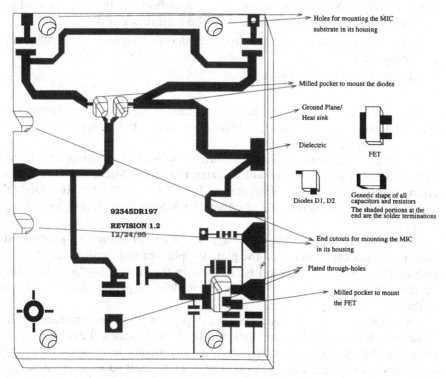

Figure 14.6 Development of mechanical features on the Mixer-IF amplifier substrate.

"Setup end-milling tool" is the next task, and, because we just started, we assume that the correct end-milling tool is not installed on the machining center. Thus, this task decomposes into "Install end-milling tool (of the appropriate size)," which is a primitive task. Assuming tight tolerances, "End-mill" decomposes into "Rough end-mill" and "Finish end-mill," both of which are primitive tasks.

"Make features" continues to decompose until a plan has been created for all five milling features and all thirteen drilling features. The next complex task is "Make artwork."

"Make artwork" decomposes into "Preclean for artwork," "Apply photoresist," "Artwork photolithography," and "Etching." In our planner, all of these tasks but "Apply photoresist" are primitive. "Apply photoresist" has several alternate methods: "Spread photoresist from a spinner" or "Spindling the photoresist" or "Spraying the photoresist." "Painting on the photoresist" is not a feasible alternative in this case because painting on the photoresist is not accurate enough for this substrate.

The rest of the plan is generated in a similar manner, and output is provided in the format shown in Figure 14.7. The output of the detailed process planner includes:

· A totally ordered sequence of process specifications that can be used to produce the finished substrate from the materials given.
· Process parameters of all the processes that are required to manufacture the device.
· Estimates of cost and cycle times.

The output can be fed back to the designers, with cycle time "hot spots" indicated. The designer can then choose to change the design elements, in order to reduce the cycle time.

When the designers and manufacturing engineers are satisfied with the design, the artwork elements will be extracted out of Microstation, and the equivalent IGES file will be generated and sent to ACADEMY. ACADEMY can then export the design file in either IGES format or Gerber format for manufacturing.

As mentioned before, because the method of application of photoresist does not affect anything else in the plan, the planner will locally decide which photoresist application method is cheapest in this instance—"Spindling the photoresist," let us say—keep only that subtask in the plan, and ignore the remainder.

The planning module constructs a set of process plans and evaluates them to see which takes the least amount of time. In some cases, it evaluates a set of incomplete process plans and discards all but the one which takes the least amount of time. For example, because the method of application for

```
Parts:
Block
  Dimensions: 7,4,1
  Ground material: Aluminum
  Substrate: Teflon
  Substrate thickness: 30 mils
  Metallized layer: Copper
  Metallized layer thickness: 7 mils
  Part number: 80280SA/2
Resistor
  Name: P1
  Part number: RNC55H2370FS
  Description: Motorola SS163
  Specification: MIL-R-55182
[...]
Processes:
```

Opn A	BC/WW	Setup	Run	LN	Description
001 A	VMC1	2.0	0.0	01	Hold substrate with flat vise jaws at 3.5,4,0.5 and 3.5,0,0.5
				02	Establish datum point at 0,0,1
001 B	VMC1	0.0	0.6	01	Drill hole: 1,4,0 depth: 1 using 0.25 radius bit
				02	Drill hole: 3,4,0 depth: 1 using 0.25 radius bit
001 C	VMC1	0.0	0.3	01	Drill hole: 3.5,6.5,0 depth: 1 using 0.125 radius bit
001 D	VMC1	0.0	5.0	01	Mill slot: 0.5,1,0 dimensions 3,1,1 using 0.5 radius end-milling tool
001 T	VMC1	2.0	5.9	01	Total time on VMC1

Figure 14.7 Part of a process plan in a standard format.

photoresist does not affect the method of application for solder paste, if the quickest method of applying photoresist is spraying it on, then there is no need to generate process plans in which some other method of application is used. If no process plans can manufacture the device—because some manufacturability constraint, such as achievable tolerance, is violated—the planner reports the failure and the reason for the failure to the designers.

14.3   TRADE-OFF ANALYSIS MODEL

The second research effort explores in more detail the trade-off issues faced during the microwave module design. It proposes a trade-off analysis model and the associated procedure that allows the designer to choose sets of alternate parts and processes that are desirable with respect to a set of metrics. This research explores multiple issues related to integrated design and planning: representing design and planning options and comparing feasible alternatives on multiple criteria.

The trade-off is performed with respect to five metrics: cost, manufacturing yield, number of suppliers, supplier lead time, and quantity discounts. The problem is formulated as a multiobjective integer program that the designer iteratively solves to search for and sort desirable solutions, as described in the following discussion.

The modeling approach exploits the following assumptions: The conceptual design for the microwave module (board) is given and is to be realized as a single assembly. The design specifies the set of required generic component types and, for each such component type, a number of specific alternatives. For each specific component, there is a list of processes that are related to the component and the alternatives (if any) for each such process. This defines an and–or tree that captures the structure of the design. Key attributes such as material costs, run times, setup times, and defect rates are known for components, processes, and component–process combinations. In addition, the supplier's lead time and the supplier's quantity discount structure are known for each component. The designer's problem is to determine a set of components (and thus suppliers) and processes that are "efficient" with respect to the five objectives mentioned earlier.

The model uses the following notation:

m = number of generic components required

P_i = generic component i, $i = 1, \ldots, m$

n = number of alternate components available

V = $\{p_1, \ldots, p_n\}$, the set of available components

V_i = alternate components for generic component P_i, $V_i \subset V$

s_j = number of generic processes required for p_j

Q_{jk} = generic process k for component p_j, $k = 1, \ldots, s_j$

r = number of alternate processes available

W = $\{q_1, \ldots, q_r\}$, the set of available processes

W_{jk} = alternate processes for generic process Q_{jk}, $W_{jk} \subset W$

The decision variables are x_j, $j = 1, \ldots, n$, and y_t, $t = 1, \ldots, r$. $x_j = 1$ if component p_j is selected and 0 otherwise. $y_t = 1$ if process q_t is used in the assembly and 0 otherwise.

The following constraints define the and–or structure of the model:

$$\sum_{p_j \in V_i} x_j = 1 \qquad \text{for all } i = 1, \ldots, m$$

$$\sum_{q_t \in W_{jk}} y_t = x_j \qquad \text{for all } j = 1, \ldots, n, \, k = 1, \ldots, s_j$$

The first set of constraints represents the design requirements: The design must contain generic components $P_1, P_2, \ldots, P_m$. Similarly, the second set of constraints represents the requirements of component p_j (if p_j is a selected component): p_j requires generic processes $Q_{j_1}, Q_{j_2}, \ldots, Q_{j_{s_j}}$. Each set V_i represents the design options: Generic component P_i requires p_1 or p_2, if both are elements of V_i. Similarly, set W_{jk} represents the process options: Generic process Q_{jk} requires q_1 or q_2, if both are elements of W_{jk}.

The model includes additional parameters and constraints necessary to measure the five objective functions, which are normalized with respect to designer-supplied limits (lower and upper bounds) and combined using designer-specified weights. In addition, feasible solutions must satisfy all of the upper bounds; thus, these upper bounds define the search space. The resulting integer program resembles an uncapacitated facility location problem, which is well structured and can be solved using the linear programming tool CPLEX[10] with reasonable computational effort.

After specifying the model parameters, the designer iteratively solves the trade-off analysis model to generate a set of designs that are "efficient" with respect to the five metrics mentioned earlier. The designer specifies, for each objective function, upper bounds and weights. The bounds limit the search space, and the optimization tool sorts the feasible solutions by their weighted performance and outputs the best solution(s). From this feedback, the designer changes the bounds to expand or contract the search space or changes the relative weights to find other good solutions in the search space. This continues until the designer has located the most desirable solutions.

14.4 PARTNER SELECTION FOR MICROWAVE MODULE MANUFACTURING

Our third effort in the area of microwave module design and planning addresses selecting partners for the joint manufacture of a new microwave module design. Specifically, we present an approach that, given a new microwave module design, generates feasible process and partner alternatives, evaluates the feasible alternatives, allows the designer to search for and sort these alternatives on multiple criteria, and selects the most efficient set of partners.

Section 14.4.1 presents an overview of the design evaluation and partner selection methods and system. Section 14.4.2 describes the necessary information models. Section 14.4.3 describes high-level process planning, the method that generates process and partner alternatives. Section 14.4.4 describes evaluating the alternatives, and Section 14.4.5 describes selecting an efficient partnership.

14.4.1 Overview

Figure 14.8 illustrates our approach. The output of the designer's CAD system is translated and stored in an integrated product model. This model uses the data definitions of STEP, the international Standard for the Exchange of Product Data (ISO 10303[21]), and thus supports the free exchange of data between the firm and its partners.

Design evaluation requires more abstract product information than that in the STEP-based product model. Concise group technology (GT) codes are used to search for and retrieve similar products, and high-level generative process planning uses some detailed data about those product attributes that the GT code includes. This information forms the object-oriented group technology (OOGT) product model. We have developed (and implemented as the Group Technology Design Processor in Fig. 14.8) algorithms that derive the OOGT product model from the STEP-based product representation. In the design retrieval step (the product search module), the designer exploits the concise nature of the GT codes to search quickly for similar products in the product databases of candidate partners.[22]

To generate and evaluate partnering alternatives, we use a high-level process planning approach. In the first step of this approach, the feasibility assess-

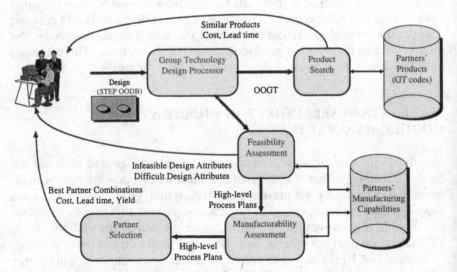

Figure 14.8 Design evaluation and partner selection approach.

ment module generates feasible manufacturing alternatives. The system uses generic data about manufacturing processes and specific information about the process capabilities of the candidate partners to construct feasible plant-specific process plans and identify features of the design that are infeasible with respect to generic or partner-specific process capabilities. The feasible process plans specify the sequence of manufacturing operations, the candidate partners who could perform these operations, and the design attributes to be realized at each operation. (Unlike the approach presented in Section 14.2, the process plans do not describe process details, process parameters, tooling, fixtures, or other specific manufacturing instructions necessary for actual production.) For infeasible processes, this step identifies for the designer the related attributes that need revision.

The manufacturability assessment module, which uses generic data about manufacturing processes and specific performance measures about the processes of the candidate partners, evaluates each feasible process and partner combination with respect to cost, quality, and cycle time. In addition, in this step the designer can determine those attributes that most affect the design's cost, quality, and cycle time. With this information, the designer can initiate redesigns that improve the product's performance within the given set of processes and partners.

Once the design evaluation is complete, the system allows the designer to sort the alternative high-level process plans on selected criteria, identify the partners that form the most desirable plan, and receive feedback on the plan's expected cost, quality, and cycle time.

Note that Figure 14.8 illustrates the entire design evaluation and partner selection system. This section describes only the portions that generate, evaluate, and compare process planning alternatives. The high-level process planning approach consists of the feasibility assessment and manufacturability assessment modules. The partner selection module allows the designer to compare alternatives and select the one that is most suitable on multiple criteria.

14.4.2 Information Models

The partner selection approach requires three general types of data: product design data, manufacturing process data, and manufacturing resource data. We identify and manage the necessary data by constructing appropriate information models (see Candadai et al.[4] for a complete description).

Product Information As described previously, the designer initially stores a product design in an integrated product model that uses STEP to support the free exchange of data between the firm and its partners. Design evaluation requires more abstract product information, however. The product information required for high-level process planning is captured in the object-oriented group technology (OOGT) product model[5, 16] shown in Figure 14.9. This model is a concise view of a product design. It stores critical design information more compactly and at a different level of abstraction than the complete product model.

The top level of this information model describes general product attributes including part number, raw material, and production quantity. The lower levels capture information about both mechanical and electrical product attributes. The mechanical information describes the product envelope in terms of enveloping faces and the product features in terms of parametric attributes such as feature volume, corner radii, minimum tolerance, and surface finish. Additional feature-related information includes thin sections, sections with abrupt thickness changes, and directions along which a feature causes an undercut. The electrical information describes the electrical product design requirements including artwork layout and tolerances, component types and mounting specifications, and soldered and nonsoldered hardware requirements.

Process Information   The generic process knowledge used in this approach is organized in a simple process information model. This information, typically

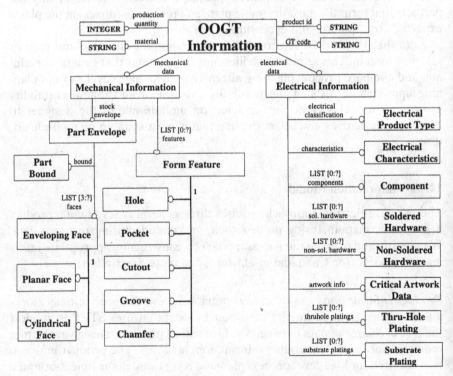

Figure 14.9   OOGT product model.

found in manufacturing handbooks, describes universal process capabilities, material–process compatibilities, and recommended production quantities. Table 14.1 shows a representative table from the generic process information model, which was populated with data from various sources including design handbooks,[2, 8, 30] manufacturing handbooks,[9, 12, 17, 35] and materials handbooks.[14] It shows the compatible material–process combinations, compatible feature–process combinations, and some global process capabilities such as the feasible design quantity range.

Manufacturing Resource Information   The manufacturing resource model includes general information about a corporation and its manufacturing facilities (plants) and also detailed data about the systems in each plant. Most important to process planning and manufacturability analysis are the data that describe the capabilities and performance of a plant's manufacturing processes and the associated resources. Information about process availability and process capabilities (such as maximum envelope size and achievable accuracy of a milling process) are used to generate the plant-specific process plans (as discussed in Section 14.4.3). The performance measures (including cost rates, queue time, capacity, process variance, and yield) is used to evaluate the plans (Section 14.4.4). Additional details are given by Candadai et al.[4]

14.4.3   Generating Partnering Alternatives: High-Level Process Planning

Supporting the high-level process planning approach, which generates feasible partnering alternatives, is a process planning data structure (PPDS) that captures information about the various process alternatives, their sequence, and the plants that perform these processes.[23]

The PPDS structure reflects the processes used to manufacture microwave modules and discussed in Section 14.2: drilling and plating conductive through-holes (vias), machining the microwave module substrate features, generating artwork (substrate etching and plating), automated or manual component assembly and soldering, and testing. (Although the same principles apply, a different product's PPDS would include a different set of processes. For example, a strictly mechanical product would include primary, secondary, and tertiary processes.) As shown in Figure 14.10, the PPDS has alternating levels of process and plant options, which represent the processes and plants that may be used to manufacture the product. The combination of a process option and a plant option represents a complete processing step in a high-level process plan: It describes the operations performed at the manufacturing plant and the remaining features that need to be manufactured at subsequent steps. A high-level process planning alternative is a sequence of process–plant combinations.

High-level process planning uses the OOGT product model to obtain critical design attributes, the process information model to relate design attributes to manufacturing processes, and the manufacturing resource model to identify the potential partners' specific manufacturing capabilities.

TABLE 14.1 Sample Table of the Process Information Model

Processes	Selected Materials										Features					Other Attributes			
	Aluminum	Brass	Copper	Cast Iron	Carbon Steel	Tungsten	LDPE	HDPE	Nylon	PEEK	Hole	Pocket	Cutout	Chamfer	Groove	Smallest env. Dimension (in.)	Largest env. Dimension (in.)	Lower Qty. Bound	Upper Qty. Bound
Die Casting	•	•	•								•	•	•	•	•	1.25	12	1,000	99,000
Inv. Casting	•	•	•	•	•						•	•	•	•	•	1.00	25	1,000	99,000
Sand Casting	•	•	•	•	•	•					•	•	•	•	•	1.25	20	1	100,000
Forging	•		•		•	•					•	•	•	•	•	0.50	1,000	2,500	25,000
Inj. Molding							•	•	•	•	•	•	•	•	•	0.035	50	7,500	99,000
Milling	•	•	•	•	•	•	•	•	•	•	•	•	•	•	•	0.25	100	1	250
Drilling	•	•	•	•	•	•	•	•	•	•	•					0.25	100	1	250
Int. Grinding	•	•	•	•	•	•	•	•	•	•	•					1.00	25	1	250
Sf. Grinding	•	•	•	•	•	•	•	•	•	•				•		0.25	100	1	250

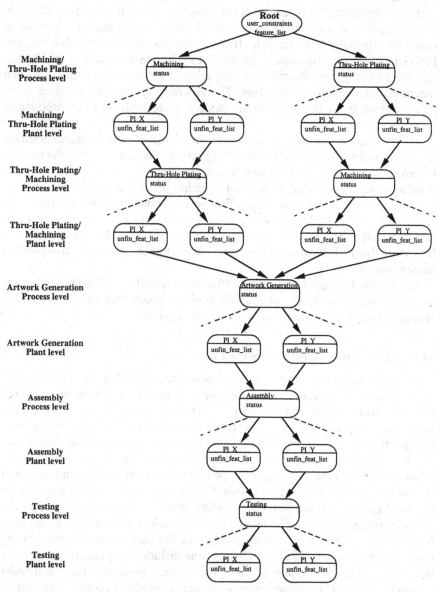

Figure 14.10 Process plan data structure (PPDS) for electromechanical parts.

The approach constructs the PPDS by selecting feasible process and plant alternatives at each step.[15] Process selection is a plant-independent procedure that retrieves all candidate processes (from the process database) associated with key design attributes and discards processes that are globally infeasible (i.e., infeasible at any plant). All required subprocesses must be feasible in order for a process to be feasible. If the design has plated through-holes, the PPDS includes two processing alternatives: One corresponds to machining and then plating; the second corresponds to through-hole plating and then machining the remaining features. The assembly process (manual or automatic) depends on the component mounting methods and the production quantity.

Plant selection uses process capability information from the manufacturing resource model to identify the candidate partners that can perform the process (or all required subprocesses) to generate the corresponding attributes of the product design. For example, a plant's plating process must be able to plate the required thickness, and the etching process must be able to achieve the required line width tolerance and line spacing tolerance. If a process or plant option is infeasible, the process planning approach identifies the reason and lists it in the PPDS, which may allow the designer to modify the product design appropriately.

Each path through the resulting PPDS corresponds to a feasible high-level process plan (a sequence of feasible process–plant combinations with no remaining features) or ends in an infeasible option.

14.4.4 Evaluating Feasible Process Plans

After the feasibility assessment module generates the PPDS, the manufacturability assessment procedure evaluates the cost, quality, and cycle time of each feasible process–plant combination.[18, 23] The procedure uses process-specific knowledge, expressed as rules and formulas, and the potential partners' process performance data, which the manufacturing resource model describes.

The cycle time associated with each process is the queue time for the process, the setup time for the entire production quantity and each batch, and the total run time of all subprocesses. For example, the milling setup time is the total recurring setup time (for loading, unloading, and cleaning) and the nonrecurring setup time. The milling run time includes the actual cutting time for all features (roughing and finishing) and the tool approach time (during rapid and slow travel). The total etching time includes the photoresist masking time, the photoresist exposure time, the etching time, and the photoresist stripping time. The manufacturing resource model provides the plant-dependent queue time. Process-specific procedures calculate the process setup and run times based on design characteristics, plant capabilities, and process knowledge. The approach includes procedures for milling, drilling, plating, etching, automated assembly, automated soldering, manual assembly, and testing. (We have also developed procedures for other mechanical processes:

sand casting, investment casting, forging, surface grinding, and internal grinding.)

The cost of the process is the setup cost and direct labor cost of the process. The costs are the plant-specific setup and labor rates multiplied by the setup and run times and a plant-specific overhead rate. The quality of a process is the process capability ratio C_p (where appropriate) and a plant-specific yield otherwise. The C_p for etching is the quotient of the minimum artwork tolerance (the minimum of the line width tolerance and the line spacing tolerance) and six times the plant's etching standard deviation. If a process consists of subprocess, the procedure determines the performance of each subprocesses and aggregates them to calculate the process performance. (In this case, C_p's are converted to yields, multiplied, and transformed again to a composite C_p.) When this step is completed, the PPDS contains the feasible processes and plants and the cost, quality, and cycle time of each combination, which is required for the comparison of high-level process plans and selection of partners.

14.4.5 Partner Selection

The partner selection approach allows the designer to compare the different high-level process plans. Partner selection follows the generation and evaluation of high-level process plans, as described previously.

An explicit enumeration technique constructs all feasible high-level process plans from the feasible process–plant pairs in the PPDS. Each feasible alternative is evaluated with respect to cost, quality, cycle time, and the transportation cost between consecutive plants in the process plan. These performance measures combine the cost, quality, and cycle time for the plan's component process–plant pairs. The transportation cost depends on the location of the candidate manufacturing plants.

The designer may search for desirable alternatives by excluding those alternatives that are dominated by some other alternative with respect to any combination of criteria and by excluding those alternatives that are inferior with respect to user-specified thresholds for one or more criteria. The designer can sort the remaining alternatives on a linear combination of some criteria. The designer provides a weight for each performance criterion, and the weighted combination of the criteria forms the new performance criterion. For example, these weights allow the designer to convert all criteria to dollars or to give relative weights to the criteria.

In addition, the designer can specify preferences in the form of natural language expressions about the importance of each performance attribute (cost, quality, cycle time). Using a fuzzy extension of the analytic hierarchy process (fuzzy-AHP),[26, 27] the partner selection approach combines these preferences with existing data (from industrial surveys and statistical analysis) to reemphasize attribute priorities. These redefined attribute priorities reflect the specific needs of the firm for this product. In the fuzzy-AHP procedure, the pairwise comparisons in the judgment matrix are fuzzy numbers that are modified by the designer's emphasis. Using fuzzy arithmetic and alpha cuts, the procedure calculates a sequence of weight vectors that will be used to combine the process plan's scores on each attribute. The procedure calculates a corresponding set of scores and determines one composite score that is the average of these fuzzy scores.

14.5 CONCLUSIONS

This chapter identifies the issues related to the integrated design and process planning of microwave modules. In addition, this chapter discusses three related research efforts that explore these issues: detailed process planning, trade-off analysis, and high-level process planning. We anticipate that our methods and results provide significant insight into concurrent engineering of other electromechanical systems. In this section, we review the specific contributions of our research efforts and discuss promising research directions.

Integrating Electrical and Mechanical Design   The ultimate solution to the problem of integrating electronic and mechanical design can be found in one of at least two ways. One possibility is the implementation of a single monolithic software system that includes both an electronic design subsystem and a solid modeling engine for mechanical design. The data structures in such an implementation would relate the solid model of each shape element in the mechanical design with its function in the schematic of the electronic design. Such a solution would allow tightly coupled interaction between the electronic design subsystem and the mechanical design subsystem—and could be used to generate sophisticated feedback to the designer, such as suggestions for how to change the proposed design to improve its manufacturability while maintaining acceptable performance. Unfortunately, such an approach requires the creation of a completely new system, which may be incompatible with the legacy systems already used in practice.

Another possibility—the approach we have taken in the detailed process planning research—is to integrate existing systems for electrical and mechanical design. In addition, this approach requires extending the electronic design system to keep track of some of the information needed for mechanical design so that this information will not be lost when users change the electrical design, and similarly extending the mechanical design system. The disadvantage of such a solution is that it may limit the interaction between the electronic design system and the solid modeler and that, in any case, translating and transferring information from one system to another takes time and work. (In our system, because our feedback is based on the process plan for manufacturing, we did not have to translate much information back to the electronic design system from the solid modeler.) However, such a solution allows companies to keep legacy systems in place; in addition, designers can change their electronic design system without changing their solid modeler or vice versa.

Representing and Analyzing Design and Planning Options   In an integrated design and planning environment, a designer needs to represent and analyze alternate design and planning options at multiple levels of detail. These options include alternate components, suppliers, manufacturing processes, and manufacturing partners. Our research explores different structures for representing these alternatives.

The trade-off analysis model specifies the set of required generic component types and, for each component type, a number of specific component alternatives. For each specific component is a list of processes that need to be performed on the component and the alternatives (if any) for each such process. This defines the basic and–or tree that captures the structure of the design.

The high-level process planning approach includes a process planning data structure (the PPDS) that captures information about the various process alternatives, their sequence, and the plants that perform these processes. Each path through the PPDS corresponds to a feasible high-level process plan or ends in an infeasible alternative. The combination of a feasible process option and a feasible plant option represents a complete processing step in a high-level process plan. A feasible high-level process plan is a sequence of feasible process–plant combinations.

Our detailed process planning procedure uses an approach from artificial intelligence called hierarchical task network (HTN) planning, which proceeds by taking a complex task to be performed and considering alternate methods for accomplishing the task. Each method provides a way to decompose the task into a set of smaller tasks. By applying other methods to decompose these tasks into even smaller tasks, the planner will eventually produce a set of primitive tasks that it can perform directly.

The trade-off analysis model's and–or tree provides a very general way to describe design and planning requirements and the associated alternatives. The PPDS uses a version of this structure to describe high-level process planning and partnering alternatives. The HTN approach, which uses methods and tasks to explore a search space that has the and–or tree structure, specifies process sequences and allows a more general process decomposition. Although externally different because they support different types of decision making that occur at different times during the design life cycle, these data structures have the same hierarchical and–or structure. It seems clear that this structure supports design and planning during the evolution from conceptual design to preliminary design and detailed design. While refining the design, the designer identifies the additional requirements and alternatives associated with the design and planning alternatives chosen earlier.

Generating Feasible Design and Process Options   In order to explore the complete search space and overcome the inertia that complex system design has (because the large number of required decisions limit the time available to develop new ideas), a designer requires tools that can generate, using a product design (at any level of detail) and appropriate manufacturing knowledge, feasible design and planning options and can identify the causes of infeasible options. Our research efforts include methods for identifying feasible manufacturing alternatives.

Most researchers have had great difficulty in developing generative process planners for complex mechanical parts, because their shape features may have complex interactions. However, generative process planning can be more easily applied to microwave modules, because the process plans use a relatively small set of operations and the mechanical features have fewer interactions.

During preliminary design, high-level process planning allows the designer to identify the most suitable processes and manufacturing facilities. The system uses generic data about manufacturing processes and specific information about the process capabilities of the candidate partners to construct feasible plant-specific process plans and identify features of the design that are infeasible with respect to generic or partner-specific process capabilities.

After the detailed design is complete, hierarchical task network planning appears to be an ideal approach for generating detailed process plans from the selected high-level process plan. The decomposition in an HTN naturally corresponds to the decomposition of a microwave module into the parts and processes required to manufacture it, and HTNs provide a unified framework that accommodates both electronic and mechanical manufacturing processes.

Evaluating Feasible Alternatives To choose the best options, a designer must know how each alternative performs. Our research explores different plan-based approaches for evaluating designs using multiple metrics. These approaches provide the designer with valuable feedback about the design during preliminary and conceptual design. This allows the designer to improve the design's manufacturability and avoid unnecessary iterations through the design and manufacturing cycle.

The detailed process planning approach estimates manufacturing cost and time based on the parameters of the required processes. The trade-off analysis model evaluates a design based on component–process combinations. The high-level process planning approach evaluates feasible process–partner combinations.

Comparing Feasible Alternatives on Multiple Criteria Faced with a large number of alternatives and the need to balance multiple criteria, a designer needs a convenient way to compare his or her performance and methods for making trade-offs according to specified criteria.

The trade-off analysis model and the partner selection approach provide tools that search for and sort alternatives (designs or process plans). In general, the designer first specifies thresholds to eliminate undesirable solutions and then weighs the different criteria to sort the remainder. An iterative approach allows the designer to change the thresholds and weights and therefore locate solutions that balance, subjectively at least, the various performance measures.

Providing Seamless Access to External Data Sources To generate and evaluate alternatives, integrated design and planning requires manufacturing knowledge that resides in a variety of sources (e.g., CAD models, parts catalogs, manufacturing process databases, and manufacturing resource models). Therefore, a designer needs seamless access to these sources so that their information can be retrieved and updated as needed. Our research identifies some required data sources and approaches for providing access to them.

The high-level process planning research described previously has identified some of the external data sources needed to support design and planning: product information models that describe the critical design information, relevant manufacturing process knowledge, the manufacturing resources' capabilities and performance (for each manufacturing facility or potential supplier), and a parts repository that has indexes for efficient searches.

Seamless access requires common data structures. The high-level process planning approach uses one data structure (the OOGT product model) to link the product design and process planning functions and another (the PPDS) to link the different modules that generate, evaluate, and compare the process planning alternatives. Similarly, the detailed process planner uses IGES files and a product information model to link the design and process planning modules.

In the detailed process planning system, a user interface written in the Tcl/Tk language provides seamless access. It allows the designer to smoothly interact with the heterogeneous modules that constitute the system.

Future Directions Although, as described previously, the research efforts described here explore many of the relevant issues and integrate portions of the design and manufacturing process, they are largely separate approaches, and one can clearly see that additional integration work remains. Our next research effort will integrate the trade-off analysis and detailed process planning approaches. The designer will generate an initial schematic based on device specifications and will simulate the schematic to test its functionality. In addition, the designer will specify the component types required. A high-level process planning procedure will determine the processes that the component types require and estimate the process performance. This provides the necessary input for the trade-off analysis model, and the designer will use this model to generate preliminary designs that are efficient with respect to multiple criteria. For each preliminary design, the designer will use electronic CAD tools to generate the artwork and mechanical CAD tools to create a solid model and add substrate features. Finally, the detailed process planner will generate and evaluate a complete process plan.

ACKNOWLEDGMENTS

Portions of the work described here were sponsored by the U.S. Army Tank-Automotive Command under Grant DAAE-07-93-C-R086. The research team was led by University of Maryland researchers and included researchers from the State University of New York at Buffalo, the National Institute for Standards and Technology, Westinghouse Electric Corporation, and Martin-Marietta. The researchers at the University of Maryland were partially supported by the Institute for Systems Research under Grant NSFD CD 8803012.

Portions of the work described here were supported in part by NSF Grants NSF EEC 94-02384, IRI-9306580, and DDM-9201779, by ARPA Grant DABT63-95-C-0037, and by in-kind contributions from Spatial Technologies and Bentley Systems.

REFERENCES

1. *ACIS Geometric Modeler,* 1993, Spatial Technology, Inc., Boulder, Co.

2. Bralla, J. G., 1986, *Handbook of Product Design for Manufacturing,* McGraw-Hill, New York.

3. Brindley, K., 1990, *Newnes Electronics Assembley Handbook,* Heinemann Newnes, Oxford.

4. Candadai, A., J. W., Herrmann, I. Minnis, and V. Ramachandran, 1994, Product and Process Information Models for Microwave Modules, in *CAE/CAD Application to Electronic Packaging,* D. Agonager and R. E. Fulton (eds.), ASME, New York, pp. 33–42.

5. Candadai, A., J. W. Herrmann, and I. Minis, 1995, A Group Technology-Based Variant Approach for Agile Manufacturing, in *Concurrent Product and Process Engineering,* A. R. Thangaraj, R. Gadh, and S. Billatos (eds.), ASME, New York, pp. 289–306.

6. Candadai, A., J. W. Herrmann, and I. Minis, 1996, Applications of Group Technology in Distributed Manufacturing, *J. Intelligent Manufacturing,* Vol. 7, pp. 271–291.

7. Chenu, J. P., and H. Muller, 1989, Manufacture of Microwave Telecommunication Equipment, *Electric. Comm.* Vol. 63, pp. 159–167.

8. Clark, D. S., and W. A. Pennington, 1962, *Casting Design Handbook,* American Society for Metals, Metals Park, OH.

9. Coombs, C. F., 1979, *Printed Circuits Handbook,* 2nd ed., McGraw-Hill New York.

10. CPLEX Optimization, Inc., Incline Village, NV.

11. Currie, K., and A. Tate, 1985, O-Plan-Control in the Open Planner Architecture, BCS Expert Systems Conference, Cambridge University Press.

12. Dym, J. B., 1979, *Injection Molds and Molding—A Practical Manual,* Van Nostrand Reinhold, New York.

13. *EEsof Series IV, Version 4,* 1992, EEsof Inc., Westlake Village, CA.

14. Farag, M., 1979, *Materials and Process Selection in Engineering,* Applied Science Publishers, London.

15. Gupta, S. K., J. W. Herrmann, G. Lami, and I. Minis, 1995, Automated High Level Process Planning for Agile Manufacturing, *Proceedings of the 1995 ASME Design*

Engineering Technical Conferences, O. M. Jadaan, A. C. Ward, S. Fukuda, E. C. Feldy, and R. Gadh ASME, New York, pp. 835–852.

16. Hebbar, K., S. J. J. Smith, I. Minis, and D. S. Nau, 1996, Plan-Based Evaluation of Designs for Microwave Modules, 1996 ASME Design Engineering Technical Conference and Computers in Engineering Conference.

17. Heine, R. W., C. R. Loper, and P. C. Rosenthal, 1967, *Principles of Metal Casting,* McGraw-Hill New York.

18. Harrmann, J. W., G. Lam, and I. Minis, 1996, Manufacturability Analysis using High-Level Process Planning, ASME Design for Manufacturing Conference, University of California, Irvine.

19. Hoffmann, R. K., 1987, Handbook of Microwave Integrated Circuits, Artech House, Norwood, MA.

20. *The Initial Graphics Exchange Specification Version 5.1,* 1991, IGES/PDES Organization, Gaithersburg, MD.

21. *ISO 10301-1, Product Data Representation and Exchange,* 1992, International Organization for Standardization.

22. Iyer, S., and R. Nagi, 1994, Identification of Similar Parts in Agile Manufacturing, in *Concurrent Product Design,* R. Gadh (ed.), ASME, New York, pp. 87–96.

23. Lam, G., 1995, Automated High Level Process Planning and Manufacturability Analysis for Agile Manufacturing, M.S. Thesis, University of Maryland, College Park.

24. *Microstation Version 5,* 1995, Bentley Systems, Inc., Exton, PA.

25. Ousterhout, J. K., 1994, *Tcl and the Tk Toolkit,* Addison-Wesley, Reading, MA.

26. Saaty, T. L., 1986, Axiomatic Foundation of the Analytic Hierarchy Process, *Management Sci.* Vol. 32, pp. 841–855.

27. Saaty, T. L., 1990, *Multicriteria Decision Making: The Analytic Hierarchy Process,* RSW Publications.

28. Sacerdoti, E. D., 1977, *A Structure for Plans and Behavior,* Elsevier North-Holland, New York.

29. Salomons, O. W., 1993, Review of Research in Feature-Based Design, *J. Mfg. Systems,* Vol. 12, pp. 113–132.

30. Sheridan, S. A., 1972, *Foraging Design Handbook,* American Society for Metals, Metals Park, OH.

31. Smith, S. J. J., D. S. Nau, and T. Throop, 1996, A planning approach to declarer play in contract bridge, Computer Intelligence, Vol. 12.

32. Tate, A., 1977, Generating Project Networks, *Proceedings of the Fifth International Joint Conference on Artificial Intelligence,* Morgan Kaufmann, San Mateo, CA, pp. 888–893.

33. Trinogga, L. A., G. Kaizhou, and I. C. Hunter, 1991, *Practical Microstrip Design,* Ellis Horwood, Chichester.

34. Wilkins, D. E., 1984, Domain Independent Planning: Representation and Plan Generation, *Artificial Intelligence,* Vol. 22, pp. 269–301.

35. Wilson, F. W., and P. D. Harvey, 1963, *Manufacturing Planning and Estimating Handbook,* McGraw-Hill, New York.

INDEX